iophysics for the Life Sciences

eries Editor
Norma Allewell

For further volumes:
http://www.springer.com/series/10230

Andres F. Oberhauser
Editor

Single-molecule Studies of Proteins

Springer

Editor
Andres F. Oberhauser
Neuroscience and Cell Biology
University of Texas Medical Branch
Galveston, Texas, USA

ISBN 978-1-4614-4920-1 ISBN 978-1-4614-4921-8 (eBook)
DOI 10.1007/978-1-4614-4921-8
Springer New York Heidelberg Dordrecht London

Library of Congress Control Number: 2012950398

Printed on acid-free paper

Springer is part of Springer Science+Business Media (www.springer.com)

Preface

Single-molecule methods have emerged as powerful tools in life science research. These techniques allow the detection and manipulation of individual biological molecules and investigate, with unprecedented resolution, their conformations and dynamics at the nanoscale level. These techniques overcome the restrictions of traditional bulk biochemical studies by focusing not on a population of molecules but on the molecule itself. During the last two decades, this field has grown at an almost exponential rate in terms of biological and biophysical applications. Single-molecule techniques have opened a new field of science, that is, at the crossroads of several disciplines, namely biology, physics, chemistry, material science, and computer science. Since the development of single-ion channel recording techniques in the 1970s, the family of single-molecule methods has expanded significantly to include, among others, optical and magnetic tweezers, atomic force microscopy (AFM), and single-molecule fluorescence.

The present volume considers a range of important and timely applications of biological problems mainly tackled by single-molecule AFM. The AFM was first described in 1986 and originally developed as a high-resolution imaging tool before it began to be used to probe and manipulate molecules. During the last two and half decades, AFM has evolved as a very powerful and versatile tool in biology that can be used, for example, to manipulate and detect single proteins, DNA, or polysaccharides; to image single molecules in physiological conditions with nanometer resolution; to measure the interaction forces between proteins and DNA; to track exocytotic fusion; to map cell surface receptors; or to visualize molecular motors in action. One of the key advantages of the AFM as a single-molecule technique is the straightforward sample preparation, the ability to conduct imaging and manipulation experiments of biomolecules under physiological relevant conditions, and the direct analysis of the dynamics of single molecules or complexes.

The chapters are aimed at the nonexpert—advanced graduate students, postdoctoral investigators or established investigators wanting to enter new fields with the knowledge needed to pursue research in various areas in which single-molecule approaches are important.

Carrión-Vazquez and coworkers contribute with a comprehensive up-to-date review on the use of AFM techniques to study the nano-mechanical properties of structured and disordered proteins. Amyloidogenic proteins are a rich and diverse group of molecules capable of forming amyloid fibrils, and many of them are implicated in severe degenerative diseases. Kellermayer and coworkers describe how single-molecule techniques can be used to study the behavior of amyloidogenic proteins and amyloid fibrils.

Three chapters are devoted to different single-molecule approaches to analyze biomolecular interactions. Hinterdorfer and coworkers discuss single-molecule force spectroscopy as well as simultaneous Topography and RECognition imaging (TREC) applied to living cells using AFM techniques. While the force spectroscopy yields information regarding the structure and the dynamics of the recognition process, the TREC method offers the possibility to map the distribution of specific molecular recognition events on the cell surface with nanometer accuracy under physiological condition. Zhu and Chen describe single-molecular interaction experiments that demonstrate catch bonds between integrin receptors and their respective ligands, intercellular adhesion molecule 1, or fibronectin. They also discuss experimental analyses and models that relate integrin catch bonds to their structures and conformations. Lyubchenko outlines the advances in AFM technology in studies of various types of protein–DNA complexes. He gives special attention to time-lapse AFM imaging in aqueous solutions, which enables direct observation of protein–DNA dynamics and interactions.

Ando and coworkers overview the use of high-speed AFM to visualize dynamic events of proteins in action under physiological conditions at subsecond temporal and subnanometer resolution. They highlight recent imaging studies of myosin V walking on an actin filament, rotary catalysis of F_1-ATPase, and the processive movement of cellulase. Perez-Jimenez and Alegre-Cebollada cover some of the most significant achievements of the single-molecule methods in elucidating enzyme catalysis. They describe recent developments of force-clamp single-molecule assays to study the chemistry of thioredoxin enzymes, a class of oxidoreductases present in all known living organisms that act as electron donors for reducing disulfide bonds.

Alsteens and Dufrêne discuss how single-protein manipulation and imaging experiments can be applied to live cells to shed new light into the highly sophisticated functions of cell surfaces. They review recent progress made in understanding the elasticity, clustering, and functions of cell surface proteins, owing to the integration of the modern tools of molecular genetics (protein design in live cells), with a powerful set of AFM techniques (live cell imaging, single-molecule manipulation, and localization).

In the last chapter, Makarov introduces essential theoretical concepts needed to understand how the dynamics and thermodynamics of proteins are affected by mechanical stress. Various computational approaches and theoretical models used to explore the relationship between protein structure and mechanical response are critically reviewed and computational and theoretical predictions are then contrasted with experimental observations.

I gratefully acknowledge the contributions by the different authors each of whom donated their unique expertise in order to further enhance the ever expanding single-molecule field.

Galveston, TX, USA Andres F. Oberhauser

Contents

1 **Nanomechanics of Proteins, Both Folded and Disordered** 1
Rubén Hervás, Albert Galera-Prat, Àngel Gómez-Sicilia, Fernando Losada-Urzáiz, María del Carmen Fernández, Débora Fernández-Bravo, Elena Santana, Clara Barrio-García, Carolina Melero, and Mariano Carrión-Vázquez

2 **Force Spectroscopy and Recognition Imaging of Cells from the Immune System** 49
Bianca L. Bozna, Memed Duman, and Peter Hinterdorfer

3 **Catch Bonds of Integrin/Ligand Interactions** 77
Cheng Zhu and Wei Chen

4 **AFM Visualization of Protein–DNA Interactions** 97
Yuri L. Lyubchenko

5 **Nanovisualization of Proteins in Action Using High-Speed AFM** 119
Takayuki Uchihashi, Noriyuki Kodera, and Toshio Ando

6 **Enzyme Catalysis at the Single-Molecule Level** 149
Raul Perez-Jimenez and Jorge Alegre-Cebollada

7 **Single-Molecule Studies of Amyloidogenic Proteins** 169
Miklós S.Z. Kellermayer, Árpád Karsai, Ünige Murvai, Szilvia Erdélyi-Bótor, József Kardos, and Ricardo H. Pires

8 **Stretching and Imaging Individual Proteins on Live Cells Using Atomic Force Microscopy** 211
David Alsteens and Yves F. Dufrêne

9 **Individual Proteins Under Mechanical Stress: Lessons from Theory and Computer Simulations** 235
Dmitrii E. Makarov

Index 269

Contents

Contributors

Jorge Alegre-Cebollada Department of Biological Sciences, Columbia University, New York, NY, USA

David Alsteens Institute of Condensed Matter and Nanosciences, Université catholique de Louvain, Louvain-La-Neuve, Belgium

Toshio Ando Department of Physics and Bio-AFM Frontier Research Center, Kanazawa University, Kanawaza, Japan

Clara Barrio-García Centro de Investigación Biomédica en Red sobre Enfermedades Neurodegenerativas (CIBERNED), Instituto Cajal, IC-CSIC, Madrid, Spain

Instituto Madrileño de Estudios Avanzados en Nanociencia (IMDEA-Nanociencia), Madrid, Spain

Bianca L. Bozna Institute for Biophysics, Johannes Kepler University Linz, Austria

Department of Pharmaceutical Sciences, University of Nebraska Medical Center, Omaha, NE, USA

Mariano Carrión-Vázquez Centro de Investigación Biomédica en Red sobre Enfermedades Neurodegenerativas (CIBERNED), Instituto Cajal, IC-CSIC, Madrid, Spain

Instituto Madrileño de Estudios Avanzados en Nanociencia (IMDEA-Nanociencia), Madrid, Spain

Wei Chen The Coulter Department of Biomedical Engineering, Georgia Institute of Technology, Atlanta, GA, USA

María del Carmen Fernández Centro de Investigación Biomédica en Red sobre Enfermedades Neurodegenerativas (CIBERNED), Instituto Cajal, IC-CSIC, Madrid, Spain

Instituto Madrileño de Estudios Avanzados en Nanociencia (IMDEA-Nanociencia), Madrid, Spain

Yves F. Dufrêne Institute of Condensed Matter and Nanosciences, Université catholique de Louvain, Louvain-La-Neuve, Belgium

Memed Duman Institute for Biophysics, Johannes Kepler University, Linz, Austria

Nanotechnology and Nanomedicine, Institute of Science, Hacettepe University, Beytepe, Ankara, Turkey

Szilvia Erdélyi-Bótor Department of Biophysics and Radiation Biology, Semmelweis University, Budapest, Hungary

University of Pécs, Pécs, Hungary

Débora Fernández-Bravo Centro de Investigación Biomédica en Red sobre Enfermedades Neurodegenerativas (CIBERNED), Instituto Cajal, IC-CSIC, Madrid, Spain

Instituto Madrileño de Estudios Avanzados en Nanociencia (IMDEA-Nanociencia), Madrid, Spain

Albert Galera-Prat Centro de Investigación Biomédica en Red sobre Enfermedades Neurodegenerativas (CIBERNED), Instituto Cajal, IC-CSIC, Madrid, Spain

Instituto Madrileño de Estudios Avanzados en Nanociencia (IMDEA-Nanociencia), Madrid, Spain

Àngel Gómez-Sicilia Centro de Investigación Biomédica en Red sobre Enfermedades Neurodegenerativas (CIBERNED), Instituto Cajal, IC-CSIC, Madrid, Spain

Instituto Madrileño de Estudios Avanzados en Nanociencia (IMDEA-Nanociencia), Madrid, Spain

Rubén Hervás Centro de Investigación Biomédica en Red sobre Enfermedades Neurodegenerativas (CIBERNED), Instituto Cajal, IC-CSIC, Madrid, Spain

Instituto Madrileño de Estudios Avanzados en Nanociencia (IMDEA-Nanociencia), Madrid, Spain

Peter Hinterdorfer Institute for Biophysics, Johannes Kepler University Linz, Austria

Christian Doppler Laboratory for Nanoscopic Methods in Biophysics, University of Linz, Linz, Austria

József Kardos Department of Biochemistry, Eötvös Loránd University, Budapest, Hungary

Árpád Karsai Department of Biophysics and Radiation Biology, Semmelweis University, Budapest, Hungary

Department of Neurobiology, Physiology, and Behavior, University of California-Davis, Davis, CA, USA

Miklós S.Z. Kellermayer Department of Biophysics and Radiation Biology, Semmelweis University, Budapest, Hungary

Noriyuki Kodera Bio-AFM Frontier Research Center, Kanazawa University, Kanawaza, Japan

Fernando Losada-Urzáiz Centro de Investigación Biomédica en Red sobre Enfermedades Neurodegenerativas (CIBERNED), Instituto Cajal, IC-CSIC, Madrid, Spain

Instituto Madrileño de Estudios Avanzados en Nanociencia (IMDEA-Nanociencia), Madrid, Spain

Yuri L. Lyubchenko Department of Pharmaceutical Sciences, University of Nebraska Medical Center, Omaha, NE, USA

Dmitrii E. Makarov Department of Chemistry and Biochemistry and Institute for Computational Engineering and Sciences, University of Texas at Austin, Austin, TX, USA

Carolina Melero Centro de Investigación Biomédica en Red sobre Enfermedades Neurodegenerativas (CIBERNED), Instituto Cajal, IC-CSIC, Madrid, Spain

Instituto Madrileño de Estudios Avanzados en Nanociencia (IMDEA-Nanociencia), Madrid, Spain

Ünige Murvai Department of Biophysics and Radiation Biology, Semmelweis University, Budapest, Hungary

Raul Perez-Jimenez Department of Biological Sciences, Columbia University, New York, NY, USA

Ricardo H. Pires Department of Biophysics and Radiation Biology, Semmelweis University, Budapest, Hungary

Elena Santana Centro de Investigación Biomédica en Red sobre Enfermedades Neurodegenerativas (CIBERNED), Instituto Cajal, IC-CSIC, Madrid, Spain

Instituto Madrileño de Estudios Avanzados en Nanociencia (IMDEA-Nanociencia), Madrid, Spain

Takayuki Uchihashi Department of Physics and Bio-AFM Frontier Research Center, Kanazawa University, Kanawaza, Japan

Cheng Zhu The Coulter Department of Biomedical Engineering, Georgia Institute of Technology, Atlanta, GA, USA

Chapter 1
Nanomechanics of Proteins, Both Folded and Disordered

Rubén Hervás, Albert Galera-Prat, Àngel Gómez-Sicilia, Fernando Losada-Urzáiz, María del Carmen Fernández, Débora Fernández-Bravo, Elena Santana, Clara Barrio-García, Carolina Melero, and Mariano Carrión-Vázquez

Abbreviations

AFM	Atomic force microscopy
Arc	Arctic
Aβ	Amyloid beta
CaM	Calmodulin
D^2	Disorder in disorder
DHF	7,8-Dihydrofolate
DHFR	Dihydrofolate reductase
DMSO	Dimethyl sulfoxide
ELP	Elastin-like-polypeptides
F_U	Unfolding force
GB1	B1 immunoglobulin binding domain of streptococcal protein G
HD	Huntington's disease
hM	Hyper-mechanostable
IDP	Intrinsically disordered protein
Ig	Immunoglobulin
L_C	Contour length
M	Mechanostable

R. Hervás • A. Galera-Prat • À. Gómez-Sicilia • F. Losada-Urzáiz • M.C. Fernández • D. Fernández-Bravo • E. Santana • C. Barrio-García • C. Melero • M. Carrión-Vázquez (✉)
Centro de Investigación Biomédica en Red sobre Enfermedades Neurodegenerativas (CIBERNED), Instituto Cajal, IC-CSIC, Madrid, Spain

Instituto Madrileño de Estudios Avanzados en Nanociencia (IMDEA-Nanociencia), Madrid, Spain
e-mail: mcarrion@cajal.csic.es

A.F. Oberhauser (ed.), *Single-molecule Studies of Proteins*, Biophysics for the Life Sciences 2, DOI 10.1007/978-1-4614-4921-8_1,

MBP	Maltose binding protein
mtx	Methotrexate
N_A	Avogadro's number
NADPH	Nicotinamide adenine dihydrogen phosphate
NM	Non-mechanostable
NMR	Nuclear magnetic resonance
p	Persistence length
PC1	Polycystin-1
pFS	Plasmid for force spectroscopy
PKA	Protein kinase A
PolyQ	Polyglutamine
QBP1	PolyQ binding peptide 1
RC	Random coil
SMD	Steered molecular dynamics
SMF	Single-molecule fluorescence
SMFS	Single-molecule force spectroscopy
SMT	Single-molecule techniques
SPM	Scanning probe microscopy
S–S	Disulfide bond
STM	Scanning tunneling microscope
THF	5,6,7,8-Tetrahydrofolate
TNfn3	Third fibronectin type III domain
TNXfn7	Seventh FnIII domain of human tenascin-X
TS	Transition state
WLC	Worm-like chain
α-syn	Alpha synuclein
ΔL_C	Contour length increase

1.1 Summary and Overview

Single-molecule techniques have recently provided a versatile tool for imaging and manipulating protein molecules one at a time, enabling us to address important biological questions in key areas of cell biology (e.g., cell adhesion and signaling, neurodegeneration) and protein science (e.g., protein folding, protein structure and stability, catalysis, protein evolution, conformational polymorphism, and amyloidogenesis). One of these techniques, single-molecule force spectroscopy (SMFS) based on atomic force microscopy (AFM), combined with theoretical/computational approaches and protein engineering, has allowed unprecedented progress in characterizing and understanding at the molecular level the mechanical properties of biomolecules, particularly those of proteins, which has recently opened the new, exciting and fast-growing research field of protein nanomechanics. The aim of this review is to describe the principles of this methodology and to discuss the main achievements in this field, with special emphasis on its emerging application to the analysis of IDPs.

1.2 Force, an Important Parameter in Biology

Since the days of Isaac Newton, we know that force is a parameter that governs the motion of the bodies. However, in biology, force only began to emerge in the 1980s as an important player in many cellular processes that involve a variety of biological systems. Since then, new instrumentation and methodology have been developed allowing to directly measure forces at the level of single molecules and/or their complexes (Carrión-Vázquez et al. 2006).

Cells have the ability to sense external mechanical stimuli—either from the extracellular matrix or neighboring cells—and react as a consequence. Indeed, they have specific protein complexes that act as mechanosensitive transducers that translate deformations and mechanical forces into biochemical responses that can modulate cellular and extracellular structure and function. This mechanosensitive feedback regulates vital cellular functions like muscle contraction, cell locomotion, cell signaling and division or transport processes, and is crucial for organ development and homeostasis (Bustamante et al. 2004; Gräter et al. 2005; Vogel and Sheetz 2006; Johnson et al. 2007). Proteins play such a pivotal role in these mechanically guided processes that any disruption in their properties that interrupts or alters this mechanical sensing pathway and its subsequent cell signaling events could jeopardize normal cellular activities. These abnormalities in cellular mechanoresponses could potentially result in diverse diseases that have a mechanical origin such as the loss of hearing, cardiovascular disorders, muscular dystrophy, osteoporosis, and cancer (Kazmierczak and Müller 2012; Vogel and Sheetz 2009).

1.3 Single-Molecule Vs. Bulk Biophysics

In classical bulk measurements, the properties of all the molecules in the sample contribute to the observed signal, thus measuring ensemble averages over N molecules (where N is typically a large number close to Avogadro's number, N_A). In this population averaging, the major contributors dominate such that the effects of minor contributions are hidden to the signal so that fluctuations are mostly canceled out, and information on transient processes is lost. Thus, these techniques, although robust, give us an idealized picture of the molecule with well-defined dynamics assuming a homogeneous population (unimodal distribution).

One of the differential features of single-molecule techniques (SMT) is that they report the distribution of values for a given property and not just the mean value of the property measured as an average over a large molecular ensemble. According to the ergodic principle, the average behavior of a single molecule at a given location accumulated over sufficiently long timescales does not offer a less accurate description of the system than that obtained from the analysis of an ensemble of independent, identical molecules, each measured once (Matsunaga and Komatsuzaki 2004). The explicit identification that arises from single-molecule approaches results in a richer and more accurate description of the behavior of the system than the ensemble approach. For instance, this approach can reveal intermediates that are not detectable by bulk techniques, resulting in a more detailed description of the reaction mechanism. In SMT, molecular fluctuations are monitored, and the distributions and the dynamics of certain properties can be obtained, not just an average behavior (Bilsel and Matthews 2006; Deniz et al. 2000, 2001; Michalet et al. 2006; Schuler 2005; Schuler et al. 2002; Tinnefeld and Sauer 2005). This allows the observation of molecules that co-exist in various states (multimodal distribution, providing direct access to molecular heterogeneity, an intrinsic feature of complex biomolecules) and even the detection of molecules found in states far from the average of the population (extreme states). These fluctuations are ultimately the ones that characterize the underlying statistical nature of each property.

One interesting example is found in the hot and growing field of amyloid proteins. The details of the structure and dynamics of early monomeric intermediates are crucial to obtain information on the molecular mechanism of amyloid formation (Jahn et al. 2006; Platt et al. 2005) and to understand the toxic gain-of-function of proteins causally involved in several conformational diseases, such as neurodegenerative ones (Nagai et al. 2007). Although solid-state nuclear magnetic resonance (NMR) and X-ray crystallography provide valuable structural information on the mature amyloid fibers (Tycko 2011; Sawaya et al. 2007), the heterogeneity and transient nature of the partially folded intermediates in the amyloidogenic pathway impede structural characterization by traditional bulk techniques. In recent years, single-molecule biophysics has proven to be very useful in studying complex systems such as disordered proteins (Ferreon et al. 2010; Hervás et al. 2012). Furthermore, combined with the information obtained in traditional bulk assays, using SMT we can obtain deeper insight into complex cellular processes such as neurodegeneration (Hervás et al. 2012).

1.4 Single-Molecule Force Spectroscopy Techniques

SMFS has emerged as a powerful tool to manipulate individual molecules and investigate the forces and motions associated with biological particles, enzymatic catalysis, and folding reactions (Galera-Prat et al. 2010). The most common force-based detection and manipulation SMT are optical tweezers, magnetic tweezers, and AFM. These three techniques have been extensively used to characterize the mechanical properties of biomolecules, each of them being appropriate for certain type of studies based on their different temporal and spatial resolutions as well as force ranges. Furthermore, a new methodology that combines force measurements and fluorescence capabilities has been reported (Lang et al. 2004). A comparative study of SMFS techniques can be found in Neuman and Nagy (2008), where the authors describe these methods and show exciting application examples, current capabilities, strength, limitations, and practical considerations for the selection of a particular technique for a specific application.

AFM-based SMFS can in principle measure distances with an angstrom-scale resolution, events with millisecond temporal resolution and forces in the range of $10–10^4$ pN. This technique is ideal to characterize the mechanical resistance of both individual polypeptides (intramolecular interactions) and protein–biomolecule bonds (intermolecular interactions) (Carrión-Vázquez et al. 2006) among other biological processes (Galera-Prat et al. 2010). However, a major limitation is the poor signal-to-noise ratio in the low-force regime (about 10 pN). However, the recent development of the so-called lock-in mode increases the resolution and the range of forces measured into the sub-pN range, making this technique applicable also to the study of protein folding and ionic coordination reactions in the low-force regime (Junker et al. 2009; Schlierf et al. 2007).

1.5 AFM-Based Single-Molecule Force Spectroscopy

1.5.1 AFM: Invention and Uses

A general feature in all scanning probe microscopy (SPM) techniques, to which AFM belongs, is the presence of a tip that directly interacts with the sample surface. This enables us to study details of the sample at the molecular and atomic scale, getting deeper insight on the nature of nanoscale systems and how they work, revealing a wealth of otherwise inaccessible information in many science fields.

The AFM was developed to overcome a basic drawback found with the Scanning Tunneling Microscope (STM, the ancestor of all SPMs and developed in 1982 by Binnig and Rohrer at IBM Zurich, who were awarded the 1986 Nobel prize in Physics for their design). While STM can only image conducting or semiconducting surfaces (Binnig and Rohrer 1986), the AFM, invented by Binnig, Quate, and Gerber in 1985, has the advantage of imaging almost any type of surface, including biological samples in liquids (Binnig et al. 1986). AFM has been most widely used in its imaging configuration to map the topography of surfaces at atomic resolution. In addition, it can be used as a force-measuring tool to estimate the bond strength between different biological molecules such as ligands and receptors or to measure the forces responsible for the mechanical resistance of the fold of a single protein.

1.5.2 AFM-Based SMFS: Physical Principle and Modes

In the so-called "force spectroscopy" or "force-measuring" configuration, a single polyprotein (a tandem repeat of the protein of interest, the periodicity of which is used as a single-molecule marker) is usually immobilized between the force sensor (a cantilever with a sharp tip of silicon nitride) and a substrate (usually gold-covered or otherwise functionalized glass coverslip), mounted on a high-precision

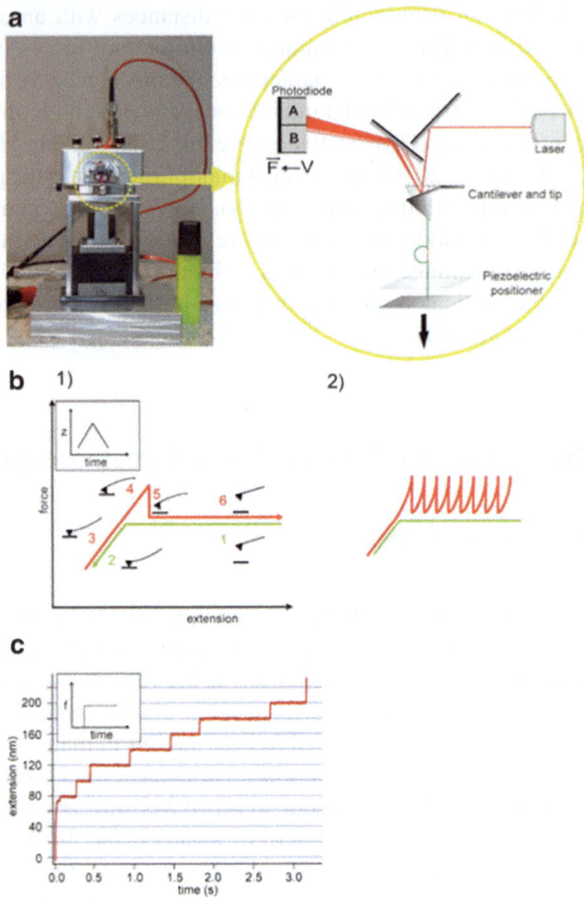

Fig. 1.1 Mechanical unfolding of proteins by SMFS. Physical principle and modes of operation. (**a**) Picture (*left*) and schematic diagram of an SMFS–AFM setup. The *line* shows the laser light path before (*dashed line*; *pale red*) and after (*solid lines*; *red*) a deflection of the force sensor (a cantilever). The protein (*green*) has to form a mechanical circuit connecting the tip and the substrate, which is attached to a piezoelectric positioner. The contraction of the piezoelectric positioner results in an elongation of the trapped stretch of the protein molecule and, in turn, a bending of the cantilever. As the distance between the substrate and the tip of the cantilever increases, the protein elongates against a restoring force originated from the reduction of its entropy (described by the WLC model) that bends the cantilever along the same axis. This bending is trackedby changes in the reflected angle of a laser beam that is reflected on the cantilever. Deviations are converted into voltage differences by a split photodiode detector and transformed into force following Hooke's law in order to obtain the final force–distance plot. The observed force continues increasing until the noncovalent interactions that hold its structure are broken (the breakpoint is usually known as a "mechanical clamp"). The average force reached at that instant indicates the mechanical resistance of that protein. (**b**1) A typical recording diagram in length-clamp mode SMFS showing the movement of the cantilever in a complete approach-retraction cycle mediated by the piezoelectric positioner: It starts with the substratenot in contact with the tip (1), then it contacts with it, which bends the cantilever (2) increasingly (3); afterwards it is withdrawnfrom the tip, which bends the cantilever the other way (4) as it adheres to the tip,

piezoelectric positioner to allow control of its position, so that a "mechanical circuit" is established between both elements (Fig. 1.1a). There are two basic modes of this methodology depending on which variable is controlled: length-clamp (at controlled pulling speed, the most frequently used) and force-clamp (at controlled force). The traces originating from the length-clamp mode of operation (Fig. 1.1b) are usually analyzed by fitting the worm-like chain (WLC) model of polymer elasticity (Bustamante et al. 2004), which relates the force acting on a polymer to its length through two parameters known as "contour length," L_C, which is the length the polymer would have under infinite force, and "persistence length," p, the maximum length along which the direction of the polymer remains correlated, which is related to its rigidity. Once the interactions involved in the mechanical resistance of a protein fold (region known as "mechanical clamp") are broken, the protein usually becomes completely unfolded. The force-hidden protein length is released and it can be measured from the distance between the two consecutive peaks, which is the contour length increase, ΔL_C. This parameter is characteristic of each protein, depends on the protein fold, and is important to identify the molecule under study (as a kind of molecular "fingerprint").

Length-clamp data are usually interpreted by means of a model based on the transition state (TS) theory (Bell 1978; Evans and Ritchie 1997). This is a two-state model where the native and the unfolded states of a protein are separated by an energy barrier, the relative height of which is lowered by the applied force as described in the following rate equation:

$$k_U(F) = A\exp[-(\Delta G^{\ddagger} - F\Delta x_U)/k_BT]$$

where $k_U(F)$ is the unfolding rate at a given force F, A is a pre-exponential factor, $\Delta G^{\ddagger}$ is the height of the energy barrier between the native state and the TS, Δx_U represents the distance over which force is applied in the reaction coordinate, and k_BT is the thermal energy. It is clear from this equation that the applied force directly reduces the height of the energy barrier, thus facilitating the unfolding of the protein.

Furthermore, according to this model, the most probable unfolding force of a protein, F_U, can be predicted as

$$F_U = \left(\frac{k_BT}{\Delta x_U}\right)\ln\left(\frac{r\Delta x_U}{k_U^0 k_BT}\right)$$

Fig. 1.1 (continued) originating a force peak (4) on "dumpingoff contact" from the tip (5) which ends with the substrateagain not in contact with the tip (6). (**b2**) Schematic force–extension recording showing a typical sawtooth pattern obtained by several cycles of stretching from a homomeric polyprotein molecule using the length-clamp mode. (**c**) Typical staircase extension-time recording obtained using force-clamp mode of SMFS. This particular example shows an N–C linked polyubiquitin protein being stretched at a constant force of 110 pN (Taken from Carrión-Vázquez at al. 2009, with permission from Springer Science and Business Media)

where k_U^0 is the unfolding rate extrapolated to 0 force and r is the loading rate that indicates the rate of change of the force, which is approximated as the product of the cantilever spring constant and the pulling speed. This implies that the average F_U depends on the logarithm of the pulling speed.

In force-clamp experiments, length vs. time traces at the selected force are obtained (Fig. 1.1c). Each unfolding event appears here as a step where the height represents the ΔL_C after the protein breakage at that specific force (as mentioned, at a given force, ΔL_C is only part of the total extended length of the molecule) while its duration, or length of the step, is related to the mechanical stability of the protein. Remarkably, the force-clamp mode allows to directly obtain the kinetic parameters of the process by applying the previous equations in order to get information on the energy landscape of the protein. A variation of this mode consists on ramping the force (force-ramp mode).

1.6 SMFS in Ordered Versus Disordered Proteins

Protein folding is still an unsolved problem. Our classical view of the protein structure–function relationship is based on the key point that the biological function obligatorily depends on a well-defined 3D structure, which is determined in turn by its amino acid sequence. Ordered proteins are composed of two main types of secondary structure: α-helix and β-sheet. These structures differ mostly in the Ramachandran Φ and Ψ dihedral angles of the peptide bond, which determines the orientation of the side chain. In general, secondary structures have been shown to be the typical determinants of mechanical stability in proteins (Carrión-Vázquez et al. 2006).

However, since the discovery of IDPs, there has been a considerable shift in this traditional protein folding paradigm. It has been shown that for many proteins and protein domains the functional state is intrinsically unstructured and it does not maintain long-range interactions. Although they do not form tertiary structures by themselves, the presence of different levels of transient secondary structure elements in this kind of proteins has been shown. In these cases, these regions exist as an ensemble of Ramachandran Φ and Ψ angles along the backbone, instead of adopting a restricted set of Φ and Ψ angles as in folded structures.

Alternative views to the structure/function paradigm have been recently proposed, such as the so-called "protein trinity" (Dunker et al. 2001). In this model, native protein structure can correspond to one of three different thermodynamic states—ordered, collapsed (or molten globule), and extended (random coil, RC)—and protein function could derive from any of these three states or from transition forms between them (Fig. 1.2).

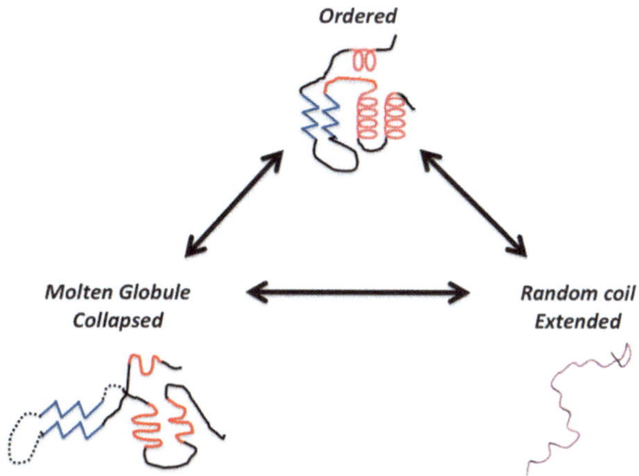

Fig. 1.2 The Protein Trinity. Native proteins may exist in one of the three thermodynamic states; ordered, collapsed, or extended

1.6.1 SMFS of Folded Proteins

One of the main advantages of SMFS is the ability to analyze the mechanical stability of proteins, which can be defined as the average force required to unfold a protein (reviewed in Galera-Prat et al. 2010). In this section, we discuss the molecular determinants of mechanical stability and novel applications derived from this methodology for folded proteins.

1.6.1.1 Mechanical Stability of Structured Proteins

From the pioneering experiments in the field in the muscle protein titin, each unfolding peak observed using the length-clamp mode in this modular protein was attributed to the rupture of a single-protein module (Rief et al. 1997), which is usually unfolded in a two-state manner. Further atomic-resolution information on the process is often obtained by complementing the experimental studies with steered molecular dynamics (SMD) simulations (see Sect. 1.7; Lu et al. 1998; Marszalek et al. 1999).

It is now widely accepted that mechanical stability is a kinetic parameter that arises mainly from localized elements of the protein structure. These elements are the ones that directly resist mechanical unfolding. Among them, the best studied is the mechanical clamp motif, which is shared among many proteins with a mechanical function (Valbuena et al. 2009). It consists of a patch of backbone hydrogen bonds from local secondary structure elements (Fig. 1.3a), which are usually β-strands that fasten the protein module in a shear configuration. When force is applied to this motif, the bonds that constitute this element have to be broken in order to unfold the complete structure. Thus, geometry and the number of hydrogen bonds present in the mechanical clamp motif are of key importance in determining the mechanical stability of a protein (Valbuena et al. 2009).

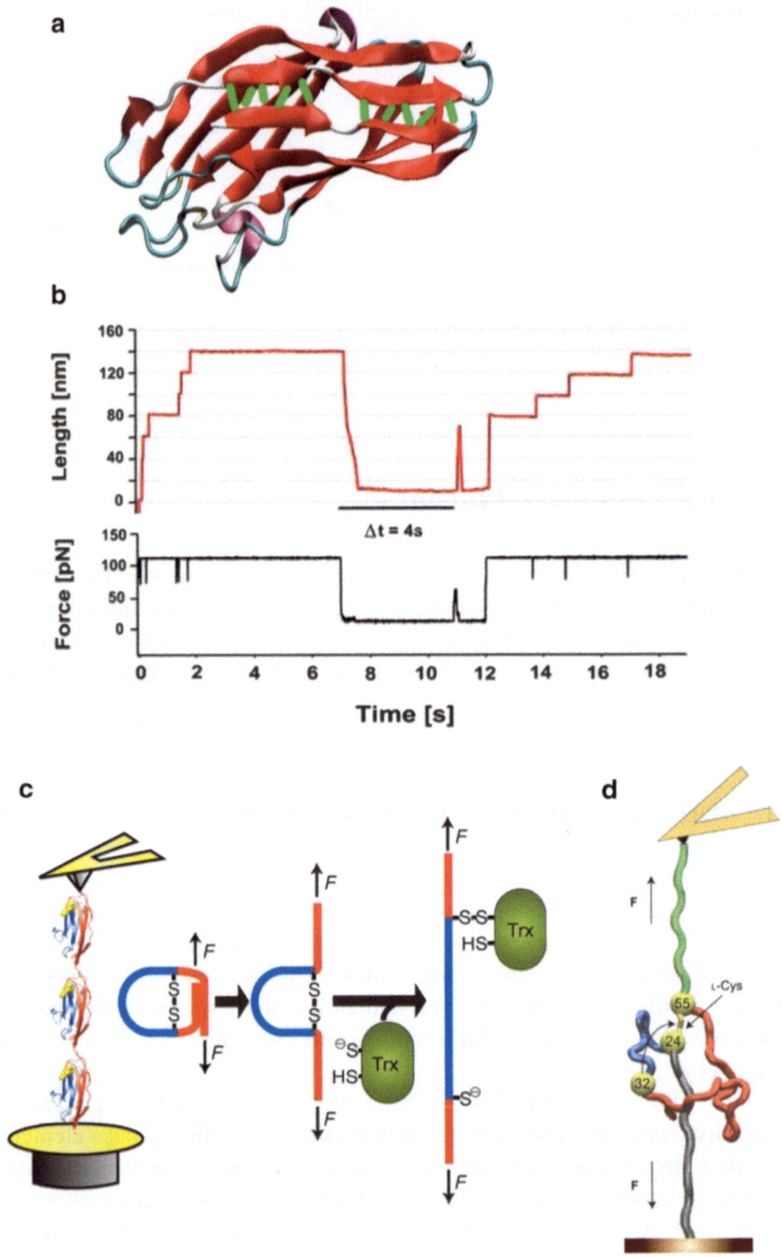

Fig. 1.3 (**a**) Cartoon representation of a cohesin module where the backbone hydrogen bonds are highlighted in *green*. This is an example of shear mechanical clamp. Figure adapted with permission from Valbuena et al. 2009. (**b**) Force-clamp trace probing the properties of the minimum energy collapsed structures. Force is initially set to a high value to unfold all protein modules (initial steps). Then, force is lowered to allow folding to occur: during this step, force is

Some proteins have evolved to function under force in different ways. SMFS is a technique that is ideally suited for the study of this kind of proteins, since force can be applied to the molecules in a controlled way. These "mechanical proteins" can be composed of β-stranded (e.g., titin Ig domains) or α-helical (e.g., spectrin) structures.

β-Structured Proteins: Mechanical Clamps of the Shear and Zipper Configurations

Since force is a vector magnitude, its effect depends also on the direction and point of application. Thus, protein topology and pulling geometry have revealed to be crucial determinants in protein mechanical stability (Carrión-Vázquez et al. 2006). Proteins subjected to mechanical stress during its physiological activity such as some Ig modules of human titin (e.g., the 27th Ig module, which is a model protein for SMFS studies) often have a mechanical clamp composed of two shearing β-strands. In this configuration, the strands are parallel with respect to the force direction and hence the backbone hydrogen bonds are perpendicular to it. When the critical β-strands are perpendicular to the pulling direction, backbone hydrogen bonds have to break sequentially, which usually results in lower mechanical stabilities compared to shearing mechanical clamps. This is known as a zipper mechanical clamp and can be found, for example, in the C2A domain from synaptotagmin I, a synaptic vesicle protein (Carrión-Vázquez et al. 2000).

α-Helical Proteins

In general, proteins composed mainly of α-helical proteins tend to be much less mechanostable than those rich in β structure (Rief et al. 1999; Carrión-Vázquez et al. 2000). For instance, calmodulin, CaM, was found to present no force peaks using the conventional technique. Still, very low force peaks (12–15 pN) were found when the lock-in configuration was used (Carrión-Vázquez et al. 2000; Junker et al. 2009). α-Helical structures are often associated in bundles or other complexes, which increases its mechanostability.

Fig. 1.3 (continued) momentarily raised to disrupt the folding precursors before their properties are finally probed. Taken with permission from Garcia-Manyes et al. 2009a. (**c**) Cartoon representation of the engineered poly-I27 substrate used in the experiments of single-molecule catalysis. In a first step, the substrate is partially unfolded and hence the S–S bond is exposed. Later the bond is reduced either by a thioredoxin enzyme or a reducing agent. Taken with permission from Wiita et al. 2007. (**d**) Representation of the engineered I27 with two S–Ss. In this case, the substrate is partially unfolded and the S–Ss exposed. After reduction of one of them, a free cysteine is capable of reducing the still formed disulfide. This results in the expected length step corresponding to the number of residues previously enclosed by the SS-bond loop (Taken with permission from Alegre-Cebollada et al. 2011)

Spectrin, a cytoskeletal protein thought to be subjected to mechanical forces, is composed by bundles of α-helices that contribute to the mechanical properties of cells. This protein was reported to show much lower unfolding forces than some β structured proteins. This was attributed to the fact that the tertiary structure of this protein is mainly stabilized by hydrophobic interactions while the hydrogen bonds in the α-helices just stabilize the helix itself (Rief et al. 1999).

A slightly more complex behavior was later observed when studying the unfolding pathway of T4 lysozyme, another protein with a high content in α-helices (Peng and Li 2008). This protein has two domains, an α/β N-terminal and an all-α C-terminal one, which have an unusual topology surprisingly coupled through helix A (this region, although formed by residues of the N terminus, is part of the C-terminal domain). Since the thermodynamic stability of the two domains is very different, the authors hypothesized that stretching it from different ends might result in a different behavior. Hence, they engineered a circular permutant (an engineered sequence with preserved fold but different connectivity where the native N- and C-termini are connected while there are artificial new N- and C-termini, such that helix A is now part of the C-terminal domain) and studied it by SMFS. Using this construct, it was found that the shuffling of helix A from the N terminus to C terminus weakened the protein, such that a shift of the mechanical unfolding barrier made it easier to unfold the mutant. Coupling helix A to the C terminus is the key also to explain the thermodynamic stability of T4 lysozyme as a whole. Furthermore, the authors reported evidence of kinetic partitioning, a phenomenon in which two or more possible states are kinetically accessible originating alternative pathways. This was concluded based on the fact that the majority of molecules unfolded in an apparent two-state way (all or none) while 13% unfolded in a three-state way that involved a partially unfolded intermediate state.

Coiled-coil structures have also been shown to resist mechanical unfolding (Bornschlögl and Rief 2008; Bornschlögl et al. 2009a; Gebhardt et al. 2010). These are double-stranded structures formed by two α-helices wrapped around each other, which typically show repeats of seven amino acids with their interaction surface generally formed by hydrophobic residues. The most studied coiled-coil structure is the leucine-zipper region CGN4 of the growth-control transcription factor. Using SMFS as well as SMD simulations under nonequilibrium conditions, the authors reported a detailed description of the hysteresis between unzipping and rezipping at different pulling speeds on a triple-GCN4-like leucine zipper (LZ26, Bornschlögl and Rief 2008). The energy landscape of this structure revealed four energy minima (states) in the unzipping pathway. Furthermore, the mechanical unzipping of coiled-coil structures was found to lead to a slow seed formation, starting at the cross-linked C terminus upon rezipping of the coiled coil.

Nonnative Mechanical Clamps

The mechanical resistance of a protein is usually determined only by its native structure and hydrogen bonding at the force-bearing region. However, several studies indicate that there are cases where native contacts may not suffice to predict the mechanical properties of proteins under a stretching force (Dembo et al. 1988; Forman et al. 2009; Marshall et al. 2003). This highlights the importance of intermediate transitions and nonnative states in determining the mechanical properties of proteins. The existence of bonds with an increased lifetime under force, i.e., "catch bonds," was first predicted (Dembo et al. 1988) and more recently it has been experimentally observed in P-selectin glycoprotein ligand-1 (Marshall et al. 2003) and PKD domains from polycystin-1 (Forman et al. 2009). In the latter study, specific residues were mutated to impede the formation of the nonnative interactions that provide stability against forced unfolding. As expected, the mechanical stability of the protein was dramatically reduced by these mutations, although their destabilization effect of the native state was just moderate. These nonnative interactions could resist mechanical unfolding even when native interactions have been broken, and they also depend on the magnitude and direction of the applied force.

Beyond the Mechanical Clamp

Knots

Recent studies have shown that pulling proteins from both termini does not always lead to a linear unfolded polypeptide chain. Several proteins seem to adopt knotted conformations in their native state, although the physiological function of these conformations is still unclear. SMFS provides a way to control the open ends of the knots. Thus, the mechanical stability of the figure-eight knot in the chromophore-binding domain has been recently explored in this way (Bornschlögl et al. 2009b). In that study, knotted monomers were covalently linked by means of disulfide bonds (S–Ss) to generate a polyprotein. Upon stretching, it yielded a main unfolding peak that was attributed to contacts between the protein and the small ligand tetrapyrrole. The measured contour length was shorter than the expected value for a completely stretched protein of its length, in agreement to the presence of the knot. It is assumed that this kind of structures is somehow beneficial for the stability and function of the proteins possessing it. For instance, in the case of phytochrome, it was proposed that, upon the absorption of light energy by the chromophore, this knot limits the movement of its domains so that it impairs the motion along trajectories other than the one that exposes the right-side chains (Bornschlögl et al. 2009b).

Other Mechanically Resistant Structures

Several additional structures capable of withstanding force have been recently predicted. These include shear structures between helical regions from two domains (e.g., cfr10I, the A-chain of the *Citrobacter freundii* restriction endonuclease), intradomain shear between strands stabilized by another domain (i.e., 20S proteasome subunit), tensile clamps stabilized by knot–loops (i.e., glucosidase), and clamps that oppose tensile stress by swinging two domains apart (i.e., D-crystallin) (Sikora and Cieplak 2011).

Intrinsic Elements That Modulate the Mechanical Stability of Proteins

Neighboring Strands

Other factors not directly involved in the force-bearing region of a protein have been described to play an important role in its mechanical resistance. The study of the mechanical properties of proteins with different topologies has stressed the importance of other elements in determining their mechanical response. In particular, neighboring strands can also participate in defining the mechanical stability as observed in a study of Top7 and Barstar using SMFS and SMD (Sharma et al. 2008). The mechanical clamp of these two proteins consists of two shearing terminal β-strands that do not directly establish hydrogen bonds since they are separated by a third β-strand. In spite of the fact that the pulling geometry and the force-bearing strands of the mechanical clamp of these two proteins are fairly similar, these proteins have been described to unfold through completely different mechanisms, which result in different unfolding forces. These findings support the idea that the stabilization effect of neighboring strands is an important factor.

Hydrophobic Core

The hydrophobic core has been shown to have also an effect on the mechanical stability of certain proteins. One of the studies used TNfn3, the third fibronectin type III domain from human tenascin C, combining protein engineering Φ-value analysis (to characterize the TS of the unfolding process) with SMFS and SMD simulations (Ng et al. 2005). Using mutational analysis, it was shown that mutations in some residues of the core lowered the force required to unfold the protein. The mutated residues were hydrophobic, buried, and placed away from the domain–domain interface, which was shown to interact. Mutations close to the C terminus and the E–F loop lowered the mechanical unfolding rate, and some of the other mutations affected the mechanical resistance.

Protein Loops

Protein loops are flexible regions usually related with protein thermodynamic stability. They offer an opportunity to evaluate the contribution of configurational entropy in the mechanical stability and unfolding kinetics. The effect of loop length on the mechanical response of GB1, the B1 immunoglobulin binding domain of streptococcal protein G, was studied by SMFS and SMD (Li et al. 2008). In this study, the authors used protein engineering to insert different number of amino acid residues into a loop of the protein. They found that by progressively increasing its length the mechanical stability of GB1 is decreased while the mechanical unfolding rate is accelerated (as reported by the significant decrease on unfolding forces of the mutants at different pulling speeds). Although the mutated loop does not connect any of the strands that form the mechanical clamp, the entropic cost of longer loops would result in less thermodynamically stable native states. Furthermore, since the distance to the TS, Δx_U, was found constant in most of the mutants in that study, the changes in the free energy barrier ($\Delta G_{T\text{-}N}$) observed account for the lower mechanical stability observed.

However, loop modification does not always affect the mechanical stability of a protein. In a recent study, a mutation on a loop of TNXfn7, the seventh FnIII domain of human tenascin-X, which is associated with Ehlers–Danlos syndrome, has been mechanically characterized (Zhuang et al. 2010). The phenotypical effect of this mutation is characterized by hyperelastic and fragile skin, easy bruising, and hypermobile joints, which suggest an abnormal mechanical behavior of the protein. However, the comparison of this mutant with the wt showed that, although the 3D structure was not altered, a small destabilizing effect on mechanical stability was observed in the mechanical unfolding. Simulations revealed that the only difference was a higher flexibility of the loop in the mutant, which may alter in turn the binding capacity to its ligand.

Thus, although loop regions can have important effects in protein mechanostability, their intrinsic flexibility might be of great importance also for biological function.

Other Factors

In addition to the intrinsic elements discussed so far, external factors have been shown to play a crucial role in the mechanical stability of proteins. These include ligand binding, solvent composition, and physical factors such as temperature.

Ligand Binding

In order to perform their physiological function, most proteins have to interact with other molecules, including ions, small molecules, and even other proteins (Alberts et al. 2002). It is well known that this process of ligand binding can modulate their activity and thermodynamic stability. Based on this, it is sensible to question whether it also affects the mechanical behavior of proteins, a problem that has been addressed in many studies.

The high resolution obtained with specific SMFS configurations has allowed the observation of the dynamic interaction of a single signaling protein with its ligand in real time. Thus, using a custom-built low-drift AFM, it has been possible to study ligand binding/unbinding in CaM, a eukaryotic calcium-dependent signal transducer formed by two structurally similar domains (Junker et al. 2009). In this study, the ligand-binding affinity of CaM was found to depend on the bound ligand. While bound calcium ions affect the folding kinetics of the individual CaM domains, the binding of ligand peptides stabilizes the folded structure. Furthermore, equilibrium fluctuations between numerous intermediate states of a single CaM molecule revealed the binding affinity of some peptides to only partially Ca^{2+}-saturated CaM form, confirming that the affinities of the target peptides are not exclusively determined by their pattern of hydrophobic anchor residues (Junker and Rief 2009). Different degrees of binding cooperativity were found and, by applying force to CaM–peptide complexes, single CaM molecules were unfolded revealing their folding energy landscape with a variety of equilibrium fluctuations between intermediate states. Finally, force was used to slow down the transition kinetics of CaM-target peptide complexes to observable time-scales and thus to clearly separate the individual steps of peptide unbinding.

The effects of engineering an artificial cation-binding site in a protein have also been examined. Thus, a bi-histidine motif (two histidine residues close in the structure) was created in the mechanical clamp of GB1 (Cao et al. 2008a). The preferential binding of divalent metal ions to the native state over the denatured and TS mechanically stabilized the protein in a reversible fashion. In addition to stressing the importance of contacts within the mechanical clamp of a protein to determine its mechanical stability, this study also showed a way to rationally tune it. Engineered metal chelation sites can also be used to get insight into the structure of the mechanical unfolding TS.

The capability of SMFS to measure different folding forces depending on ion binding was used also to determine the binding affinity of metal ions to proteins at the single-molecule level (Cao et al. 2009). This method has set up novel applications in the field of metal-binding systems, particularly in cases where traditional assays were difficult to apply, and has expanded the applications of SMFS. One obvious advantage is the minimum amount of protein required compared with traditional bulk binding assays. Nevertheless, it must be noted that this method is not applicable to metal ion binding systems in which this interaction does not alter the protein's mechanical stability and is also limited by the relatively low-force resolution of the standard AFM technique.

The nanomechanics of cadherins, a family of calcium-dependent cell–cell adhesion protein receptors, has been examined (Oroz et al. 2011) and the mechanical behavior and rigidity of their extracellular domain was directly related with the Ca^{2+} concentration. In the absence of this ion, the rigidity of the molecule is diminished and its mechanical unfolding response is clearly altered, displaying a highly variable unfolding pathway. This seemed to be due to the electrostatic repulsion between the negative charges of the chelating residues, which destabilize the cadherin modules in the absence of the ion. In the presence of Ca^{2+}, the mechanical unfolding followed a defined ("canalized") pathway, as assessed by a constant ΔL_C, which was consistent with the rupture of a single mechanical clamp per cadherin domain. Importantly, Ca^{2+} was shown not only to provide indirect stabilization to the molecule by balancing the electrostatic charges but also to generate a second and autonomous mechanical element, which was termed the "calcium rivet." The rupture of this element was identified in SMFS recordings as a force peak with a shorter ΔL_C that preceded the rupture of a cadherin module.

Further studies suggest that protein mechanical stability is enhanced by binding larger molecules such as enzyme substrates or inhibitors. This has been observed in the case of dihydrofolate reductase (DHFR); an enzyme that converts 7,8-dihydrofolate (DHF) to 5,6,7,8-tetrahydrofolate (THF) in the presence of the cofactor nicotinamide adenine dihydrogen phosphate (NADPH). Some drugs like methotrexate (MTX) have higher affinity for the binding site of DHFR than the original substrate. In the absence of any ligand (both the inhibitor and the cofactor), the unfolding of DHFR results in no detectable mechanical stability, while when the ligand is bound an F_U peak from DHFR appears (Ainavarapu et al. 2005). In another study, different results were found for the same protein. Thus, it was found that under mechanical force DHFR shows a metastable intermediate with a lifetime of about 8 ms, which upon binding of two substrates (an analog of MTX and the cofactor NADPH) was increased by a factor of 2 (Junker et al. 2005).

The mechanical stabilization of a protein by ligand binding is not a general phenomenon, even if a thermodynamic stabilization is observed. This has been clearly shown for the case of maltose binding protein (MBP) (Bertz and Rief 2009). MBP consists of two lobes, connected by a hinge region that switches between open and closed conformations upon maltose binding to the central cleft located between the lobes. By changing the direction of pulling from N–C to others (using cysteine replacement to construct polyproteins), it was shown that the direction of the applied force not only affected the mechanostability of the protein itself but also the effect of ligand binding on protein mechanostability. This approach changed the pulling geometry to probe barriers that are stabilized by ligand binding. Thus, when force is applied in the axis of the conformational movement that separates the two lobes of the hinge, the mechanical stability of MBP increases and it adopts an open conformation prior to unfolding when its ligand is bound. The mechanical stabilization effect was shown to be due to a few interactions between the protein and its ligand, which was demonstrated by introducing mutations that disrupted some of these interactions between the ligand and the force-bearing region of MBP.

The effects of protein–protein interactions on the mechanical stability of proteins have also been demonstrated using SMFS. A functional assay has been established in which the protein under study has different mechanical stability depending on whether the interacting protein is bound or not (Cao et al. 2008b). In comparison with the interaction of a protein with a small ligand, protein–protein interactions usually have higher affinities and larger interaction surfaces. In particular, the binding of IgG fragments (Fc and Fab) to GB1 was used as a proof-of-concept that protein–protein interaction constitutes an efficient mechanism to modulate the mechanical stability of a protein in a controlled and reversible fashion. Strikingly, the authors demonstrated that upon binding of IgG fragments, the mechanical stability of GB1 was enhanced, although those ligands did not directly interact with the mechanical clamp of GB1.

Electrostatic Forces

Electrostatic interactions have been largely studied due to its important contributions to protein conformational stability. Nevertheless, research addressing their role in the mechanical stability of proteins has just begun. To investigate this, it is necessary to distinguish the effect of electrostatic interactions due to the protonation/deprotonation of the selected residues and from other charged residues. In a recent study, the effect of pH on the mechanical stability of GB1 was investigated using an engineered pH-sensitive bi-histidine motif into its mechanical clamp. By varying the pH, histidine residues can be switched between protonated and deprotonated states, which leads to a change on the electrostatic interactions between the two histidine residues. Decreasing the pH from 8.5 to 3 resulted in a reduced mechanical stability of the engineered protein, presumably due to the increased electrostatic repulsion between the two positively charged histidine residues at this acidic pH (Zheng et al. 2011).

Solvents

Some studies have characterized the influence of solvent molecules on the TS and mechanical stability both experimentally and computationally (Guzmán et al. 2010; Li et al. 2010). These studies indicate that water insertion into the mechanical clamp is a key event in the unfolding process through the formation of a bridge that replaces an internal hydrogen bond, thus facilitating the breakage of the mechanical clamp and the unfolding of the molecule. This is consistent with the observation that a protein is more mechanostable when its mechanical clamp is solvent-hidden than when it is exposed (Guzmán et al. 2010). Also, in another line of evidence, substitutions of small residues near the mechanical clamp have been shown by simulations to facilitate water insertion resulting in lowering the mechanical resistance (Li et al. 2010).

The nanomechanical properties of proteins can also be modulated by the osmolytes present in the environment. In polycystin-1 (PC1), a large multimodular membrane protein located in the renal tubules, it has been shown that the presence of naturally occurring osmolytes can either increase or decrease its mechanical stability. The effect of urea and protective osmolytes on the first PKD domain of human PC1 has recently been studied, and it was found that a progressive increase in urea concentration (a common denaturing agent) weakened the protein, while protective osmolytes such as sorbitol and trimethylamine *N*-oxide had the opposite effect and enhanced the refolding rate. Two possible explanations have been proposed to explain these results: firstly, urea can act "indirectly" by altering hydrophobic interactions via water molecules redistribution and, secondly, it can act "directly" on the hydrogen bonds from both the backbone and side chains, resulting in a decrease of the free-energy barrier. The authors favored a direct weakening of the force-bearing hydrogen bonds, thus leading to the stabilization of the unfolded state and slowing down the refolding rate due to the binding of urea to the protein backbone (Ma et al. 2010).

In the cell, different species of macromolecules occupy a significant fraction of the volume (Minton 2001) facilitating nonspecific interactions between macrosolutes and therefore affecting the total free energy of the medium. In these environments, referred to as "crowded," excluded volume effects should be taken into account. In the case of IDPs, crowding in solutions may trigger stable structures and can prevent stress-induced aggregation and misfolding by modifying the folding kinetics. Furthermore, some proteins (i.e., DHFR, enolase, and GFP) were reported to fold spontaneously in dilute but not in crowded solutions (Martin 2002).

Recently, the effect of macromolecular crowding on protein mechanostability has been examined using SMFS. The mechanical stability of ubiquitin was found to be enhanced as dextran (a crowding agent) concentration increased in the solution. The process of force-induced unfolding of this protein could be well described by a two-state model at each dextran concentration where crowding reduces ${k_U}^0$, whereas Δx_U remains unchanged. The reduction in the unfolding rate could be explained by an increase in the free-energy barrier of the process. Thus, the presence of crowding agents might entropically discourage the expanded conformations promoting the collapsed states and hence increasing the unfolding free energy barrier height and limiting the diffusion motion of the protein molecule. Considering the solution viscosity and the internal viscosity of ubiquitin from the movement of its atoms as potential factors modulating progress to the TS during its unfolding pathway, the rate-limiting step has been proposed to involve residues shielded from the solvent rather than large-scale conformational changes (Yuan et al. 2008).

Several studies have also investigated the effect of physical parameters such as temperature on the nanomechanical properties of proteins (Schlierf and Rief 2005; Popa et al. 2011). In these publications, the mechanical stability of a protein was found to decrease as the temperature was raised. Nevertheless, it should be noted that this phenomenon does not occur through the same mechanism in all proteins.

In the case of filamin, the increase in temperature does not affect the height of the energy barrier but readily increases the distance to the TS (Schlierf and Rief 2005). However, in the case of ubiquitin, increasing the temperature significantly lowered the height of the energy barrier $\Delta G^{\ddagger}$, whereas distance to the TS, Δx_U, increased only slightly within the range of temperatures tested (Popa et al. 2011). The shift in TS position to the higher values might indicate that different molecular interactions, like hydrophobic ones, contribute mainly to the structure of the TS at higher temperatures.

By comparing the effect of different precisely controlled temperatures on the mechanical unfolding of ubiquitin, it has been also possible to accurately measure the pre-exponential factor (A in the rate equation. Sect. 5.2) of the reaction (~4 $\times$ 10^9 s^{-1}), which is three orders of magnitude smaller than that predicted by the TS theory (Popa et al. 2011). The pre-exponential factor is a key parameter that reflects the frequency of attempts to convert the reactants into products and is crucial to correctly estimate the height of the activation energy barrier.

Table 1.1 summarizes the molecular determinants of mechanical stability in folded proteins.

1.6.1.2 Protein Folding

Since the initial experiments of SMFS, special interest has been placed in applying this methodology to study protein folding. Typically the protocol, using the length-clamp mode, consisted in unfolding the protein first and, prior to detachment,

Table 1.1 Molecular determinants of mechanical stability of folded proteins

Molecular determinants of mechanical stability	Description		Underlying mechanism	Example	References
Intrinsic factors	Mechanical clamp		Patch of backbone hydrogen bonds established between two local secondary structure elements, most commonly β-strands that fasten the protein module		
	β-strands	Shear	Formed when the applied force is parallel to the force-bearing β-strands, and therefore perpendicular to the direction of the backbone hydrogen bonds of this region. Usually results in higher forces since all bonds have to be broken at a time before unfolding proceeds	I27 Ubiquitin Cohesin I	Carrión-Vázquez et al. (1999, 2003) Valbuena et al. (2009) Lu and Schulten. (1999)
		Zipper	Formed when the applied force is perpendicular to the force-bearing β-strands, and therefore parallel to the direction of the backbone hydrogen bonds. Tends to show lower unfolding forces than the shear mechanical clamp since hydrogen bonds break sequentially	C2A	Carrión-Vázquez et al. (2000)
	α-helices		Backbone hydrogen bonds majorly stabilize the helices themselves while the tertiary structure is maintained by hydrophobic interactions. This usually results in lower mechanical stabilities compared to shear and zipper mechanical clamps	Lysozyme Spectrin Coiled-coils	Peng and Li (2008) Rief et al. (1999) Bornschlögl and Rief (2008)
	Loop length		The entropic cost of closing longer loops can destabilize compact conformations such as the native state thus reducing the mechanical stability of the protein and increasing its unfolding rate.	GB1 mutants with increased loop length	Li et al. (2008)
	Hydrophobic core		Highly packed hydrophobic residues in the core of a protein can increase the mechanical stability of a protein	Module Fibronectin type3 of tenascin	Ng et al. (2005)

	Electrostatic interactions	Electrostatic interactions can increase the number of contacts that have to be broken before mechanical unfolding thus increasing the mechanical stability of the protein	Engineered GB1	Zheng et al. (2011)
	Nonnative interactions	Force-induced deformation of the native state of a protein can induce a rearrangement of its structure to form an intermediate state with new nonnative hydrogen bonds capable of resisting unfolding	PSGL-1 Polycystin-1 PKD	Forman et al. (2009)
	Knots	Some proteins acquire a knotted conformation, with open ends. When these proteins are stretched, the knot imposes a restriction to the complete unfolding of the protein	Figure-eight knot of phytochrome CBD	Bornschlögl et al. (2009a, b)
Ligands	Ions	Ion binding can modify the mechanical behavior of a protein by allowing more contacts to be established and even canalizing the unfolding pathway	CaM–Ca^{2+}	Junker et al. (2009)
		In particular cases, ion binding can result in an independent mechanical element known as calcium rivet	Cadherins–Ca^{2+}	Oroz et al. (2011)
	Small molecules	When a small molecule is bound to a protein new contacts between the protein and its ligand are established which can alter the mechanical stability of the protein	CaM–peptide DHFR–NADPH + methotrexate	Junker et al. (2009) Junker et al. (2005)
	Other proteins	Protein–protein interaction usually involves large interaction surfaces among which many additional contacts can be formed. Furthermore, long range interactions can modulate mechanical stability even if the force-bearing region is distant and not directly associated with the binding site	GB1–Fc GB1–Fab	Cao et al. (2008a)
Media	Water	Water molecules can stabilize the broken hydrogen bonds thus facilitating the unfolding of the molecule. Therefore, water insertion into the mechanical clamp is a key event in protein unfolding. The presence of large	Af1521 macrodomain Ubiquitin	Guzmán et al. (2010) Li et al. (2010)

(continued)

Table 1.1 (continued)

Molecular determinants of mechanical stability	Description	Underlying mechanism	Example	References
		groups that impede this have been identified to increase the mechanical stability of a protein		
	Osmolytes	Osmolytes can alter the hydrophobic interactions via water molecules redistribution or weakening of the force-bearing hydrogen bonds, thus preferentially stabilizing the unfolded state	Polycistin (PKD domains)	Ma et al. (2010)
		Protecting osmolytes might reduce the solvent quality for the backbone-solvent hydrogen bonds thus preferentially stabilizing the folded state and hence increasing the mechanical stability of proteins	GB1 in DMSO and glycerol solution	Aioanei et al. (2011a, b)
	Crowders	Macromolecular crowding is a common phenomenon on cellular environments which imposes excluded volume effects. In these conditions, extended conformations such as the unfolded state are disfavored against the more compact folded state. This results in higher mechanical unfolding forces	Ubiquitin in dextran solution	Yuan et al. (2008)
Physical factors	Temperature	Higher temperatures can accelerate the unfolding rate of a protein resulting in lower unfolding forces	Ubiquitin at 5–45 °C	Popa et al. (2011), Schlierf and Rief (2005)
Artificial elements		Engineering an artificial bi-histidine motif onto the mechanical clamp of a protein can chelate metal ions resulting in additional contacts between the force-bearing elements which reversibly modify the mechanical stability of a protein	Engineered GB1	Cao et al. (2008b)

reducing the tip-surface distance to relax the protein to a predefined value. Then, after a defined time period, the molecule was re-stretched and successful folding was determined according to the recovery of the unfolding pattern (namely a force peak of similar force and ΔL_C). This strategy allowed to study folding efficiency at the single-molecule level and emphasized the high resolution of the technique which could even detect rare misfolding events occurring at low frequencies (Oberhauser et al. 1999). Since then, different studies have been published describing the folding landscape of different proteins (Ainavarapu et al. 2007; Bullard et al. 2006; Ma et al. 2010; Qian et al. 2005; Schwaiger et al. 2005; Serquera et al. 2010) and the folding rate of aggregation-prone proteins for which previous classical studies have failed, such as endoxylanase from *Bacillus circulans* (Jolleymore and Li 2010).

In length-clamp experiments conducted at constant velocity, the protein length, the force, and the loading rate are dynamically changing (over wide ranges on a short time scale) during the force–extension cycle. This makes it necessary to utilize Monte Carlo models to interpret the results. The force-clamp mode of SMFS (including a variant of it, namely force-ramp, in which the force is changed linearly with time) has also been used to study protein folding and has the advantage of directly extracting the kinetic parameters. By applying a constant force, a protein can be first unfolded, then force can be lowered to allow it to re-fold as its end-to-end distance is being measured. Finally, the folding success can be assessed by checking the recovery of the unfolding pattern. In a pioneer work using ubiquitin as a model, after the initial recoil, the formation of a collapsed state with highly fluctuating end-to-end length was observed (Fernandez and Li 2004). This state lasted until the protein was completely folded, which was marked by a final length drop, which no longer showed large-length fluctuations. Although initially controversial (Sosnick 2004; Best and Hummer 2005; Brujic and Fernandez 2005; Fernandez et al. 2004), this procedure has allowed the folding of different proteins to be studied in detail and has revealed the existence of different stages during this process.

From the first experiments on polyubiquitin folding, when force was quenched to lower values, fast recoil that resulted in a collapsed state of reduced end-to-end length was observed to be the first step in protein folding (Fernandez and Li 2004). Further studies showed that ethanol, which can reduce the strength of the hydrophobic interactions, severely affects the extent of the recoil, thus resulting in more extended conformations, which failed to re-fold. These results, which were reproduced by SMD simulations where hydrophobic interactions were artificially removed, suggest that hydrophobicity is the major force driving extended proteins to the recoiled state, with entropic forces playing only a minor role (Walther et al. 2007).

According to the statistical theories of protein folding, the protein is thought to explore a progressively smaller ensemble of conformations along a rough and funneled energy surface that leads to a stable folded state (Bryngelson et al. 1995). In order to experimentally test these theories, the mechanical properties of the collapsed state were studied. Thus, force was used to probe the collapsed states of ubiquitin at different times after the force quench. Two phases could be identified from these experiments according to their unfolding rate constant and length steps (Garcia-Manyes et al. 2009a). The first phase was observed as a fast initial extension characterized by a broad step length distribution including steps even longer than the length of a single ubiquitin module, which have recently been suggested to correspond to domain swapped structures (Xia et al. 2011). These events were followed by a slower staircase phase showing the typical step size of the folded protein and a rate constant similar to the one obtained when unfolding the native protein. Importantly, the rate constants of both phases remained constant at all the studied relaxation times but the proportion of molecules in the second phase was seen to increase exponentially over time. Thus, during the folding process, the authors experimentally verified the existence of an ensemble of collapsed states that were able to convert to folded states through a barrier-activated process. To test whether these structures represent either folding intermediates or kinetic traps, the authors designed a different protocol to apply a pulse of force (low and brief enough to disrupt the ensemble of collapsed conformations but not the folded ones, Fig. 1.3b). With this approach, it was shown that disturbing the ensemble of collapsed states slows down the folding kinetics, thus indicating that these states were true folding intermediates. Taken together, these results provide evidence of the existence of the minimum energy collapsed structures that were predicted by the statistical theories of protein folding. It should be noted that due to the restriction imposed by anchoring the protein ends to surfaces, folding occurs through different pathways (Cieplak and Szymczak 2006), which still may be physiologically relevant at least for certain proteins.

The capability of SMFS to monitor the details of the folding process of proteins has been exploited in several studies to explore the mechanism by which protecting osmolytes stabilize most proteins. These small organic molecules, which have been seen to stabilize the folded state of most proteins, are thought to affect protein folding through a general mechanism known as the osmophobic effect. The underlying principle arises from a decrease of the solvent quality for the protein backbone, thus disfavoring backbone-solvent hydrogen bonding. Protecting osmolytes are expected to affect all conformations of a protein while the magnitude of their effect should be correlated to the number of backbone-solvent hydrogen bonds. Therefore, it is expected to preferentially destabilize the unfolded state (as most of its backbone is solvent exposed), barely destabilize the folded state, and destabilize the TS by an intermediate amount, thus resulting in a global thermodynamic stabilization of the folded state. In two recent studies using two protecting osmolytes, dimethyl sulfoxide (DMSO) (Aioanei et al. 2011a) and glycerol (Aioanei et al. 2011b), the stabilization of GB1 with increasing osmolyte concentrations was monitored as an increase in the mean F_U (length-clamp) or longer steps (force-clamp). Further analysis revealed a negative correlation between the unfolding rate constant and the osmolyte concentration and an increase of the unfolding energy barrier while the distance to the TS Δx_U remained constant at a similar value to that obtained in the absence of cosolvent. These results indicate that the stabilization of GB1 was due to slower unfolding and enhanced folding kinetics, although the osmolyte did not directly interact with the molecule at the TS, as indicated by the invariable Δx_U. These experiments provide strong single-molecule evidence that supports the osmophobic effect.

A completely different approach to study protein folding relies on the use of a lock-in AFM setup in which the force sensitivity of the AFM was reduced to sub-picoN range by combining a low-frequency oscillation of the sample and a slow pulling velocity. This setup allowed the authors to study the folding process of a filamin domain, for which a hopping pattern was observed and interpreted as an apparent equilibrium between the native, intermediate, and unfolded states (Schlierf et al. 2007).

1.6.1.3 Studying Chemical Reactions at the Single-Bond Level

Quantifying the effect of a mechanical force on a chemical reaction has proven to be difficult. This is mainly due to the fact that force is a vector magnitude which usually results in its heterogeneous application over uncontrolled reaction orientations. AFM–SMFS has opened a door to fully quantify the effect of a calibrated force on a well-defined system with known orientation. These experiments rely on a clever substrate design which includes engineered cysteines capable of forming a disulfide bridge. This covalent bond is not broken at the typical forces applied in an SMFS experiment and hence effectively hides the trapped amino acids between both residues to the applied force. If these cysteines are introduced in the region protected by the mechanical clamp of the protein, it is possible to obtain a substrate that can be unfolded in two steps: first the protein is partially unfolded (up to the S–S bridge) by means of force and, in a second stage, the covalent bond can be reduced, resulting in a second step that liberates the remaining trapped sequence (Fig. 1.3c). By following the characteristic length increase, one can measure the reduction of the S–S bridge as a function of force. Furthermore, by applying an initial force pulse, the mechanical unfolding can be temporally separated from the reduction events in a force-clamp experiment (Wiita et al. 2006).

Applying this approach to a polyprotein of engineered I27 modules containing a pair of solvent-hidden cysteine residues, the reduction of an S–S bridge by multiple reducing agents (DTT, TCEP, OH^-, and SH) was studied in depth (Ainavarapu et al. 2008a). The rate of reduction of the S–S has a first-order dependence on the reducing agent concentration, which indicates that the reaction is bimolecular, while force dependence is exponential. Further analysis of the data allowed the calculation of the distance to the TS of the reaction, which corresponds to the elongation of the S–S bond during the S_N2 (bimolecular nucleophilic substitution) reaction and which qualitatively correlated with quantum chemical calculations. Furthermore, this parameter seems to be comparable when using different thiols as reducing agents (Ainavarapu et al. 2008a) but it differs significantly when using other reducing agents such as phosphines, which proceed through a different and irreversible mechanism.

It is interesting to note that when forces exceeding 500 pN are applied to an S–S, an abrupt shift in the TS of the reaction can be observed (at least with TCEP and hydroxide anions), which probably reflects a conformational change in the substrate (Garcia-Manyes et al. 2009b). Although length-clamp has not allowed information to be obtained on the reduction of S–S bridges, this mode has also been applied in order to study the bond exposure to solvent according to the characteristic length patterns generated by the substrates (Ainavarapu et al. 2008b).

By comparing the effect of different reducing agents at different precisely controlled temperatures, it has been possible to accurately measure the pre-exponential factor of the reactivity of the protein-embedded S–S (~10^7 M^{-1} s^{-1}). Surprisingly, the obtained value was significantly lower than the theoretical one, which implies that, apart from simple collisions, other factors such as desolvation play a role in this reaction (Liang and Fernandez 2011).

All these results show that AFM–SMFS can readily provide important information on the mechanism of a reaction even at the single-bond level. However, this application is not limited to the study of simple chemical reactions but can also be easily extended to get a deeper insight into the catalytic mechanism of biological catalysts (Fig. 1.3c). Thus, the force dependence of the S–S reduction by thioredoxins (broadly distributed enzymes that catalyze this reaction) has been also studied by substituting the reducing agent by the enzyme. *Escherichia coli* and human thioredoxins were originally studied. Unlike chemical reducing agents, the reaction rate of both thioredoxins was seen to decrease as a function of the applied force. This dependence was modeled according to a Michaelis–Menten mechanism with an intermediate state where the enzyme is bound to the substrate before the S_N2 reaction occurs (Wiita et al. 2007).

Strikingly, when the applied force is raised above 200 pN, a completely different force dependence could be observed. For the human thioredoxin, the reaction rate became force-independent, which was later interpreted as a switch toward a single-electron transfer mechanism (Perez-Jimenez et al. 2009). On the other hand, the *E. coli* thioredoxin reduction rate was accelerated as a function of the applied force, as in the S_N2 mechanism observed when using chemical reducing agents. By studying thioredoxins from multiple organisms, the observed mechanisms in human and *E. coli* thioredoxins were later generalized to the behavior of this type of enzymes in eukaryotes and bacteria, respectively. More recently, the sequence of seven precambrian thioredoxins were reconstructed and their catalytic properties probed by SMFS. This allowed a broader generalization of the different mechanisms of these enzymes, associated to two evolutionary lines of organisms: one for bacteria and a second for archaea and eukaryotes (Perez-Jimenez et al. 2011).

Finally, in a recent study with a substrate design that includes a second pair of engineered cysteins (Fig. 1.3d), the disulfide isomerization in a single protein has been studied (Alegre-Cebollada et al. 2011). This is of great biological importance since it is a highly dynamic process in physiological conditions that plays a key role in the folding process of many proteins. The results obtained showed that this isomerization is very region-specific and occurs at a rate that could interfere with other redox processes in the cell.

1.6.2 SMFS of IDPs

1.6.2.1 What Are IDPs?

Various kingdoms of life have biologically active proteins that lack, in vitro, a well-defined 3D structure under physiological conditions (Dunker et al. 2000a), but they often adopt a highly structured conformation upon binding to their biological partners to perform their cellular functions (Wright and Dyson 1999). These are IDPs, proteins with a distinctly different amino acid composition relative to that of folded proteins, which allows their identification by computational algorithms even at the proteome level. Recent computational predictions based on 29 genomes indicate that more than 30% of eukaryotic proteins contain disordered regions of 50 consecutive amino acid residues or more, and this percentage is much higher than those found in either bacteria or archaea. The fraction of the proteome that includes all IDPs, "unfoldome," provides essential biological functions and is thus a critical part of the whole proteome (Dunker et al. 2000b), which complements the functions of ordered proteins including crucial roles in eukaryotic signaling (Uversky et al. 2005). Furthermore, numerous IDPs have been shown to be associated with various human conformational diseases, such as neurodegenerative diseases, cancer, cardiovascular disease, or diabetes, a link that motivated the so-called "disorder in disorders" or D^2 concept (Uversky et al. 2008). According to this idea, predicting and analyzing such disorder should help us understand protein functions and conformational diseases.

1.6.2.2 How to Study IDPs?

IDPs share several distinctive structural properties that can be used to identify them by means of biochemical and traditional structural methods such as the sensitivity to protease digestion, insensitivity to denaturing conditions, X-ray crystallography, NMR spectroscopy, circular dichroism spectroscopy, and Stoke's radius determination using either small angle X-ray scattering or size exclusion chromatography. Interestingly, IDPs could also be easily discriminated from ordered proteins using just computational tools based on plotting the average hydrophobicity of the residues in the sequence against the net charge of the residues (Uversky et al. 2009).

For the case of IDPs implicated in amyloid diseases (aka amyloidoses: disorders in which amyloid aggregates are formed), the conformational fluctuations in the monomer play an essential role in promoting the formation of early aggregates, which are critically associated to disease. The characterization of these conformational fluctuations in amyloidogenic proteins is fiendishly difficult for ensemble biophysical techniques. On the other hand, SMT are well suited for the task as they provide an unprecedented resolution that is extremely useful for investigating structural distributions and dynamics of complex biological systems. Two of them have been particularly useful in investigating monomeric IDPs (a) single-molecule fluorescence (SMF) methods (single-molecule fluorescence resonance energy transfer and fluorescence correlation spectroscopy), used to investigate the structure and dynamics of monomeric IDPs (Ferreon et al. 2010), and (b) SMFS, specifically the AFM-based one, which has been used to investigate the (mechanical) conformational polymorphism of these species (Hervás et al. 2012).

1.6.2.3 SMFS Results on IDPs

This section summarizes the SMFS studies on both nonamyloidogenic and amyloidogenic IDPs.

Nonamyloydogenic IDPs

Tropoelastin and Elastin

Elastin is a protein that enables the reversible deformation of elastic tissues in all mammals (Li et al. 1998). Its soluble precursor, tropoelastin, is frequently described as an unstructured protein, specifically an "asymmetric" coil, with an appended C-terminal cell-binding foot.

By stretching and relaxing single tropoelastin molecules several times by AFM–SMFS, its ability to return to its original state was studied. Tropoelastin showed a single-chain elasticity curve pattern that could be fit to the WLC model, and since no hysteresis was observed after repeated cycles this structure was suggested to act as a highly efficient molecular nanospring (Baldock et al. 2011).

Elementary components of α-elastin [synthetic penta-polypeptides $(GVGVP)_{251}$ and $(GVGIP)_{260}$] were also studied using SMFS observing single-chain force–extension curves from which these polypeptides were described as ideal elastomers behaving also as an entropic spring in equilibrium with no hysteresis below certain temperature for each composition (Urry et al. 2002).

The mechanochemical and hydration properties of elastin-like-polypeptides (ELPs) have also been studied at the single-molecule level. In this work, the effect of solvent quality, temperature, and type of the so-called "guest" residue (X is any residue except proline in the VPGXG sequence) on the force–extension behavior was measured for three different ELP polypeptides (with different repeats and guest residues): ELP1-180 (180 pentapeptide repeats), ELP4-120, and ELP2-160. The authors concluded that the hydrophobic hydration modulates the molecular elasticity for the different ELPs studied, in qualitative agreement with theoretical predictions (Valiaev et al. 2008).

PEVK

The elastic region of titin (responsible for the passive elasticity of muscle) is the so-called I-band, which consists of a region of immunoglobulin module repeats and two unstructured regions: PEVK and N2B. The PEVK domain contains coiled structures that are elongated when the muscle is stretched.

To analyze the nanomechanical properties of the cardiac 188-residue PEVK, heteromeric polyproteins containing a mechanical marker (I27 module) were engineered. In contrast to other random coils, PEVK displayed a range of different persistence lengths from 0.3 to 2.3 nm (average of 0.91 nm). This variability was suggested to be originated from varying degrees of proline isomerization (Li et al. 2001). Furthermore, since different muscles have different alternatively spliced forms of the PEVK region (made out of 116 exons), there was the possibility that the elasticity of each titin form in each muscle was due to the different content in proline (flexibility cassette hypothesis). This hypothesis was tested by analyzing the nanomechanical properties of thee different PEVK exons that varied significantly in their proline content, and which surprisingly showed similar p values. Hence, it seems that alternative splicing only serves to adjust the length of the PEVK domain of titin (Sarkar et al. 2005).

In a different study, the distribution of p was found to be multimodal, with three p modes, 1.4, 0.8, and 0.45 nm, which were interpreted as originated from 1, 2, and 3 molecules, respectively (Watanabe et al. 2002).

Despite which the interpretation of the results was, this analysis has revealed that the PEVK region behaves as an entropic spring.

N2B

The N2B-unique sequence from titin I band is found exclusively in the cardiac muscle and was shown to extend toward the upper limit of the physiological sarcomere length range, contributing to the passive-tension generation of cardiac myofibrils (Helmes et al. 1999; Linke et al. 1999). This region was characterized by SMFS using heteromeric polyproteins that contained this region of 572 residues. Stretching the N2B-construct resulted in a single p (0.66 nm; Li et al. 2002). Still, in a different study, two values of p were found, 0.6 and 0.3 nm, which were interpreted as originating from 1 or 2 molecules, respectively (Watanabe et al. 2002). N2B showed the absence of force peaks in the force–extension curves and the stretch-release cycles were found to be largely reversible following a WLC behavior that suggests that mechanically N2B behaves like an entropic spring.

The nanomechanics of phosphorylated N2B was also analyzed, based on the fact that N2B can be phosphorylated by protein kinase A (PKA), which causes a drop in myocardial passive stiffness (Yamasaki et al. 2002). However, no mechanical effect was found (Leake et al. 2006).

Furthermore, it has been postulated that additional modulatory effects on titin stiffness may arise from disulfide bonding under oxidant stress, as many Ig domains in titin's I-band region can potentially form S–S bonds. In this context, the cardiac-specific N2B was analyzed using SMFS in oxidizing conditions. These results revealed the existence of molecules of a shorter L_C, consistent with the presence of intramolecular S–S bond (s), which is (are) not typically broken in SMFS experiments (Grützner et al. 2009).

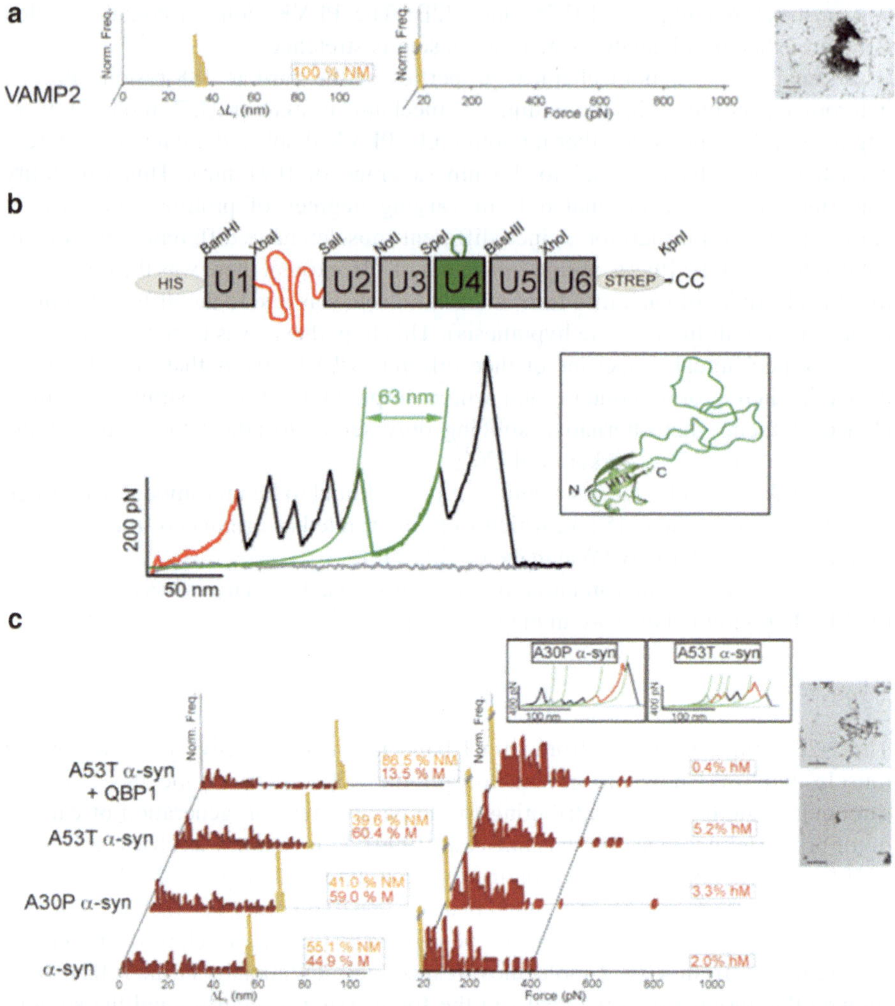

Fig. 1.4 Nanomechanics of IDPs by AFM–SMFS. (**a**) Nanomechanical analysis of pFS-2 +VAMP2. ΔL_C (*left*) and *F* (*right*) histograms of pFS-2 polyproteins carrying VAMP2. This nonamyloidogenic IDP does not show conformational polymorphism. The scale bar from the TEM image on the *right* corresponds to 0.6 μm. Figure taken with permission from Hervás et al. 2012. (**b**) Schematic representation of the pFS-2 protein used in the carrier-guest strategy. The ubiquitin repeats are represented as *gray boxes* (U), followed by the number of the repeat in the construct. A N2B fragment is represented in *red* as a nonfolded polypeptide. A multicloning site (MCS) inserted into loop A–B of the cDNA coding for ubiquitin (between residues T9 and G10), is represented in *green* (*upper panel*). Typical force–extension recording of the pFS-2 + VAMP2 protein carrying (in this order) the N2B region (*red*), unfolding peaks from four ubiquitin repeats (*black*) and a peak from the "carrier" ubiquitin (*green*) that shows a larger ΔL_C due to the inclusion in its fold of the MCS and the VAMP2 protein. Using this strategy, SMFS data obtained from the IDP are always registered far from the proximal region (*lower panel*). Figure taken with permission from Oroz et al. 2012. (**c**) Nanomechanical analysis of α-syn using the pFS-2 strategy. ΔL_C (*left*) and *F* (*right*) histograms of pFS-2 polyproteins carrying α-syn. The wild-type protein

VAMP2

The disordered cytoplasmic region of the synaptic vesicle protein VAMP2 (synaptobrevin) has been also recently studied by SMFS using a new strategy that overcomes the noise of the proximal region (i.e., the carrier-guest strategy based on the pFS-2 vectors; see below). This region showed an RC-like behavior (Fig. 1.4a), with no detectable force peaks and a single value of p around 0.4 nm, which was similar in two different carriers used (I27 module and ubiquitin; Hervás et al. 2012).

Amyloid-Related IDPs

α-Synuclein

α-Synuclein (α-syn) is a 140-amino acid protein, expressed mainly in the presynaptic terminals of the central nervous system, the function of which is not yet well known, although it is thought to participate in the endoplasmic reticulum-Golgi vesicle trafficking. Missense mutations or overexpression of α-syn has been related to the pathogenesis of several neurodegenerative diseases, termed synucleopathies, like Parkinson disease and multiple system atrophy, which are usually characterized by the presence of intracellular aggregates (Lewy Bodies) rich in β-structure. There is evidence that the monomeric α-syn, at least in vitro under physiological conditions, populates an ensemble of conformations that includes extended conformers and structures that are more compact than those expected for a completely unfolded chain (Kim et al. 2007; Lee et al. 2004).

The initial SMFS studies on α-syn were done with the standard approach of heteromeric polyproteins (i.e., a single α-syn flanked on either side by three tandem I27 domains), which has proved to be a valid strategy for nonamyloidogenic IDPs. However, amyloidogenic proteins like α-syn seemed to be able or form a variety of structures evidenced by mechanical conformational polymorphism (Sandal et al. 2008; Brucale et al. 2009). This polymorphism is likely to be contaminated by the noise of the proximal region of the AFM (a region prone to nonspecific interactions) which compromises the reliability of the data obtained. In order to get data free of this noise, a new strategy was later used to revisit the nanomechanics of α-syn. This strategy was termed the "carrier-guest" strategy (Fig. 1.4b, Oroz et al. 2012) and is based on a mechanical protection of the protein of interest inside the force-hidden region of a mechanical marker. By using this strategy, the SMFS data obtained from

Fig. 1.4 (continued) (*first row*) exhibits a wide-range polymorphism with NM conformers (*orange bars*) and M conformers (*red bars*). Familial-disease mutations A30P and A53T (*middle rows*) increase the frequency of M and hM. Treatment with QBP1 peptide (*last row*) reduces the frequency of M and hM conformers in A53T α-syn. Transmission electron microscopy images of the amyloid fibers are shown on the *right*. Amyloid fibers are clearly not formed in the presence of QBP1 (*top*). From *bottom* to *top*, the *scale bars* correspond to 0.45 and 0.6 μm, respectively. Examples of hM conformers of A30P and A53T α-syn are shown in the insets (Taken with permission from Hervás et al. 2012)

the neurotoxic protein are always read far from the unreliable proximal region. In a recent study, the authors overcome also two additional limitations of previous studies: the absence of a control to demonstrate the amyloidogenic behavior of the neurotoxic protein fused to the polyprotein and the absence of a control to show that the data originate from intramolecular interactions in the monomer instead of intermolecular interactions between monomers (Hervás et al. 2012).

Thus, using AFM–SMFS in the length-clamp mode, it has been possible to describe unequivocally the conformational polymorphism of wild-type and several mutants of α-syn, as well as other neurotoxic proteins. The SMFS analysis showed that the wild-type α-syn exhibited a rich conformational polymorphism (in force and length), which included two main classes of events: nonmechanostable conformers (NM) and mechanostable (M) conformers (Fig. 1.4c). Furthermore, a subset of the M conformers presented an extremely high resistance to stretching (these conformers were termed hyper-mechanostable (hM) conformers, and were operatively defined as those M conformers showing a mechanical stability higher than 400 pN, at 0.4 nm ms^{-1}). Frequently, more than one force peak was found for each molecule, suggesting the presence of more than one structured region in the M conformers. The work also shows that all studied familial-disease mutations of the α-syn (A30P and A53T) promoted an increase in the frequency of M conformers, an effect that is correlated with the aggregation/fibrillogenesis capabilities of the neurotoxic protein (Fig. 1.4c). Finally, it was demonstrated that QBP1 (polyQ binding peptide 1), an inhibitor peptide originally developed for expanded polyQ containing proteins (Nagai et al. 2007), strongly diminished the formation of M conformers over WT α-syn and over its mutants, with concomitant reduction of aggregation and fibrillization capabilities (Fig. 1.4c). Strikingly, the same effect was also observed for other neurotoxic proteins (polyQs and Sup35NM), which suggests common molecular mechanisms at the monomer level in the initial stages of amyloidogenesis (Hervás et al. 2012).

Polyglutamine Tracts

Homopolypeptide repeats are very common in eukaryotes, with the amino acids glutamine (Q) and alanine (A) as the most common uninterrupted repeats (Faux et al. 2005). Huntington's disease (HD) and other diseases (e.g., spinocerebellar ataxias) are linked to the insertion of glutamines into a specific protein, such as huntingtin in HD, which results in polyglutamine (polyQ) expansions that self-associate to form aggregates (Cummings and Zoghbi 2000).

Recent experimental and computational studies have shown that polyQ chains are very insoluble in water and form collapsed structures (Crick et al. 2006; Vitalis et al. 2008a, b). The conformational stability of these collapsed structures was examined by SMFS using standard heteromeric polyproteins and in the absence of the two additional aforementioned controls, obtaining some puzzling results (Dougan et al. 2009). The use of the carrier-guest strategy and the appropriate controls allowed examining unambiguously the nanomechanics of this protein. Thus, monomeric polyQ tracts of three different lengths, i.e., sub-, near-, and super-threshold, for triggering Huntington disease, Q_{19}, Q_{35}, and Q_{62}, were examined. A positive correlation between the number of Q repeats, the level of amyloidogenesis observed, and the frequency of the formation of M conformers (including the hM ones) was found. QBP1 peptide inhibited the formation of M species, as well as over β-structure acquisition and fibrillation process, which was demonstrated by structural techniques (Hervás et al. 2012).

Sup35NM

The *Saccharomyces cerevisiae* epigenetic element [PSI$^+$] is the prion form of the translation termination factor subunit Sup35p (Cox 1965), which arises from conversion of soluble active monomers to an inactive amyloid (King et al. 1997) leading to a change in the yeast metabolic phenotype. This prion has been extensively used as a model system to study human prions.

Using the length-clamp mode of SMFS and the carrier-guest strategy, a high conformational polymorphism in Sup35NM has been found, with 64% of the M conformers. Again, QBP1 strongly inhibited aggregation and fibrillization as well as the formation of M and hM conformers (Hervás et al. 2012).

Amyloid β

Amyloid β (Aβ) is a peptide of variable length (36–43 amino acids) associated with Alzheimer's disease that is processed by the cellular machinery from the amyloid precursor protein. Aβ is the main component of amyloid plaques, and recent research suggests that soluble oligomeric forms of the peptide may be causative agents in the development of Alzheimer's disease (Shankar et al. 2008).

Aβ has not been crystallized. Structural knowledge on this peptide comes from NMR-guided simulations and MD studies, which suggest that Aβ peptides are IDPs that display a high conformational polymorphism (Sgourakis et al. 2007; Yang and Teplow 2008; Zhang et al. 2000). The SMFS study, using the length-clamp mode and the carrier-guest strategy, of the conformational polymorphism of the Aβ42 peptide showed the existence of M and hM conformations that coexisted with RC conformers. This neurotoxic protein was found not to be inhibited by QBP1. Two mutants were also studied by this technique. The Arctic mutant (Arc; E22G), which has been implicated in familial Alzheimer's disease, showed an increased polymorphism. On the other hand, the fibrillogenesis-defective mutant F19S/L34P Aβ behaved as an RC. Furthermore, the effect of SV111, which inhibits oligomerization and fibrillogenesis of Aβ42 but does not affect the conformational fluctuations of the unbound monomers (Arslan et al. 2010), was also studied as a control to rule out a significant contribution of intermolecular interactions to the SMFS data (Hervás et al. 2012).

Tau

Tau is a microtubule-associated IDP that stabilizes neuronal microtubules under normal physiological conditions. Defective Tau proteins may no longer stabilize microtubules properly, resulting in dementias that are termed taupathies (Kidd 1963). In these diseases, both binding to microtubules and amyloid fibril assembly are mediated through the Tau repeat domain (Hirokawa et al. 1988).

SMFS was used to examine this protein. Although we must be aware that most data in this study were obtained without single-molecule markers (only an external polyprotein was used for comparison) and that the noise of the proximal region could not be ruled out, the alternative use of careful controls and comparative studies seems to allow the authors to draw some conclusions, namely, that the protein seems to display three conformations that can be modulated by proaggregant and antiaggregant mutants as well as ionic strength and heparin (Wegmann et al. 2011). However, no RC conformations were identified. As with the previous neurotoxic proteins, these data should be revisited using the carrier-guest strategy to confirm or refute them, mostly considering that they depart so much from the behavior of the other neurotoxic proteins studied so far by the carrier-guest strategy.

In Table 1.2 we summarize the nanomechanics of IDPs, both nonamyloidogenic and amyloidogenic.

1.7 *In Silico* Force Spectroscopy

Although experimental work gives us a lot of information, as seen in this review, it fails when it comes to providing the atomic details of the processes taking place in the experiments. Computer simulations, especially MD, are used in order to get this important information.

The predictive power of these simulations in the field (even of those coarse-grained) of protein nanomechanics has been truly remarkable. To name a few, the list of predictions includes the following: the existence of both the mechanical clamp and the I27 intermediate were predicted before and independently from the SMFS experiments (Lu et al. 1998), relative mechanical stabilities of the α-helix and the β-sheet (both in shear and zipper configurations) (Sikora et al. 2009), or relative mechanostabilities of cohesins, the most mechanostable protein structures reported to date (Valbuena et al. 2009).

The idea behind this kind of simulation is simple: in order to obtain the information on the whole system, the interactions between each pair of its components are studied, as well as their interactions with the environment, thus computing the movement of each particle. In the case of proteins, different models can be adopted, ranging from simulating each and every atom in the system (including not only the ones that belong to the protein but also solvent molecules and other species that might be present) to considering a group of atoms as a single entity.

Table 1.2 The nanomechanics of disordered proteins

Protein	Cellular Function	Amyloid disease related?	Nanomechanical behavior	References
Tropoelastin	Precursor to the elastin molecule		Random coil behavior. Individual tropoelastin molecules are highly extensible yet elastic without hysteresis. *p*: 0.36 ± 0.14 nm	Baldock et al. (2011)
α-elastin Elastin-like polypeptides (ELPs)	Source of elasticity in vertebrate tissues (i.e., arteries, vocal cords, skin and certain cartilages)		Random coil behavior. This protein behaves as an entropic spring in equilibrium, showing no hysteresis below certain temperature. *p*: 0.4 at lower forces and 0.6 at higher forces for $(GVGVP)_{251}$0.7 with 20 % error for $(GVGIP)_{260}$	Urry et al. (2002)
			Hydrophobic hydration modulates the molecular elasticity of EPLs	Valiaev et al. (2008)
PEVK	Entropic spring in titin (I band). Responsible for the global elastic properties of the molecule		Random coil behavior. *p*: 0.3–2.3 nm. Average: 0.91 nm	Li et al. (2001)
			Random coil behavior. Different values of *p*: 0.89 ± 0.42 nm (exon 120) 0.92 ± 0.38 nm (exon 161) 0.98 ± 0.4 nm (exon 184)	Sarkar et al. (2005)
			Random coil behavior. Multimodal *p*: 1.4, 0.8 and 0.45 nm	Watanabe et al. (2002)
N2B	Entropic spring in cardiac titin (I band). Responsible for the global elastic properties of the molecule		Random coil behavior. *p*: 0.66 nm	Li et al. 2001
			Random coil behavior. *p*: 0.6 and 0.3 nm	Watanabe et al. (2002)
VAMP2 (cytoplasmatic region)	Component member of a protein complex involved in the docking and/or fusion of synaptic vesicles with the presynaptic membrane		Random coil behavior. It shows a single *p* around 0.4 nm	Hervás et al. (2012)

(continued)

Table 1.2 (continued)

Protein	Cellular Function	Amyloid disease related?	Nanomechanical behavior	References
α-synuclein and variants	Not well defined: it is assumed to be involved in the functioning of the neuronal Golgi apparatus and vesicle trafficking, act as a molecular chaperone in the formation of SNARE complex and may have an activity as potential microtubule-associated protein	Synucleopathies, like Parkinson disease and Multiple System Atrophy	Rich conformational polymorphism including two main classes of events: non mechanostable conformers (NM) and mechanostable (M) conformers. Polymorphism increased by familial mutations and strongly diminished by QBP1	Hervás et al. (2012)
PolyQ tracts	Unclear: i.e., Huntingtin is essential for development, upregulates the expression of BDNF, mediate endocytosis and may play a role in cytoskeletal anchoring or transport of mitochondria	Huntington's disease, Spinocerebellar Ataxias, Kennedy disease or Spinobulbar Muscular Atrophy and Dentatorubropallidoluysian Atrophy	Rich conformational polymorphism in expanded PolyQ tracts (>35 Qs), strongly diminish by QBP1. Q_{19} Behaves as a random coil	Hervás et al. (2012)
Sup35NM	Translation termination factor (eRF3) in yeast (C-terminal domain)		Rich conformational polymorphism, strongly diminish by QBP1	Hervás et al. (2012)
Amyloid-β and variants	Several potential activities including activation of kinases, protection against oxidative stress, regulation of cholesterol transport or functioning as a transcription factor	Alzheimer's disease	Rich conformational polymorphism: no effect by QBP1 and increased by the Arctic familial mutation	Hervás et al. (2012)
Tau	Interact with tubulin to stabilize microtubules and promote tubulin assembly into microtubules	Taupathies as Alzheimer's disease, frontotemporal Dementia and frontotemporal lobar degeneration	Conformational polymorphism, with three main folds. Modulated by mutations, electrolyte concentration and Heparin	Wegmann et al. (2011)

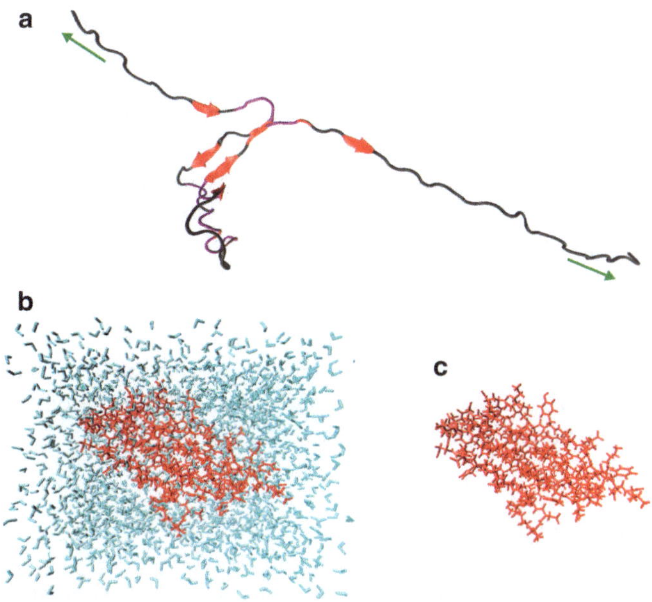

Fig. 1.5 *Top*: representation of a Steered Molecular Dynamics simulation (**a**). *Green arrows* represent the pulling force involved in the process of mechanical unfolding. *Bottom*: representation of the I27 module of human cardiac titin (PDB code: 1tit) with explicit water molecules (**b**) and with implicit solvent (**c**). This figure was done using Visual Molecular Dynamics, VMD (Humphrey et al. 1996)

In order to reproduce the experiments *in silico*, a methodology termed steered molecular dynamics (SMD, Fig. 1.5a) was conceived early in the single-molecule field (Lu and Schulten 2000; Sotomayor and Schulten 2007). This development of MD involves applying an external force on the molecule under study and monitoring the behavior produced. SMD can reproduce both length-clamp and force-clamp experiments, by computationally attaching a spring on two of the atoms in the system and either pulling them apart at a constant speed or maintaining a controlled force between them, respectively.

1.7.1 All-Atom MD

The most precise way to describe the properties of a system *in silico* is to divide it down to the smallest possible entities. Therefore, given that the physical properties of each of the atoms present in a sample are known, the forces between each of them can be computed so as to know the movements that take place inside the molecule and also the movement of each of the molecules in the system as a whole.

This can be achieved by performing all-atom MD, which, as the name indicates, simulates all the atoms present in the system. Thus, in a minimal cubic box with a protein module such as I27 surrounded by water molecules, there are around 7,000 atoms (Fig. 1.5b) for all of which the position, velocity, and interaction with the others need to be computed at every step (typically of the order of 1 fs). Therefore, this system, although very precise, involves many calculations and makes the simulation run very slowly.

In an attempt to gain some velocity at the expense of a low information loss, the solvent molecules were firstly treated as single entities with established properties concentrated at the center of mass of the water molecules. In a similar approach, since hydrogen atoms have such a low mass and therefore have very fast fluctuations, their position in the molecule was computed through other algorithms (i.e., SHAKE), which allowed calculations to speed up at a low informational cost. Both of these simplifications are nowadays widely used in the field, and they indeed made an improvement in the computational speed, but simulation timescales are still limited to the nanosecond time scale.

1.7.1.1 Implicit Solvent Models

In a further attempt to reduce the computational cost of these simulations, and only in cases where the solvent is not a crucial part of the study, the solvent molecules can be eliminated from the system, thus permitting a serious reduction of the number of atoms in the system (from 7,000 to 1,000 in a molecule like I27, Fig. 1.5c). To do this, a model needs to be used to account for the effect of the solvent molecules in the electrostatics of the system [i.e., generalized Born surface area (Still et al. 1990)]. Given that the calculation speed scales like the square of the number of particles, this reduction improved the performance of the simulations and permits the study of much bigger molecules (Oroz et al. 2011).

In spite of this increase in speed, the computations with implicit solvent are not yet fast enough to reach experimental time, and SMD needs to be carried out at a very high speed [typically 1 Å ps^{-1}, about 10^9 times faster than experiments (Sotomayor and Schulten 2007)], so experimental and computational results still cannot be directly compared.

1.7.2 Structure-Based or Go Models

Yet another simplification that can be done to study this kind of systems is to reduce the complexity of the force fields used in the computations to a model based on a simple assumption: those contacts that are found in the native structure are valid (attractive), while any other contact is not valid (repulsive). This kind of approach leads to much faster calculation, but provides less dynamic information on the system and can lead to wrong results if the molecule under study forms nonnative contacts upon stretching (Galera-Prat et al. 2010), unless they are properly taken into account (Wallin et al. 2007). However, SMD simulations using Go modeling was very successful in identifying mechanical-resistant elements in the structure of proteins such as shearing β-sheets, cystine slipknots (Sikora et al. 2009), and tensile mechanical clamps (Sikora and Cieplak 2011).

Structural models, as mentioned, do not take into account the real properties of the system, but the contacts present in the native structure. Thus, the dynamics observed is, in general, less realistic than those in all-atom MD. However, the gain in time is so substantial that SMD has been carried out at speeds comparable to experiments (Sikora et al. 2009—results are stored in http://info.ifpan.edu.pl/BSDB/) and it has even been possible to study protein refolding (i.e., unfolding and recovery of a module's structure), which is still far beyond the reach of all-atom MD (Cieplak and Szymczak 2006).

1.7.3 Molecular Dynamics of IDPs

The study of molecules *in silico* typically begins with the native structure as a starting point. Therefore, the study of IDPs using MD is not easy, since these molecules fluctuate in a complex equilibrium of structures and therefore they do not have a single native state to start from. This forces us to exclude the use of Go models for IDPs, and, as noted earlier, the use of all-atom MD significantly slows down the calculation time.

However, simulations have indeed been carried out with IDPs. Polyglutamine models have been developed to test whether they aggregate (Marchut and Hall 2006), as they are known to behave in vitro and in vivo; and also to check their structural transition before aggregation (Khare et al. 2005). Furthermore, the N terminus of p53 has recently been studied *in silico* using all-atom models with explicit water (Xiong et al. 2011), and a coarse-grained potential model for IDP simulations has been developed and tested on this protein (Terakawa and Takada 2011).

1.8 Future Perspectives

Since its invention, AFM–SMFS has emerged into a powerful tool for the understanding of many properties of proteins, which now include mechanical properties, interactions with other biomolecules, protein folding, and catalysis. An area where this technique holds considerable potential is to investigate the dynamics of IDPs, which still poses nontrivial challenges for current approaches.

An exciting new challenge for the single-molecule researchers is to perform in vivo single-molecule assays, with promising initial results (Brandenburg and Zhuang 2007; Xie et al. 2008). To meet this challenge, some technological developments are needed, which involve the combination of force-spectroscopy techniques with other SMT, in particular SMF detection approaches (Ishijima et al. 1998; Lang et al. 2004; Li et al. 2007). Finally, instrumentation development with enhanced spatial and temporal resolution and further development in computational biology are critical technical improvements that will grow in importance in the future, as the complexity of the biological questions increases. Also, to improve the efficiency of data acquisition as well as to get stronger attachment points, more controlled bio-functionalization protocols are needed.

Through future research and with the information unveiled by these techniques in combination with the information derived from bulk and computational experiments, it will be possible to obtain new insight on protein dynamics and structure as well as on the interactions that occur at the cellular level making life possible.

Acknowledgments We thank Douglas Laurents (Instituto Química-Física Rocasolano, IQFR-CSIC), Andrés Oberhauser (University of Texas Medical Branch), Marek Cieplak (Polish Academy of Science), Javier Oroz (Centro de Investigaciones Biológicas, CIB-CSIC) and Andrés M. Vera (Instituto Cajal, IC-CSIC) for the critical reading of the manuscript. This work was funded by grants from the Ministerio de Ciencia e Innovación (MICINN-BIO2007-67116), the Consejería de Educación de la Comunidad de Madrid (S-0505/MAT/0283), and the Consejo Superior de Investigaciones Científicas (CSIC- 200620F00) to MCV. We apologize to all researchers whose pioneering work was not cited due to limitations of space.

References

Ainavarapu SRK, Li L, Badilla CL, Fernandez JM (2005) Ligand binding modulates the mechanical stability of dihydrofolate reductase. Biophys J 89:3337–3344

Ainavarapu SRK, Brujic J, Huang HH, Wiita AP, Lu H, Li L, Walther KA, Carrión-Vázquez M, Li H, Fernandez JM (2007) Contour length and refolding rate of a small protein controlled by engineering disulfide bonds. Biophys J 92:22–233

Ainavarapu SRK, Wiita AP, Dougan L, Uggerud E, Fernandez JM (2008a) Single-molecule force spectroscopy measurements of bond elongation during a bimolecular reaction. J Am Chem Soc 130:6479–6487

Ainavarapu SRK, Wiita AP, Huang HH, Fernandez JM (2008b) A single-molecule assay to directly identify solvent-accessible disulfide bonds and probe their effect on protein folding. J Am Chem Soc 130:436–437

Aioanei D, Lv S, Tessari I, Rampioni A, Bubacco L, Li H, Samorì B, Brucale M (2011a) Single-molecule-level evidence for the osmophobic effect. Angew Chem Int Ed 50:4394–4397

Aioanei D, Tessari I, Bubacco L, Samorì B, Brucale M (2011b) Observing the osmophobic effect in action at the single molecule level. Proteins 79:2214–2223

Alberts B, Johnson A, Lewis J, Raff M, Roberts K, Walter P (2002) Molecular biology of the cell, 4th edn. Garland Science, New York, NY

Alegre-Cebollada J, Kosuri P, Rivas-Pardo JA, Fernández JM (2011) Direct observation of disulfide isomerization in a single protein. Nat Chem 3:882–887

Arslan PE, Mulligan VK, Ho S, Chakrabartty A (2010) Conversion of Aβ42 into a folded soluble native-like protein using a semi-random library of amphipathic helices. J Mol Biol 396:1284–1294

Baldock C, Oberhauser AF, Ma L, Lammie D, Siegler V, Mithieux SM, Tu Y, Chow JY, Suleman F, Malfois M, Rogers S, Guo L, Irving TC, Wess TJ, Weiss AS (2011) Shape of tropoelastin, the highly extensible protein that controls human tissue elasticity. Proc Natl Acad Sci USA 108:4322–4327

Bell GI (1978) Models for the specific adhesion of cells to cells. Science 200:618–627
Bertz M, Rief M (2009) Ligand binding mechanics of maltose binding protein. J Mol Biol 393:1097–1105
Best RB, Hummer G (2005) Comment on "Force-clamp spectroscopy monitors the folding trajectory of a single protein". Science 308:498b
Bilsel O, Matthews CR (2006) Molecular dimensions and their distributions in early folding intermediates. Curr Opin Struct Biol 16:86–93
Binnig G, Rohrer H (1986) Scanning tunneling microscopy. IBM J Res Devel 30:355–369
Binnig G, Quate CF, Gerber C (1986) Atomic force microscope. Phys Rev Lett 56:930–933
Bornschlögl T, Rief M (2008) Single-Molecule dynamics of mechanical coiled-coil unzipping. Langmuir 24:1338–1342
Bornschlögl T, Gebhardt JCM, Rief M (2009a) Designing the folding mechanics of coiled coils. Chemphyschem 10:2800–2804
Bornschlögl T, Anstrom DM, Mey E, Dzubiella J, Rief M, Forest KT (2009b) Tightening the knot in phytochrome by single-molecule atomic force microscopy. Biophys J 96:1508–1514
Brandenburg B, Zhuang X (2007) Virus trafficking-learning from single-virus tracking. Nat Rev Microbiol 5:197–208
Brucale M, Sandal M, Di Maio S, Rampioni A, Tessari I, Tosatto L, Bisaglia M, Bubacco L, Samorì B (2009) Pathogenic mutations shift the equilibria of α-synuclein single molecules towards structured conformers. Chembiochem 10:176–183
Brujic J, Fernandez JM (2005) Response to comment on "Force-clamp spectroscopy monitors the folding trajectory of a single protein". Science 308:498c
Bryngelson JD, Onuchic JN, Socci ND, Wolynes PG (1995) Funnels, pathways, and the energy landscape of protein folding: a synthesis. Proteins 21:167–195
Bullard B, Garcia T, Benes V, Leake MC, Linke WA, Oberhauser AF (2006) The molecular elasticity of the insect flight muscle proteins projectin and kettin. Proc Natl Acad Sci USA 103:4451–4456
Bustamante C, Chemla Y, Forde N, Izhaky D (2004) Mechanical processes in biochemistry. Annu Rev Biochem 73:705–748
Cao Y, Yoo T, Li H (2008a) Single molecule force spectroscopy reveals engineered metal chelation is a general approach to enhance mechanical stability of proteins. Proc Natl Acad Sci USA 105:11152–11157
Cao Y, Yoo T, Zhuang S, Li H (2008b) Protein–protein interaction regulates proteins mechanical stability. J Mol Biol 378:1132–1141
Cao Y, Er KS, Parhar R, Li H (2009) A force-spectroscopy-based single-molecule metal-binding assay. Chemphyschem 10:1450–1454
Carrión-Vázquez M, Oberhauser AF, Fowler SB, Marszalek PE, Broedel SE, Clarke J, Fernandez JM (1999) Mechanical and chemical unfolding of a single protein: a comparison. Proc Natl Acad Sci USA 96:3694–3699
Carrión-Vázquez M, Oberhauser AF, Fisher TE, Marszalek PE, Li H, Fernandez JM (2000) Mechanical design of proteins studied by single-molecule force spectroscopy and protein engineering. Prog Biophys Mol Biol 74:63–91
Carrión-Vázquez M, Li H, Lu H, Marszalek PE, Oberhauser AF, Fernandez JM (2003) The mechanical stability of ubiquitin is linkage dependent. Nat Struct Biol 10:738–743
Carrión-Vázquez M, Oberhauser AF, Díez H, Hervás R, Oroz J, Fernández J, Marínez-Martín D (2006) Protein nanomechanics—as studied by AFM single-molecule force spectroscopy. In: Arrondo JLR, Alonso A (eds) Advanced techniques in biophysics. Springer, Heidelberg, pp 163–245
Carrión-Vázquez M, Cieplak M, Oberhauser AF (2009) Protein mechanics at the single-molecule level. In: Meyers RA (ed) Encyclopedia of complexity and systems science. Springer, New York, NY, pp 7026–7051
Cieplak M, Szymczak P (2006) Protein folding in a force clamp. J Chem Phys 124:194901–194904

Cox BS (1965) [PSI+], a cytoplasmic suppressor of super-suppressors in yeast. Heredity 20:505–521

Crick SL, Jayaraman M, Frieden C, Wetzel R, Pappu RV (2006) Fluorescence correlation spectroscopy shows that monomeric polyglutamine molecules form collapsed structures in aqueous solutions. Proc Natl Acad Sci USA 103:16764–16769

Cummings CJ, Zoghbi HY (2000) Trinucleotide repeats: mechanisms and pathophysiology. Annu Rev Genomics Hum Genet 1:281–328

Dembo M, Torney DC, Saxman K, Hammer D (1988) The reaction-limited kinetics of membrane-to-surface adhesion and detachment. Proc R Soc Lond B Biol Sci 234:55–83

Deniz AA, Laurence TA, Beligere GS, Dahan M, Martin AB, Chemla DS, Dawson PE, Schultz PG, Weiss S (2000) Single-molecule protein folding: diffusion fluorescence resonance energy transfer studies of the denaturation of chymotrypsin inhibitor 2. Proc Natl Acad Sci USA 97:5179–5184

Deniz AA, Laurence TA, Dahan M, Chemla DS, Schultz PG, Weiss S (2001) Ratiometric single-molecule studies of freely diffusing biomolecules. Annu Rev Phys Chem 52:233–253

Dougan L, Li J, Badilla CL, Berne BJ, Fernandez JM (2009) Single homopolypeptide chains collapse into mechanically rigid conformations. Proc Natl Acad Sci USA 106:12605–12610

Dunker AK, Brown CJ, Lawson JD, Iakoucheva LM, Obradovic Z (2000a) Intrinsic disorder and protein function. Biochemistry 41:6573–6582

Dunker AK, Obradovic Z, Romero P, Garner EC, Brown CJ (2000b) Intrinsic protein disorder in complete genomes. Genome Inform Ser Workshop Genome Inform 11:161–171

Dunker AK, Lawson JD, Brown CJ, Williams RM, Romero P, Oh JS, Oldfield CJ, Campen AM, Ratliff CM, Hipps KW, Ausio J, Nissen MS, Reeves R, Kang CH, Kissinger CR, Bailey RW, Griswold MD, Chiu W, Garner EC, Obradovic Z (2001) Intrinsically disordered protein. J Mol Graph Model 19:26–59

Evans E, Ritchie K (1997) Dynamic strength of molecular adhesion bonds. Biophys J 72:1541–1555

Faux NG, Bottomley SP, Lesk AM, Irving JA, Morrison JR, de la Banda MG, Whisstock JC (2005) Functional insights from the distribution and role of homopeptide repeat-containing proteins. Genome Res 15:537–551

Fernandez JM, Li H (2004) Force-clamp spectroscopy monitors the folding trajectory of a single protein. Science 303:1674–1678

Fernandez JM, Li H, Brujic J (2004) Response to comment on "Force-clamp spectroscopy monitors the folding trajectory of a single protein". Science 306:411c

Ferreon AC, Moran CR, Gambin Y, Deniz AA (2010) Single-molecule fluorescence studies of intrinsically disordered proteins. Methods Enzymol 472:179–204

Forman JR, Yew ZT, Qamar S, Sandford RN, Paci E, Clarke J (2009) Non-native interactions are critical for mechanical strength in PKD domains. Structure 17:1582–1590

Galera-Prat A, Gómez-Sicilia A, Oberhauser AF, Cieplak M, Carrión-Vázquez M (2010) Understanding biology by stretching proteins: recent progress. Curr Opin Struct Biol 20:63–69

Garcia-Manyes S, Dougan L, Badilla CL, Brujic J, Fernandez JM (2009a) Direct observation of an ensemble of stable collapsed states in the mechanical folding of ubiquitin. Proc Natl Acad Sci USA 106:10534–10539

Garcia-Manyes S, Liang J, Szoszkiewicz KTL, Fernández JM (2009b) Force-activated reactivity switch in a bimolecular chemical reaction. Nat Chem 1:236–242

Gebhardt JCM, Bornsclögl T, Rief M (2010) Full distance-resolved folding energy landscape of one single protein molecule. Proc Natl Acad Sci USA 107:2013–2018

Gräter F, Shen J, Jiang H, Gautel M, Grubmüller H (2005) Mechanicallyinduced titin kinase activation studied by force-probe molecular dynamicssimulations. Biophys J 88:790–804

Grützner A, Garcia-Manyes S, Kotter S, Badilla CL, Fernandez JM, Linke WA (2009) Modulation of titin-based stiffness by disulfide bonding in the cardiac titin N2-B unique sequence. Biophys J 97:825–834

Guzmán DL, Randall A, Baldi P, Guan Z (2010) Computational and single-molecule force studies of a macro domain protein reveal a key molecular determinant for mechanical stability. Proc Natl Acad Sci USA 107:1989–1994
Helmes M, Trombitás K, Centner T, Kellermayer M, Labeit S, Linke WA, Granzier H (1999) Mechanically driven contour-length adjustment in rat cardiac titin's unique N2B sequence: titin is an adjustable spring. Circ Res 84:1339–1352
Hervás R, Oroz J, Galera-Prat A, Goñi O, Valbuena A, Vera AM, Gómez-Sicilia A, Losada-Urzáiz F, Uversky VN, Menéndez M, Laurents DV, Bruix M, Carrión-Vázquez M (2012) Common features at the start of the neurodegeneration cascade. PLoS Biol 10(5):e1001335. doi:10.1371/journal.pbio.1001335
Hirokawa N, Shiomura Y, Okabe S (1988) Tau proteins: the molecular structure and mode of binding on microtubules. J Cell Biol 107:1449–1459
Humphrey W, Dalke A, Schulten K (1996) VMD: visual molecular dynamics. J Mol Graph 14:33–38
Ishijima A, Kojima H, Funatsu T, Tokunaga M, Higuchi H, Tanaka H, Yanagida T (1998) Simultaneous observation of individual ATPase and mechanical events by a single myosin molecule during interaction with actin. Cell 92:161–171
Jahn TR, Parker MJ, Homans SW, Radford SE (2006) Amyloid formation under physiological conditions proceeds via a native-like folding intermediate. Nat Struct Mol Biol 13:195–201
Johnson CP, Tang HY, Carag C, Speicher DW, Discher DE (2007) Forced unfolding of proteins within cells. Science 317:663–666
Jolleymore A, Li H (2010) Measuring "unmeasurable" folding kinetics of proteins by single-molecule force spectroscopy. J Mol Biol 402:610–617
Junker JP, Rief M (2009) Single-molecule force spectroscopy distinguishes target binding modes of calmodulin. Proc Natl Acad Sci USA 106:14361–14366
Junker JP, Hell K, Schlierf M, Neupert W, Rief M (2005) Influence of substrate binding on the mechanical stability of mouse dihydrofolate reductase. Biophys J 89:45–48
Junker JP, Ziegler F, Rief M (2009) Ligand-dependent equilibrium fluctuations of single calmodulin molecules. Science 323:633–637
Kazmierczak P, Müller U (2012) Sensing sound: molecules that orchestrate mechanotransduction by hair cells. Trends Neurosci 35:220–229
Khare SD, Ding F, Gwanmesia KN, Dokholyan NV (2005) Molecular origin of polyglutamine aggregation in neurodegenerative diseases. PLoS Comput Biol 1:230–235
Kidd M (1963) Paired helical filaments in electron microscopy of Alzheimer's disease. Nature 197:192–193
Kim HY, Heise H, Fernandez CO, Baldus M, Zweckstetter M (2007) Correlation of amyloid fibril β-structure with the unfolded state of α-synuclein. Chembiochem 8:1671–1674
King CY, Tittmann P, Gross H, Gebert R, Aebi M, Wüthrich K (1997) Prion-inducing domain 2–114 of yeast Sup35 protein transforms *in vitro* into amyloid-like filaments. Proc Natl Acad Sci USA 94:6618–6622
Lang MJ, Fordyce PM, Engh AM, Neuman KC, Block SM (2004) Simultaneous, coincident optical trapping and single-molecule fluorescence. Nat Methods 1:133–139
Leake MC, Grutzner A, Kruger M, Linke WA (2006) Mechanical properties of cardiac titin's N2B-region by single-molecule atomic force spectroscopy. J Struct Biol 155:263–272
Lee JC, Langen R, Hummel PA, Gray HB, Winkler JR (2004) α-synuclein structures from fluorescence energy-transfer kinetics: implications for the role of the protein in Parkinson's disease. Proc Natl Acad Sci USA 101:16466–16471
Li DY, Brooke B, Davis EC, Mecham RP, Sorensen LK, Boak BB, Eichwald E, Keating MT (1998) Elastin is an essential determinant of arterial morphogenesis. Nature 393:276–280
Li H, Oberhauser AF, Redick SD, Carrion-Vazquez M, Erickson HP, Fernandez JM (2001) Multiple conformations of PEVK proteins detected by single-molecule techniques. Proc Natl Acad Sci USA 98:10682–10686

Li H, Linke WA, Oberhauser AF, Carrion-Vazquez M, Kerkvliet JG, Lu H, Marszalek PE, Fernandez JM (2002) Reverse engineering of the giant muscle protein titin. Nature 418:998–1002

Li H, Wang H-C, Cao Y, Sharma D, Wang M (2008) Configurational entropy modulates the mechanical stability of protein GB1. J Mol Biol 379:871–880

Li J, Fernandez JM, Berne J (2010) Water's role in the force-induced unfolding of ubiquitin. Proc Natl Acad Sci USA 107:19284–19289

Li PT, Bustamante C, Tinoco I Jr (2007) Real-time control of the energy landscape by force directs the folding of RNA molecules. Proc Natl Acad Sci USA 104:7039–7044

Liang J, Fernandez JM (2011) Kinetic measurements on single-molecule disulfide bond cleavage. J Am Chem Soc 133:3528–3534

Linke WA, Rudy DE, Centner T, Gautel M, Witt C, Labeit S, Gregorio CC (1999) I-band titin in cardiac muscle is a three-element molecular spring and is critical for maintaining thin filament structure. J Cell Biol 146:631–644

Lu H, Schulten K (1999) Steered molecular dynamics simulations of force-induced protein domain unfolding. Proteins 35:453–463

Lu H, Schulten K (2000) The key event in force-induced unfolding of titin's inmunoglobulin domains. Biophys J 79:51–65

Lu H, Isralewitz B, Krammer A, Vogel V, Schulten K (1998) Unfolding of titin immunoglobulin domains by steered molecular dynamics simulation. Biophys J 75:662–671

Ma L, Xu M, Oberhauser A (2010) Naturally occurring osmolytes modulate the nanomechanical properties of polycystic kidney disease domains. J Biol Chem 285:38438–38443

Marchut AJ, Hall CK (2006) Effects of chain length on the aggregation of model polyglutamine peptides: molecular dynamics simulations. Proteins 66:96–109

Marshall BT, Long M, Piper JW, Yago T, McEver RP, Zhu C (2003) Direct observation of catch bonds involving cell-adhesion molecules. Nature 423:190–193

Marszalek PE, Lu H, Li H, Carrion-Vazquez M, Oberhauser AF, Schulten K, Fernandez JM (1999) Mechanical unfolding intermediates in titin modules. Nature 402:100–103

Martin J (2002) Requirement for GroEL/GroES-dependent protein folding under nonpermissive conditions of macromolecular crowding. Biochemistry 41:5050–5055

Matsunaga Y, Komatsuzaki T (2004) Protein folding dynamics: ergodic behavior in principal component space. Aip Conf Proc 708:342–343

Michalet X, Weiss S, Jager M (2006) Single-molecule fluorescence studies of protein folding and conformational dynamics. Chem Rev 106:1785–1813

Minton AP (2001) The influence of macromolecular crowding and macromolecular confinement on biochemical reactions in physiological media. J Biol Chem 276:10577–10580

Nagai Y, Inui T, Popiel HA, Fujikake N, Hasegawa K, Urade Y, Goto Y, Naiki H, Toda T (2007) A toxic monomeric conformer of the polyglutamine protein. Nat Struct Biol 14:332–340

Neuman KC, Nagy A (2008) Single-moleculeforcespectroscopy: optical tweezers, magnetictweezers and atomic force microscopy. Nat Methods 5:491–505

Ng SP, Rounsevell RWS, Steward A, Geierhaas CD, Williams PM, Paci E, Clarke J (2005) Mechanical unfolding of TNfn3: the unfolding pathway of a fnIII domain probed by protein engineering, AFM and MD simulation. J Mol Biol 350:776–789

Oberhauser AF, Marszalek PE, Carrión-Vázquez M, Fernandez JM (1999) Single protein misfolding events captured by atomic force microscopy. Nat Struct Biol 6:1025–1028

Oroz J, Valbuena A, Vera AM, Mendieta J, Gomez-Puertas P, Carrion-Vazquez M (2011) Nanomechanics of the cadherin ectodomain: canalization by Ca2+ binding results in a new mechanical element. J Biol Chem 286:9405–9418

Oroz J, Hervás R, Carrión-Vázquez M (2012) Unequivocal single-molecule force spectroscopy of proteins by AFM using pFS vectors. Biophys J 102:682–690

Peng Q, Li H (2008) Atomic force microscopy reveals parallel mechanical unfolding pathways of T4 lysozyme: evidence for a kinetic partitioning mechanism. Proc Natl Acad Sci USA 105:1885–1890

Perez-Jimenez R, Li J, Kosuri P, Sanchez-Romero I, Wiita AP, Rodriguez-Larrea D, Chueca A, Holmgren A, Miranda-Vizuete A, Becker K, Cho SH, Beckwith J, Gelhaye E, Jacquot JP, Gaucher EA, Sanchez-Ruiz JM, Berne BJ, Fernandez JM (2009) Diversity of chemical mechanisms in thioredoxin catalysis revealed by single-molecule force-spectroscopy. Nat Struct Mol Biol 16:890–897

Perez-Jimenez R, Inglés-Prieto A, Zhao ZM, Sanchez-Romero I, Alegre-Cebollada J, Kosuri P, Garcia-Manyes S, Kappock TJ, Tanokura M, Holmgren A, Sanchez-Ruiz JM, Gaucher EA, Fernandez JM (2011) Single-molecule paleoenzymology probes the chemistry of resurrected enzymes. Nat Struct Mol Biol 18:592–596

Platt GW, McParland VJ, Kalverda AP, Homans SW, Radford SE (2005) Dynamics in the unfolded state of β_2-microglobulin studied by NMR. J Mol Biol 346:279–294

Popa I, Fernández LM, Garcia-Manyes S (2011) Direct quantification of the attempt frequency determining the mechanical unfolding of ubiquitin protein. J Biol Chem 286:31072–31079

Qian F, Wei W, Germino G, Oberhauser AF (2005) The nanomechanics of polycystin-1 extracellular region. J Biol Chem 280:40723–40730

Rief M, Gautel M, Oesterhelt M, Fernandez JM, Gaub HE (1997) Reversible unfolding of individual titin immunoglobulin domains by AFM. Science 276:1109–1112

Rief M, Pascual J, Saraste M, Gaub HE (1999) Single molecule force spectroscopy of spectrin repeats: low unfolding forces in helix bundles. J Mol Biol 286:553–561

Sandal M, Valle F, Tessari I, Mammi S, Bergantino E, Musiani F, Brucale M, Bubacco L, Samorì B (2008) Conformational equilibria in monomeric α-synuclein at the single-molecule level. PLoS Biol 6:99–108

Sarkar A, Caamano S, Fernandez JM (2005) The elasticity of individual titin PEVK exons measured by single molecule atomic force microscopy. J Biol Chem 280:6261–6264

Sawaya MR, Sambashivan S, Nelson R, Ivanova MI, Sievers SA, Apostol MI, Thompson MJ, Balbirnie M, Wiltzius JJ, McFarlane HT, Madsen AØ, Riekel C, Eisenberg D (2007) Atomic structures of amyloid cross-β spines reveal varied steric zippers. Nature 447:453–457

Schlierf M, Rief M (2005) Temperature softening of a protein in single-molecule experiments. J Mol Biol 354:497–503

Schlierf M, Berkemeier F, Rief M (2007) Direct observation of active protein folding using lock-in force spectroscopy. Biophys J 93:3989–3998

Schuler B (2005) Single-molecule fluorescence spectroscopy of protein folding. Chem Phys Chem 6:1206–1220

Schuler B, Lipman EA, Eaton WA (2002) Probing the free-energy surface for protein folding with single-molecule fluorescence spectroscopy. Nature 419:743–747

Schwaiger I, Schleicher M, Noegel AA, Rief M (2005) The folding pathway of a fast-folding immunoglobulin domain revealed by single-molecule mechanical experiments. EMBO Rep 6:46–51

Sgourakis NG, Yan Y, McCallum SA, Wang C, Garcia AE (2007) The Alzheimer's peptides Aβ40 and 42 adopt distinct conformations in water: a combined MD/NMR study. J Mol Biol 368:1448–1457

Serquera D, Lee W, Settanni G, Marszalek PE, Paci E, Itzhaki LS (2010) Mechanical unfolding of an ankyrin repeat protein. Biophys J 98:1294–1301

Shankar GM, Li S, Mehta TH, Garcia-Munoz A, Shepardson NE, Smith I, Brett FM, Farrell MA, Rowan MJ, Lemere CA, Regan CM, Walsh DM, Sabatini BL, Selkoe DJ (2008) Amyloid β-protein dimers isolated directly from Alzheimer brains impair synaptic plasticity and memory. Nat Med 14:837–842

Sharma D, Feng G, Khor D, Genchev GZ, Lu H, Li H (2008) Stabilization provided by neighboring strands is critical for the mechanical stability of proteins. Biophys J 95:3935–3942

Sikora M, Cieplak M (2011) Mechanical stability of multidomain proteins and novel mechanical clamps. Proteins 79:1786–1799

Sikora M, Sulowska JI, Cieplak M (2009) Mechanical strength of 17134 model proteins and cystein slipknots. PLoS Comput Biol 5(10):e1000547. doi:10.1371/journal.pcbi.1000547

Sosnick TR (2004) Comment on “Force-clamp spectroscopy monitors the folding trajectory of a single protein”. Science 306:411b

Sotomayor M, Schulten K (2007) Single-molecule experiments *in vitro* and *in silico*. Science 316:1144–1148

Still WC, Tempczyk A, Hawley RC, Hendrickson T (1990) Semianalytical treatment of solvation for molecular mechanics and dynamics. J Am Chem Soc 112:6127–6129

Terakawa T, Takada S (2011) Multiscale ensemble modeling of intrinsically disordered proteins: p53 N-terminal domain. Biophys J 101:1450–1458

Tycko R (2011) Solid-state NMR studies of amyloid fibril structure. Annu Rev Phys Chem 62:279–299

Tinnefeld P, Sauer M (2005) Branching out of single-molecule fluorescence spectroscopy: challenges for chemistry and influence on biology. Angew Chem Int Ed 44:2642–2671

Urry DW, Hugel T, Seitz M, Gaub HE, Sheiba L, Dea J, Xu J, Parker T (2002) Elastin: a representative ideal protein elastomer. Philos Trans R Soc Lond B Biol Sci 357:169–184

Uversky VN, Oldfield CJ, Dunker AK (2005) Showing your ID: intrinsic disorder as an ID for recognition, regulation and cell signaling. J Mol Recognit 18:343–384

Uversky VN, Oldfield CJ, Dunker AK (2008) Intrinsically disordered proteins in human diseases: introducing the D2 concept. Annu Rev Biophys Biomol Struct 37:215–246

Uversky VN, Oldfield CJ, Midic U, Xie H, Xue B, Vucetic S, Iakoucheva LM, Obradovic Z, Dunker AK (2009) Unfoldomics of human diseases: linking protein intrinsic disorder with diseases. BMC Genomics 10(Suppl 1):S7

Valbuena A, Oroz J, Hervás R, Vera AM, Rodríguez D, Menéndez M, Sulkowska JI, Cieplak M, Carrión-Vázquez M (2009) On the remarkable mechanostability of scaffoldins and the mechanical clamp motif. Proc Natl Acad Sci USA 106:13791–13796

Valiaev A, Lim DW, Schmidler S, Clark RL, Chilkoti A, Zauscher S (2008) Hydration and conformational mechanics of single, end-tethered elastin-like polypeptides. J Am Chem Soc 130:10939–10946

Vitalis A, Wang X, Pappu RV (2008a) Atomistic simulations of the effects of polyglutamine chain length and solvent quality on conformational equilibria and spontaneous homodimerization. J Mol Biol 384:279–297

Vitalis A, Wang X, Pappu RV (2008b) Quantitative characterization of intrinsic disorder in polyglutamine: insights from analysis based on polymer theories. Biophys J 93:1923–1937

Vogel V, Sheetz MP (2006) Local force and geometry sensing regulate cell functions. Nat Rev Mol Cell Biol 7:265–275

Vogel V, Sheetz MP (2009) Mechanical forces matter in health and disease: from cancer to tissue engineering. In: Vogel V (ed) Nanomedicine. Wiley-VCH, Weinheim, pp 235–303

Wallin S, Zeldovich KB, Shakhnovich EI (2007) Folding mechanics of a knotted protein. J Mol Biol 368:884–893

Walther KA, Gräter F, Dougan L, Badilla CL, Berne BJ, Fernandez JM (2007) Signatures of hydrophobic collapse in extended proteins captured with force spectroscopy. Proc Natl Acad Sci USA 104:7916–7921

Watanabe K, Muhle-Goll C, Kellermayer MS, Labeit S, Granzier H (2002) Molecular mechanics of cardiac titin's PEVK and N2B spring elements. J Biol Chem 277:11549–11558

Wegmann S, Schöler J, Bippes CA, Mandelkow E, Muller DJ (2011) Competing interactions stabilize pro- and anti-aggregant conformations of human tau. J Biol Chem 286:20512–20524

Wiita AP, Ainavarapu SRK, Huang HH, Fernandez JM (2006) Force-dependent chemical kinetics of disulfide bond reduction observed with single-molecule techniques. Proc Natl Acad Sci USA 103:7222–7227

Wiita AP, Perez-Jimenez R, Walther KA, Gräter F, Berne BJ, Holmgren A, Sanchez-Ruiz JM, Fernandez JM (2007) Probing the chemistry of thioredoxin catalysis with force. Nature 450:124–127

Wright PE, Dyson HJ (1999) Intrinsically unstructured proteins: re-assessing the protein structure-function paradigm. J Mol Biol 293:321–331

Xia F, Thirumalai D, Gräter F (2011) Minimum energy compact structures in force-quench polyubiquitin folding are domain swapped. Proc Natl Acad Sci USA 108:6963–6968

Xie XS, Choi PJ, Li GW, Lee NK, Lia G (2008) Single-molecule approach to molecular biology in living bacterial cells. Annu Rev Biophys 37:417–444

Xiong K, Zwier MC, Myshakina NS, Burger VM, Asher SA, Chong LT (2011) Direct observations of conformational distributions of intrinsically disordered p53 peptides using UV Raman and explicit solvent simulations. J Phys Chem 115:9520–9527

Yamasaki R, Wu Y, McNabb M, Greaser M, Labeit S, Granzier H (2002) Protein kinase A phosphorylates titin's cardiac-specific N2B domain and reduces passive tension in rat cardiac myocytes. Circ Res 90:1181–1188

Yang M, Teplow D (2008) Amyloid β-protein monomer folding: free-energy surfaces reveal alloform-specific differences. J Mol Biol 384:450–464

Yuan J, Chyan C, Zhou H, Chung T, Peng H, Ping G, Yang G (2008) The effects of macromolecular crowding on the mechanical stability of protein molecules. Protein Sci 17:2156–2166

Zhang S, Iwata K, Lachenmann M, Peng J, Li S, Stimson ER, Lu Y, Felix AM, Maggio JE, Lee JP (2000) The Alzheimer's peptide Aβ adopts a collapsed coil structure in water. J Struct Biol 130:130–141

Zheng P, Cao Y, Bu T, Straus SK, Li H (2011) Single molecule force spectroscopy reveals that electrostatic interactions affect the mechanical stability of proteins. Biophys J 100:1534–1541

Zhuang S, Linhananta A, Li H (2010) Phenotypic effects of Ehlers–Danlos syndrome-associated mutation on the FnIII domain of tenascin-X. Protein Sci 19:2231–2239

Chapter 2
Force Spectroscopy and Recognition Imaging of Cells from the Immune System

Bianca L. Bozna, Memed Duman, and Peter Hinterdorfer

2.1 Introduction

Several techniques have emerged in the last decade to measure biological interactions at single molecular level (Svoboda et al. 1993; Kellermayer 1997; Evans et al. 1995; Leckband and Israelachvili 1993). Among these, surface force apparatus (Leckband et al. 1992), biomembrane force probe (Merkel et al. 1999), and optical tweezers (Ashkin 1997) allow to investigate the interaction forces of systems, from small molecules to living cells. The atomic force microscope employs one of the smallest force sensors that, besides high-resolution imaging, allows to measure single molecular receptor/ligand forces down to subnanometer level.

This chapter focuses on single-molecule force spectroscopy as well as simultaneous Topography and RECognition imaging (TREC) applied to THP1 living cells using the atomic force microscope (Bozna et al. 2011). While the force

B.L. Bozna
Institute for Biophysics, Johannes Kepler University Linz, Gruberstraße 40, A-4020 Linz, Austria

Department of Pharmaceutical Sciences, University of Nebraska Medical Center, Omaha, NE, USA

M. Duman
Institute for Biophysics, Johannes Kepler University Linz, Gruberstraße 40, A-4020 Linz, Austria

Nanotechnology and Nanomedicine, Institute of Science, Hacettepe University, Beytepe, Ankara, Turkey

P. Hinterdorfer (✉)
Institute for Biophysics, Johannes Kepler University Linz, Gruberstraße 40, A-4020 Linz, Austria

Christian Doppler Laboratory for Nanoscopic Methods in Biophysics, University of Linz, Linz, Austria
e-mail: peter.hinterdorfer@jku.at

A.F. Oberhauser (ed.), *Single-molecule Studies of Proteins*,
Biophysics for the Life Sciences 2, DOI 10.1007/978-1-4614-4921-8_2,

spectroscopy yields information regarding the structure and the dynamics of the recognition process (Florin et al. 1994; Lee et al. 1994; Hinterdorfer et al. 1996; Hinterdorfer and Dufrene 2006), the TREC method offers the possibility to map the distribution of specific molecular recognition events on the cell surface with nanometer accuracy under physiological condition (Ahmad et al. 2011; Preiner et al. 2009; Stroh et al. 2004a, b).

By using single-molecule force spectroscopy, specific recognition events between Invariant Natural Killer T (iNKT) T cell Receptor (TCR) and hCD1d–glycosphingolipids (GSL) complexes were analyzed to reveal the interaction forces and the kinetic rate constants that govern the bond stability. In order to characterize the interaction forces between the iNKT-TCR and hCD1d complexes owing to different affinities for the iNKT TCR (i.e., loaded with either α-GalCer or OCH12), the receptor was attached to the AFM tip via a flexible polyethylene glycol linker, enabling force measurements to be done in vitro and in vivo. The same iNKT TCR-coated tip was used in each particular experiment for probing the specific interactions of both complexes, CD1d–α-GalCer and CD1d–OCH12.

2.2 Sample and Tip Preparation Methods

2.2.1 iNKT TCR Tip Functionalization

To allow measuring of the binding interactions between single molecule on an AFM tip and molecules on the cell membrane, the functionalization of the AFM tips with the molecule of interests is needed.

A soluble biotinylated iNKT TCR is tethered to the AFM tip via a flexible heterobifunctional PEG [poly(ethylene glycol)]-cross-linker with an extended length of 6 nm, containing an aldehyde and an *N*-hydroxysuccinimide ester (NHS-ester) group at its ends (Fig. 2.1a). The silicon nitride AFM tips (Veeco Instruments) are first functionalized with an amine group using APTES coating (3-aminoprobyl triethoxysilane) procedure. To the aminofunctionalized tip is then attached an aldehyde-PEG-NHS linker by incubation for 2 h at room temperature in 0.5 mL of chloroform containing 3.3 mg aldehyde-PEG-NHS and 0.5% triethylamine.

The tips are subsequently rinsed in chloroform and dried with nitrogen before their incubation for 50 min in a mixture of 65 μL of streptavidin with a concentration of 0.2 mg mL^{-1} in a PBS buffer (150 mM NaCl, 5 mM NaH_2PO_4, pH 7.5) and 2 μL of 1 M $NaCNBH_3$ (freshly prepared by dissolving 32 mg of solid $NaCNBH_3$ in 500 μL of 10 mM NaOH). In order to block unreacted aldehyde groups, 5 μL of 1 M ethanolamine hydrochloride (adjusted to pH 9.6 with 20% NaOH) is added while the incubation continues for 10 min. After washing the tips in PBS, they are treated with 20 μg mL^{-1} biotinylated iNKT TCR in TRIS buffer (50 mM Tris, 100 mM NaCl, Glycerol 1%, pH 7.5) and the incubation at 4°C is allowed to continue overnight. The tips are finally washed and stored in TRIS buffer in cold environment.

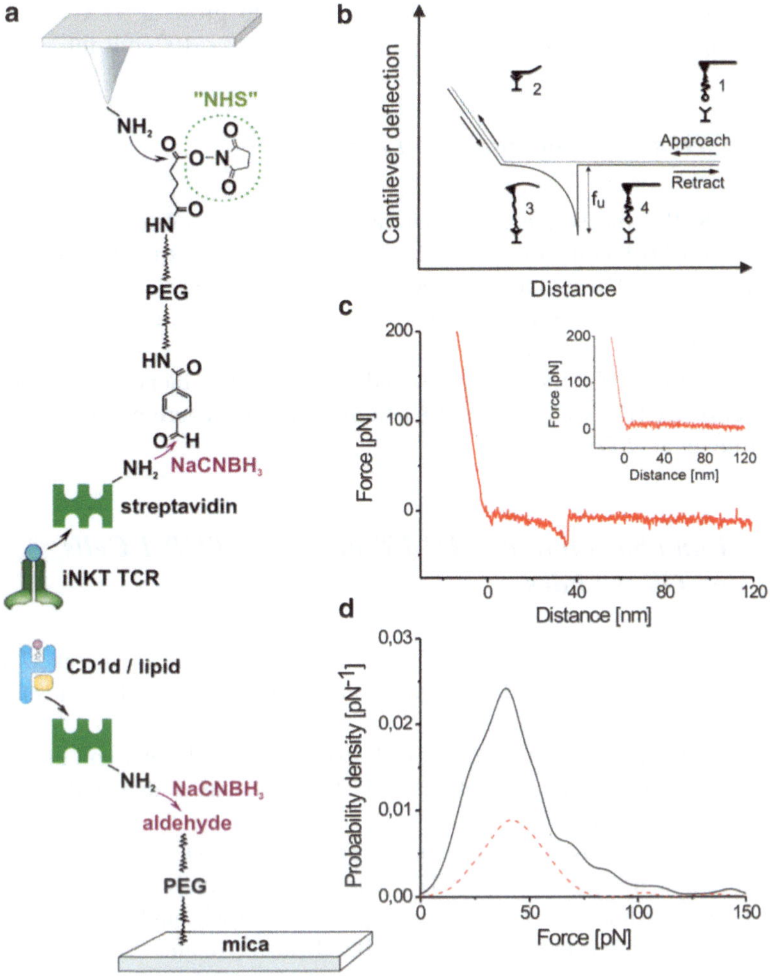

Fig. 2.1 Specific recognition of isolated CD1d–GSL molecules by an iNKT TCR coated tip. (**a**) Schematic representation of the immobilization strategy; (**b**) diagram of the force–distance cycle; (**c**) typical force curve showing an iNKT TCR-unbinding event from CD1d-α-GalCer upon tip retraction. *Inset*: the specific interaction blocked by injecting of free anti-CD1d mAb into the bath solution; (**d**) probability density function showing the distribution of unbinding force events occurring between the iNKT TCR and CD1d-α-GalCer (*black line*). The effect of the anti-CD1d blocking antibody on the iNKT TCR/CD1d interaction (*red dashed line*)

2.2.2 *THP-1 Cell Culture*

The THP1 cells [American Type Culture Collection (ATCC)], which were transfected with a lentiviral vector encoding CD1d/YFP fusion protein, were grown in RPMI 1640 medium supplemented with 10% fetal bovine serum, 2 mM Glutamax, 1% sodium pyruvate, 1% nonessential amino acids, and penicillin-streptomycin

(5,000 U penicillin/mL). The cells were allowed to grow at 37°C in 5% CO_2 maintaining their concentration between 2×10^5 and 9×10^5 cells mL^{-1}.

2.2.3 *CD1d Immobilization on Mica*

To allow SMFS measurements of interaction between isolated iNKT TCR and biotinylated CD1d–GSL complexes, freshly cleaved mica were functionalized with 20 μg mL^{-1} of CD1d-α-GalCer or OCH12, using the same chemical procedure as thoroughly described in Sect. 2.2.1. The surface modified mica substrates were then stored in TRIS buffer at 4°C.

Soluble iNKT TCR heterodimers and human CD1d monomers loaded with α-GalCer or OCH12 were prepared by following a procedure previously reported (McCarthy et al. 2007).

2.2.4 *Immobilization of CD1d Transfected THP-1 Cells on Glass Slides*

The THP1 CD1d cells were incubated overnight in a 48-well plate at 37°C with α-GalCer or OCH12 (1 μg mL^{-1}), and the next day were washed and resuspended in 500 μL RPMI 1640 w/o FBS.

After washing the glass coverslips with isopropanol and water and then dried, a Poly-L-Lysine (PLL) solution of 500 μL (0.01%) was added and allowed to incubate for 30 min at 37°C.

After rinsing the glass coverslips three times with PBS (Dulbecco), cells (500 μL) were added and incubated for 1 h at 37°C and 5% CO_2. Finally, the glass slides were washed twice with HBSS (Dulbecco) containing Ca^{2+} and Mg^{2+}. The glass coverslips were rinsed three times with PBS (Dulbecco), and cells (500 μL) were subsequently added and incubated for 1 h at 37°C and 5% CO_2. Finally, the glass slides were washed twice with HBSS (Dulbecco) containing Ca^{2+} and Mg^{2+}.

For TREC measurements, the cells were fixed with 2% paraformaldehyde for 20 min at room temperature and carefully washed with PBS buffer (150 mM NaCl, pH 7.4).

2.2.5 *Simultaneous Topography and Recognition Imaging and Instrumentation*

Both AFM topography and recognition data were recorded in the MAC (magnetic alternating current) mode by using a PicoPlus AFM (Agilent Technologies, AZ, USA). Magnetically coated Olympus cantilevers having a nominal spring constant of 80 pN nm^{-1} with a quality factor (Q) of ~1 in liquid were used for bottom-MAC (magnetic field excitation below the sample) mode. All images were taken by closed loop large-scan size scanner (100 × 100 μm^2) in TRIS buffer solution at room temperature. During TREC measurements, integral and proportional gains were adjusted to optimize the sensitivity of the feedback loop and oscillation amplitude was chosen at the optimum driving frequencies of each cantilever (~3 kHz). The TREC data were acquired by scanning ~1 × 1 μm^2 area of the cell surface and recorded by using a commercially available electronic unit (PicoTREC, Agilent, AZ, USA). The scan speed for imaging was 2 line/s at 256 data points per line. Specificity of iNKT-TCR modified AFM tips to glycolipids loaded CD1d molecule was proven by injecting free anti-CD1d antibody into the fluid cell of the AFM during scanning.

2.3 Single-Molecule Force Spectroscopy of iNKT TCR with hCD1d–GSL Complexes

iNKT cells are a heterogeneous population of lymphocytes that share properties of both T cells and natural killer (NK) cells with the ability to regulate the immune system in response to a broad range of diseases (Kronenberg 2005; Kinjo Yuki et al. 2006; Mattner et al. 2005). These cells express an invariant TCR by which they recognize glycolipids bound to or presented by the CD1d molecule, a nonpolymorphic major histocompatibility complex (MHC) class I-like molecule. The presence of CD1d-lipid complexes on several types of cells including antigen-presenting cells allows the engagement of iNKT TCR, leading to a rapid activation of iNKT cells and secretion of significant levels of inflammatory cytokines such as pro-inflammatory T helper type 1 (Th1) [interferon-γ (INF-γ) and tumor-necrosis factor-α (TNF-α)] and anti-inflammatory Th2 cytokines [interleukin-4 (IL-4), IL-10 and IL-13], which allows these cells to coordinate both innate and adaptive immunity and the development of autoimmune, antimicrobial, antitumor, antitransplant, and allergic immune responses (Spada et al. 1998; Kawano et al. 1997; Hermans et al. 2003; Fujii et al. 2003).

The affinity of the iNKT TCR for CD1d-glycolipid complexes plays an important role in evaluating the biological effects of iNKT cell agonists (Cerundolo et al. 2009). One of the most potent iNKT cell agonists is alpha-galactosylceramide (α-GalCer) that contains a galactose connected to a ceramide lipid through an α glycosidic-linkage. As revealed by the CD1d crystal structure, the ceramide lipid containing acyl and phytosphingosine chains is embedded in the groove of CD1d molecules (Koch et al. 2005; Zajonc et al. 2005). Several synthesized analogues of α-GalCer have been reported to activate iNKT cells when presented by CD1d expressing cells. Among them is OCH12, which differs from α-GalCer by a shorter phytosphingosine chain (C12 instead of C18).

The affinity of binding of the iNKT TCR to hCD1d–GSL complexes has been previously studied by ensemble-averaged methods, such as surface-plasmon resonance and flow-cytometry (FACS), suggesting that the length of the phytosphingosine chain influences the affinity of the iNKT TCR for CD1d/lipid complexes (McCarthy et al. 2007). However, the interaction forces that govern the bonds stability have not been determined. To address this issue, binding strength measurements between the iNKT TCR and hCD1d molecules loaded with GSL were performed using the atomic force microscope (AFM). A particular advantage of the AFM, besides high-resolution imaging, is the possibility to measure the intra-molecular (i.e., unfolding and refolding patterns of complementary DNA strands and proteins) and inter-molecular forces (i.e., receptor–ligand interactions) at the molecular level, yielding information regarding the dynamics of the recognition process. The dynamic aspects of molecular recognition are addressed in force spectroscopy experiments, where the unbinding forces between ligands and receptors, either on isolated molecules or on cellular surfaces, are measured. To this end, the AFM tips and solid substrates are functionalized with relevant biomolecules or cells. The low interaction force between a molecule tethered to the tip and its target molecule immobilized to the sample surface is measured in force–distance cycle by monitoring the cantilever deflection.

2.3.1 Principle of Single-Molecule Force Spectroscopy

The molecular recognition of many biological systems can be approximated by a simple configuration consisting of only two states, bound and unbound, which are separated by a transition state characterized by a single energy barrier (Fig. 2.2).

In the thermally activated model (Bell 1978; Evans and Ritchie 1997), the rate of bond dissociation under an increasing applied force is expressed by

$$k_{\mathrm{d}}(F) = k_{\mathrm{d}}(0)e^{Fx_\beta/k_{\mathrm{B}}T}, \tag{2.1}$$

where $k_{\mathrm{d}}(F)$ is the dissociation rate under applied force, and $k_{\mathrm{d}}(0)$ stands for the dissociation rate for zero external force. When a constant force is applied to the bond, the energy barrier is linearly decreased, resulting in a characteristic length scale x which signifies the distance from the bound state to the transition state (Fig. 2.2).

Force-induced dissociation of receptor–ligand complexes using AFM can be regarded as an irreversible process, because following dissociation the two binding partners are further separated. Therefore, rebinding of receptor and ligand can be neglected. Due to the small size of the system, the surrounding heat bath causes significant energy fluctuation, resulting in a stochastic escape process. Thus, the probability $N(t)$ to be in the bound state under a linearly increasing force $F = rt$ can be obtained by solving the master equation:

$$\mathrm{d}N(t) = -k_{\mathrm{off}}(rt)N(t), \tag{2.2}$$

where k_{off} is the dissociation rate constant (Strunz et al. 2000).

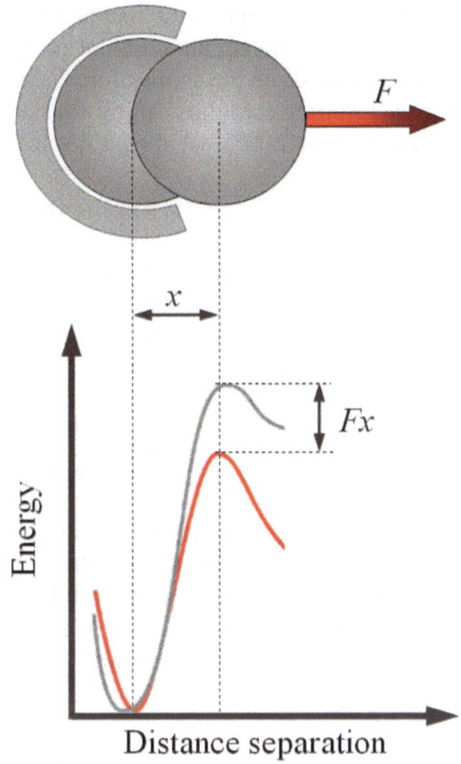

Fig. 2.2 Schematic representation of a single barrier potential under applied force. The unbinding process takes place via a transition state with a characteristic energy barrier. An external force lowers the energy barrier and facilitates the dissociation caused by thermal energy fluctuations

In force spectroscopy experiments, a force is applied on a binding complex which lowers the activation energy barrier and deforms the interaction energy landscape. The lifetime of noncovalent bonds in the absence of external forces is considerably high. By varying the dynamics of pulling on the specific receptor–ligand bonds, detailed structural and kinetic information of the bond rupture can be determined. This approach is very useful to assess kinetic parameters of the unbinding process, including the length and relative heights of the energy barriers.

By extrapolation to zero forces, the kinetic off-rate constant for the dissociation of the complex in solution can be estimated (Fritz et al. 1998).

In typical force spectroscopy experiments, the cantilever is moved upward at a constant pulling speed, which results in a linear force ramp:

$$F(t) = k_c vt, \tag{2.3}$$

where k_c is the spring constant of the cantilever, and v is the retraction speed. When a linker (e.g., PEG) is used, k_c should be substituted with the resulting spring constant of the serial combination of the cantilever and tether $(k_c{}^{-1} + k_{\text{tether}}^{-1})^{-1}$, where k_{tether} is the spring constant of the linker molecule. As the spring constant usually depends on the applied force, a simple but powerful approximation is to use for the resulting combined spring constant at rupture, k_{eff} (Evans and Ritchie 1999; Friedsam et al. 2003), yielding a linear force ramp:

$$F(t) = k_{\text{eff}} v t, \tag{2.4}$$

where $k_{\text{eff}} v = r$ represents the so-called *loading rate*.

From Eq. (2.2), the distribution of the unbinding forces $p(F)$ can be derived with $p(F) = -\,\mathrm{d}N(F)/\mathrm{d}F = -\,\mathrm{d}N(t)/(r\mathrm{d}t)$, which yields $p(F) = k_\mathrm{d}/r.\,\mathrm{N}(F/r)$. Considering Eq. (2.1) it results in

$$p(F) = \frac{k_{\text{eff}}}{r}\exp\left(\frac{Fx_\beta}{k_\mathrm{B}T} - \frac{k_{\text{off}}k_\mathrm{B}T}{rx_\beta}\left(e^{Fx_\beta/k_\mathrm{B}T} - 1\right)\right), \tag{2.5}$$

where k_{off} is the dissociation rate in the absence of an applied force, and r is the loading rate defined as $k_{\text{eff}}\ v$. The applied force decreases the energy barrier, facilitating the dissociation induced by thermal energy fluctuations, which results in a distribution $p(F)$ of the measured unbinding forces.

The most probable unbinding force $F^*(r)$ for the respective loading rate is expressed by

$$F^*(r) = \frac{k_\mathrm{B}T}{x_\beta}\ln\left(\frac{x_\beta r}{k_\mathrm{B}Tk_{\text{off}}}\right). \tag{2.6}$$

The maximum of each force distribution $F^*\ (r)$ is found to scale linearly with the logarithm of the loading rate. Hence, considering a single energy barrier, the unbinding force versus logarithm of the loading rate is characterized by a simple, linear dependence.

To gain an estimate of the rupture force distribution, hundreds of force curves at the same loading rate are typically acquired during a dynamic force spectroscopy experiment (Baumgartner et al. 2000a). The thermal off-rate constant k_{off} and the distance of the barrier from the energy minimum along the pulling coordinate x can be obtained from a linear fit of these data.

2.3.1.1 Other Models Describing the Bond Rupture

When more barriers are involved during dissociation, the dependence follows a sequence of linear regimes, each of which marking a particular barrier (Merkel et al. 1999).

Different approaches have been used for better characterizing the energy landscape of dissociation under applied external force. In particular, analytical expressions for $k(F)$ have resulted from calculations based on the free-energy surfaces model proposed by Hummer and Szabo (2003), in which the most probable unbinding force is dependent on the loading rate through $F^* \sim (\ln r)^{1/2}$. Here, a harmonic potential with a cusp-like feature at x_β was used:

$$U_0(F) = k_\mathrm{B}T \cdot \Delta G_\beta \left(\frac{x}{x_\beta}\right)^2 \tag{2.7}$$

for ($x < x_\beta$) and otherwise $-\infty$. Dudko and co-workers (Dudko et al. 2003) used a linear-cubic surface:

$$U_0(F) = \left(\frac{3}{2}\right) \Delta G_\beta \left(\frac{x}{x_\beta}\right) - 2\,\Delta G_\beta \left(\frac{x}{x_\beta}\right)^3. \tag{2.8}$$

By applying Kramers' theory of escape from a potential well, the expressions for $k(F)$ were determined, allowing the distribution of rupture forces to be obtained. These models in combination with the model described by Evans et al. (1995) can be combined within a single theoretical framework (Dudko et al. 2006):

$$k(F) = k_\mathrm{off}\left(1 - \frac{\mu F x_\beta}{\Delta G_\beta}\right)^{1/\mu - 1} \exp\left(\frac{\Delta G_\beta}{k_\mathrm{B}T}\left[1 - \left\{1 - \mu F x_\beta/\,\Delta G_\beta\right\}^{1/\mu}\right]\right) \tag{2.9}$$

$$\begin{aligned} p(F) = {} & \frac{k(F)}{r} \exp(k_\mathrm{B}Tk_\mathrm{off}/x_\beta r) \\ & \times \exp\left(-\left[k_\mathrm{B}Tk(F)/x_\beta r\right]\left[1 - (\mu F x_\beta/\,\Delta G_\beta)\right]^{1-1/\mu}\right), \end{aligned} \tag{2.10}$$

where $\mu = 2/3$ and $1/2$ correspond to the linear-cubic and quadratic free-energy surfaces, respectively. For $\mu = 1$ and for $\Delta G_\beta \to \infty$ independent of μ, the expression reduces to the result of Evans and Ritchie (1997). When $\mu \neq 1$, permissible values of force F are limited from above by the value of the critical force $F_\mathrm{c} = \Delta G_\beta/(\mu \cdot x_\beta)$ at which the barrier disappears, leading to inaccurate results for $k(F)$ as the Kramers' theory does not apply within this limit.

The most probable rupture force F^* and the variance of the force distributions σ_F can be estimated by

$$F^* \cong \frac{\Delta G_\beta}{\mu}\left\{1 - \left[\frac{k_\mathrm{B}T}{\Delta G_\beta} \ln \frac{k_\mathrm{B}Tk_\mathrm{off}e^{\Delta G_\beta/k_\mathrm{B}T}}{x_\beta r}\right]^\mu\right\} \tag{2.11}$$

$$\sigma_\mathrm{F}{}^2 \cong \frac{(k_\mathrm{B}T\pi)^2}{6x_\beta{}^2}\left[\frac{k_\mathrm{B}T}{\Delta G_\beta} \ln \frac{k_\mathrm{B}Tk_\mathrm{off}e^{\Delta G_\beta/k_\mathrm{B}T + \tilde{\gamma}}}{x_\beta r}\right]^{2\mu - 2}, \tag{2.12}$$

where $\tilde{\gamma} = \gamma^2 - 3/\pi^2\psi''(1) \approx 1.064$, with $\gamma = 0.577$ being the Euler–Mascheroni constant and $\psi''(1) = -2.404$ a particular value of the tetragamma function (Abramowitz and Stegun 1964). As concluded in Eq. (2.6), the most probable rupture force is proportional to $(\ln r)^{\mu}$.

2.3.2 Force Spectroscopy Measurements on Isolated Molecules and Living THP-1

All spectroscopy measurements were performed in force–distance cycle at room temperature using Pico Plus setup (Agilent Technologies, Tempe, USA) equipped with an optical microscope and a CCD camera. To probe the iNKT TCR/CD1d–GSL complex interaction, TCR-coated cantilevers (Veeco Instruments) with spring constants in the range of 0.01–0.03 N m^{-1} were utilized. The spring constants of the cantilevers were determined using the thermal noise method (Butt and Jaschke 1995; Hutter and Bechhoefer 1993), which is a critical parameter for single force spectroscopy measurements. In order to accurately determine the spring constant, it is first necessary to measure the cantilever sensitivity. Usually the cantilever sensitivity (i.e., cantilever deflection detected by the photodiode) is measured by bringing the tip into contact with a hard surface (such as cleaved mica) while the bending of the cantilever is monitored by continuously recording the photodiode signal. This was achieved by acquiring of few force–distance cycles with z-range of 100 nm and a frequency of 1 Hz. Subsequently, the cantilever was withdrawn far away from the surface and the free cantilever movement was recorded. From this complex signal and the already determined cantilever sensitivity, a power density spectrum is calculated by using a Fourier transformation. The spring constant of the cantilever was obtained by using the MATLAB program.

Force measurements on isolated molecules were achieved by recording a 1,000 force–distance cycles, at the same lateral position, varying the z-range (100–300 nm) and duration times (0.2–4 s) to attain different pulling speeds, which resulted in different loading rates. The acquired data were analyzed using a MATLAB program as described by Baumgartner and others (Hinterdorfer et al. 1996; Baumgartner et al. 2000b) to obtain the probability density function (pdf) of the unbinding force and unbinding length. That is achieved by fitting with a Gaussian function, calculated from the mean and the variance of each unbinding event, yielding the most likely unbinding force (maximum of the distribution). A Gaussian of unitary area with the width representing its measuring uncertainty is positioned for each data point of a measured unbinding force value. Subsequently, all Gaussians representing the measured data points are simply summed up to give the final pdf. Hence, pdfs are "continuous" histograms and benefit from the fact that the data accuracy is considered and binning artifacts can be excluded. The loading rates for the iNKT TCR/hCD1d–GSL interaction were calculated by multiplying the tip pulling velocity v

with the effective spring constant $k_{\text{eff}}(r = v \cdot k_{\text{eff}})$, resulting in values of 50–10,000 p Ns^{-1}.

The study was initially focused on the interaction forces between the iNKT TCR and hCD1d–GSL (α-GalCer and OCH12) in vitro using SMFS. The soluble biotinylated iNKT TCR was linked to AFM tips via a heterobifunctional PEG cross-linker (Haselgrübler et al. 1995), which carried an aldehyde group on its free end that chemically coupled streptavidin. The strong streptavidin–biotin bond ensures that the iNKT TCR remains firmly attached to the tip in SMFS experiments. Similarly, the biotinylated hCD1d monomer loaded either with α-GalCer or with OCH12 was immobilized on amino-functionalized mica by the same strategy. A schematic representation of the tip and surface chemistry is depicted in Fig. 2.1a.

In force spectroscopy, the unbinding force measurement is achieved through force–distance cycle by approaching and withdrawing the functionalized tip from the sample surface (Fig. 2.1b). During cantilever approaching, the signal recorded by the photo-detector is constant, as there is no bending of the cantilever (Fig. 2.1b, step 1). Once the tip reaches the surface, the cantilever bends upward and the corresponding deflection is typically represented by the ascending slope (Fig. 2.1b, step 2). When the tip is withdrawn, the cantilever first regains its initial shape. Continuing the tip retraction, if a specific interaction occurs between the receptor and glycolipid, the PEG get stretched and the cantilever bends downward (Fig. 2.1b, step 3) until the bond breaks at a critical force (Fig. 2.1b, step 4), which is a direct measure of the binding strength. This force is termed the "unbinding force" f_u which can be calculated based on Hooke's Law ($F = k \cdot \Delta x$), in which k represents the experimentally obtained spring constant (pN/nm) and Δx is the measured cantilever deflection (nm).

Figure 2.1c shows a typical force–distance curve in which specific interaction between isolated iNKT TCR and CD1d-α-GalCer is monitored. The unbinding force traces are characterized by a particular parabolic-like shape caused by the stretching of the tip-coupled PEG linker before unbinding. To check the specificity of these interactions, 25 μg mL^{-1} of free anti-CD1d monoclonal antibody (mAb) were by injected into solution which largely resulted in a disappearance of the unbinding events because of blockage of the CD1d-α-GalCer molecules on the surface (Fig. 2.1c, inset).

Empirical pdfs displaying the most probable unbinding forces were calculated based on unbinding force analysis. Figure 2.1d shows a representative example of a pdf for iNKT TCR/CD1d-α–GalCer interaction obtained from more than 100 unbinding events in which the maximal probable force was 39 ± 2 pN for a loading rate of 1,260 pN s^{-1} (Fig. 2.1d, black curve). The binding probability, which represents the frequency of occurrence of specific interaction events in force distance cycles, was 17.4%. Upon injecting the blocking antibody, the binding probability decreased to 5.7% (Fig. 2.1d, red dotted curve). The same experimental procedure and analysis sequence was applied to the CD1d-OCH12 complex (data not shown) for which the binding probability decreased from 8.6 to 2.6% after injection of blocking antibodies. To further check the interaction specificity, a bare

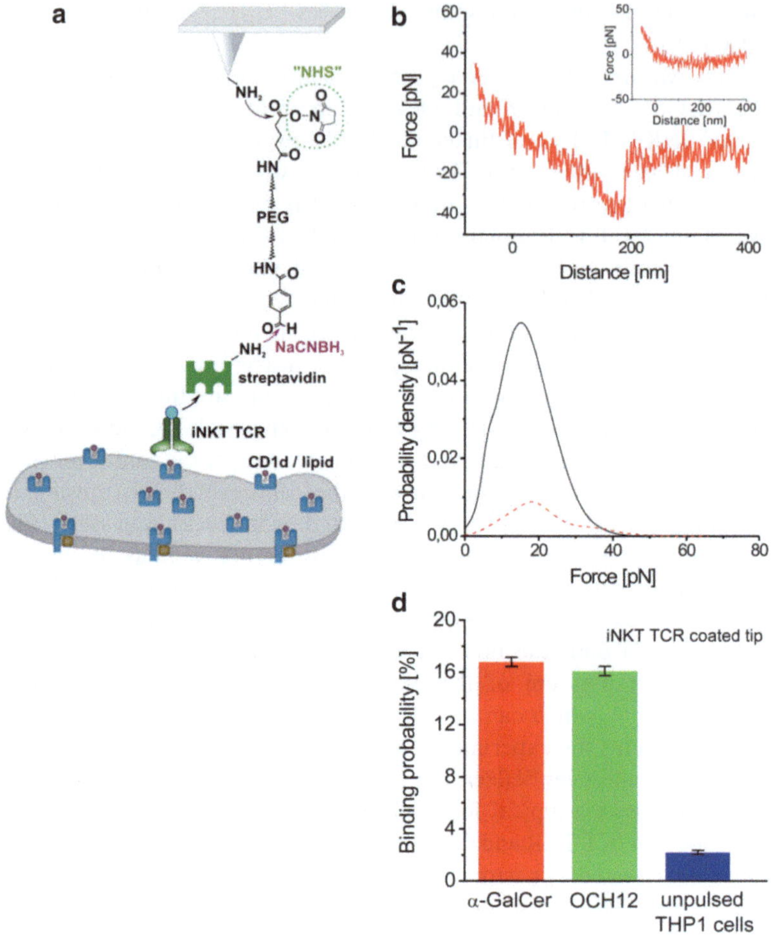

Fig. 2.3 Schematic diagram of single-molecule force measurements on live cells. The THP1 CD1d cells pulsed with iNKT agonist α-GalCer or OCH12 attached onto poly-L-lysin coated glass coverslips. (**b**) Typical force–distance cycle on living cells showing a receptor–lipid interaction. (*Inset*) Blocking of the specific interaction by free anti-CD1d monoclonal antibody. (**c**) Distribution of unbinding forces of iNKT TCR/CD1d-α-GalCer (*black line*) based on 150 unbinding events out of 1,000 force–distance cycles. (**d**) Comparison of binding probabilities of iNKT TCR coated tip on live CD1d-transfected THP1 cells either pulsed with α-GalCer (*red*) or OCH12 (*green*) or unpulsed (*blue*)

tip was used instead of a receptor-coated tip. Under these conditions, only very few rupture events were observed.

In the second part of the study, we assessed the iNKT TCR affinity for hCD1d–GSL complexes at the THP1 cell surface. SMFS measurements on living cells were carried out with iNKT TCR modified tips probing CD1d molecules loaded with α-GalCer or OCH12 (Fig. 2.3a).

A typical force curve recorded on living cells during AFM tip withdrawal exhibits a particular slope characteristic of soft substrates (Fig. 2.3b). As anticipated from previously experiments performed on isolated molecules, when the lipids were embedded in the cell membrane, significant recognition events of CD1d-lipids by the iNKT TCR modified AFM tips occurred as well (Fig. 2.3c, black line). The pdf diagram (Fig. 2.3c) reveals the maximum of the force distribution, which was found to be 15 ± 3 pN for CD1d–α-GalCer. The specificity of this interaction was proven by using CD1d mAb, which dramatically decreased the binding probability (Fig. 2.3c, red dotted line). For CD1d-OCH12 a similar behavior was found.

The specificity of the interaction was further confirmed by additional force–distance cycles performing on live CD1d transfected THP1 cells pulsed with lipids or unpulsed, using the same receptor coated tip for reference. Significant number of unbinding events was only observed with lipid pulsed CD1d transduced THP1 cells for which the binding probability values were 16.8 ± 0.37% for CD1d-αGalCer and 16.1 ± 0.36% for CD1d-OCH12, respectively. In contrast, the binding probability was only 2.2 ± 0.15% for unpulsed CD1d THP1 cells (Fig. 2.3d). These results clearly support the assumption that the iNKT TCR binds specifically to CD1d transfected cells only when loaded with lipids.

The unbinding force pdfs for living cells were compared with the ones obtained for isolated molecules, revealing a similar trend. Significantly higher mean values of the unbinding force for α-GalCer with respect to OCH12 were observed. By varying the dynamics of the force-pulling experiments, the nature of these interactions was further analyzed.

As previously discussed, the dissociation process in SMFS is driven by an external force applied to the complex that makes the bond more susceptible to reaching the unbound state by overcoming the activation barrier. Consequently, when the external force is slowly applied to the complex (slow loading rate), dissociation will occur at low forces and vice versa.

In agreement with the single-barrier model (Bell 1978; Evans and Ritchie 1997), the unbinding forces measured at different loading rates for iNKT TCR–CD1d-lipids complexes were found to increase linearly with the logarithm of the loading rate. The unbinding force dependence on the loading rate for iNKT TCR–CD1d-lipid interactions determined for both isolated molecules and for living cells is shown in Fig. 2.4. It was found that for both systems the unbinding forces for iNKT TCR–CD1d-α-GalCer were significantly higher than for iNKT TCR–CD1d-OCH12 for all loading rates.

The dynamic parameter such as kinetic off-rate, k_{off}, as well as the distance from the energy minimum to the transition state, x_β, was estimated by fitting the force spectroscopy plot (Fig. 2.4a, b) to Eq. (2.6), which estimates the most probable unbinding force at a certain loading rate.

The kinetic off-rate from the living cell measurements for α-GalCer and OCH12 was found to be $k_{off} = 0.50 \pm 0.52\ s^{-1}$ and $1.04 \pm 1\ s^{-1}$, respectively. These values compare very favorably with SPR data: $0.39 \pm 0.01\ s^{-1}$ for α-GalCer and $1.00 \pm 0.12\ s^{-1}$ for OCH12 (McCarthy et al. 2007).

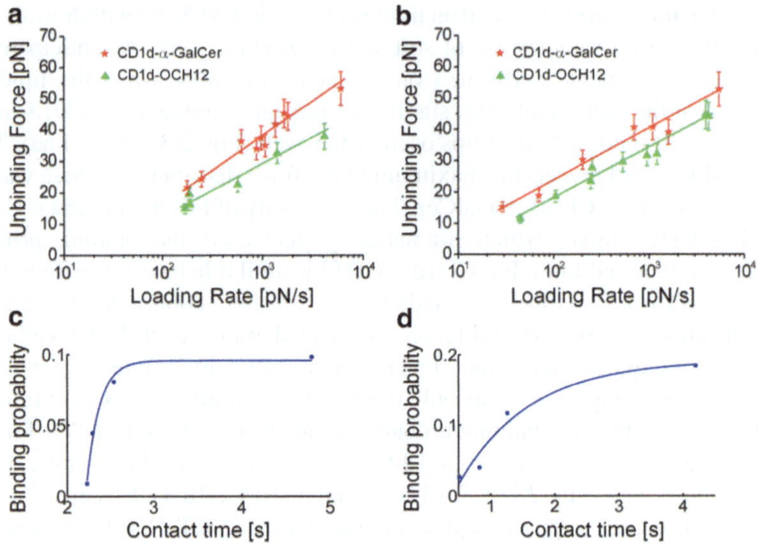

Fig. 2.4 Loading rate dependence of unbinding force. Unbinding force spectra of iNKT TCR bound to CD1d-α-GalCer (*red stars*) and CD1d-OCH12 (*green triangles*) on isolated molecules (**a**) and on living THP1 cells (**b**). The uncertainty in the determination of the unbinding force is represented by error bars. Binding probability as a function of contact time for CD1d-α-GalCer (**c**) and OCH12 (**d**) on living THP1 cells

These results indicate that the kinetic rate constant for the iNKT TCR–CD1d-α GalCer and iNKT TCR–CD1d-OCH12 complexes, k_{off}, is in good agreement with the data previously derived from surface-plasmon resonance experiments.

2.3.3 Kinetic On-Rate Measurements

In addition to determining the kinetic off-rate constant value of these complexes, we estimated the kinetic on-rate parameter k_{on} on living cells by varying the receptor–CD1d-lipid interaction time. The experimentally obtained values for CD1d–αGalCer (Fig. 2.4c) and CD1d-OCH12 (Fig. 2.4d) reveal an exponential increase. Longer dwell times resulted in a higher binding probability until reaching a saturation plateau.

Fitting the binding probability versus dwell time to $P = A\ (1-\exp(-(t-t_0)/\tau))$ in which A represents the maximum observable binding probability and t_0 the lag time, the characteristic interaction time τ was estimated (Atkins 1998). Subsequently, the kinetic on-rate was derived from τ using the equation $k_{on}(\tau c_{eff})^{-1}$ (Baumgartner et al. 2000b; Rankl et al. 2008), where c_{eff} is the effective concentration of iNKT TCR on the AFM tip expressed by the inverse of the effective volume of a half sphere with effective radius r_{eff}, in which the tip-bound iNKT TCR molecule can move freely. By summing the cross-linker length in equilibrium

(3 nm) and the diameter of streptavidin plus iNKT TCR (6 nm), the effective radius was determined.

An estimate of the dissociation constant $K_d = k_{off}/k_{on}$ was calculated, based on the obtained values for dissociation and association rate constants. For CD1d–αGalCer, the dissociation constant was found to be about 40.32 μM, whereas for CD1d−OCH12 the value was 630 μM. These values were put into perspective by comparing them to the data obtained from surface-plasmon resonance measurements (McCarthy et al. 2007).

From data analysis, one can conclude that the dissociation rate of CD1d−OCH12 from the receptor is about twofold larger than CD1d–α-GalCer. In addition, the kinetic off-rate values from the isolated molecule measurements are relatively close when compared to the living cell experiments. The larger dissociation rate on isolated molecules can be attributed to complex instability in vitro, allowing us to conclude that CD1d–GSL complexes are more stable on living cells under physiological conditions than when isolated in solution.

Unlike the dissociation rate which was close to that obtained from surface-plasmon resonance, the values for k_{on} determined by SMFS measurements appear to be scaled down by about one order of magnitude with respect to the values reported from SPR (Table 2.1), which can be explained on the basis of the rough estimate of the critical parameter represented by the effective volume (i.e., effective radius) leading to a source of errors for determining the true kinetic on-rate.

However, for both force SMFS and SPR, the k_{on} ratio between CD1d–α-GalCer and CD1d-OCH12 is comparable (~8) so that it appears evident that a similar trend exists for dissociation constant K_d (cf. Table 2.1). For both isolated molecules and living cells, higher unbinding forces are required to dissociate the iNKT TCR from the CD1d–α-GalCer complex than from CD1d-OCH12 at the same loading rate, suggesting that unlike OCH12, the CD1d–α-GalCer is more resistant to the external forces.

These results are consistent with previous findings (McCarthy et al. 2007) demonstrating that shortening the phytosphingosine chain reduced the affinity of binding to the iNKT TCR, leading to changes in the iNKT cell immunological synapse, polarization of the iNKT cell cytotoxic granules and iNKT cell activation.

Also, this study emphasizes the advantage of single-molecule force spectroscopy technique by which critical parameters can be determined directly on living cells.

2.4 TREC Imaging of CD1d-Glycolipid Complex on THP1 Cells

Simultaneous recognition, visualization, and quantifications of receptor binding over cell surfaces with high spatial accuracy are important tasks in the life sciences and especially in molecular cell biology. There are many different microscopy techniques, such as epi-fluorescence microcopy, photo-activated localization

Table 2.1 Kinetics measurements

	Force spectroscopy measurements on isolated molecules		Force spectroscopy measurements on live THP1 cells				SPR [8]		
hCD1d–GSL complex	x_β (nm)	k_{off} (s^{-1})	x_β (nm)	k_{off} (s^{-1})	k_{on} ($M^{-1}s^{-1}$)	K_d (μM)	k_{off} (s^{-1})	k_{on} ($M^{-1}s^{-1}$)	K_d (μM)
hCD1d–αGalCer	0.43 ± 0.10	1.94 ± 1.44	0.57 ± 0.14	0.50 ± 0.52	1.24×10^4	40.32	0.39 ± 0.01	3.31×10^5	1.29 ± 0.08
hCD1d–OCH12	0.56 ± 0.15	2.49 ± 1.47	0.59 ± 0.18	1.04 ± 1.0	1.65×10^3	630	1.00 ± 0.12	3.70×10^4	23.3 ± 1.41

Comparison of obtained kinetic parameters with respect to previous results by SPR

microscopy (PALM), stimulated emission depletion microscopy (STED), single-particle tracking, single dye tracing, or scanning electron microscopy, that might be used for these purposes. However, they have the drawbacks of limited resolution, lack of topographic information, and/or inapplicability under physiological conditions. On the other hand, AFM, which represents a nonoptical microscopy, offers a unique solution to obtain possible high-resolution topographical images at the nanometer scale and single-molecule interaction forces of biological specimens (e.g., proteins, DNA, membranes, cells, etc.) under ambient conditions and without the need for rigorous sample preparation or labeling (Horber and Miles 2003). With the recent development in AFM, a new fastest and most straightforward method, called "simultaneous Topography and RECognition imaging" (TREC) (Stroh et al. 2004a, b; Ebner et al. 2005), it becomes possible to quickly obtain the local distribution of receptors on cell surface with a lateral resolution of 5 nm (Chtcheglova et al. 2007). In this AFM mode, the surface of a biological specimen is scanned with a biofunctionalized tip at regular imaging speed, yielding a map of specific ligand-binding sites together with a topographic image (Hinterdorfer and Dufrene 2006; Stroh et al. 2004a, b).

2.4.1 Principle of Simultaneous Topography and Recognition Imaging

The operating principle of TREC is based on the MAC mode, in combination with a ligand attached to the AFM tip on the end of an elastic PEG linker with defined length (~6 nm). The flexible PEG linker allows the ligand molecule to easily bind the receptor on the cell surface. During scanning, the functionalized cantilever is oscillated close to its resonance frequency. When the specific recognition occurs between the ligand on the tip and the receptor site on the cell surface, oscillation amplitude of the cantilever decreases, which is evident of the binding sites. In TREC, the cantilever oscillation amplitude is divided into two parts (i.e., lower and upper parts with respect to the baseline of the oscillation) and processed in different paths by using a specially designed electronic circuit (PicoTREC, Agilent). While the lower part of the signal is used for generation of the topography image, the upper part reflects recognition events and gives the recognition image (Fig. 2.5).

The starting point for a successful TREC experiment is a cantilever with a low (~1) quality (Q)-factor. The Q-factor represents the "memory" of the cantilever. A low Q-factor (and therefore a low "memory" ability) ensures that an amplitude reduction in the lower part of the oscillation (originating from a change in the topography) is sufficiently separated in time from amplitude reductions in the upper part of the oscillation (originating from molecular recognition between the ligand and the receptor).

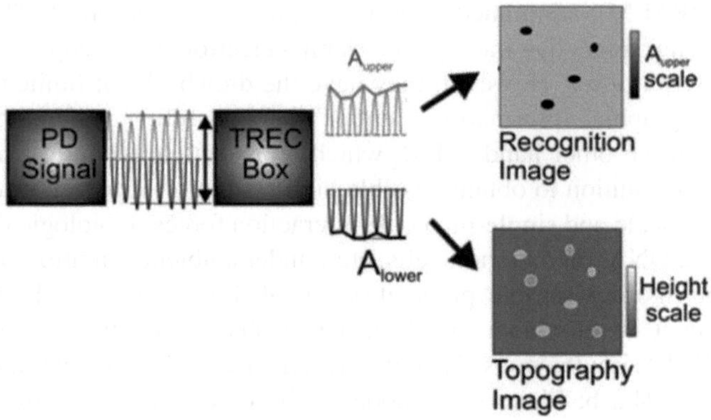

Fig. 2.5 Principle of TREC. The cantilever oscillation is divided into two parts in the TREC box. While the envelope of the *upper part* yields the recognition image, the *lower part* provides for the topography image

Consequently, only the lower part of the sinusoidal oscillation is fed into the feedback loop and is thereby held constant to obtain the unbiased surface topography. The upper part of the oscillation, solely containing information on recognition between ligand and receptor, is recorded to generate a recognition image simultaneously to the topography image.

2.4.2 Adjustment of Imaging Parameters

In order to reveal reliable recognition sites on cell membrane, imaging parameters such as feedback loop, oscillation amplitude, and driving frequency should be adjusted properly.

First, in conventional MAC mode AFM, the peak-to-peak value of oscillating amplitude of bare (i.e., without any tip functionalization) AFM tip is utilized as feedback parameter and called "full amplitude feedback." Using feedback loop, the full amplitude (FA) is held constant during scanning by adjusting the voltage applied to the piezo actuator which controls the z distance between the tip and the surface. However, during TREC imaging in which the ligand functionalized AFM cantilever tip is used, the modified "half amplitude (HA) feedback loop" has to be used to obtain the true surface topography. When molecular recognition occurs between tip-tethered ligand and its receptor on cell surface, the flexible PEG linker complex is stretched due to the upward movement of the cantilever oscillation and the top peaks of the oscillation is reduced. Thus, both topographical features (at the bottom peaks) and molecular recognition (at the top peaks) affect the value of the FA (Preiner et al. 2009). Therefore, using the conventional feedback loop for TREC leads to errors in the height value of the topographical

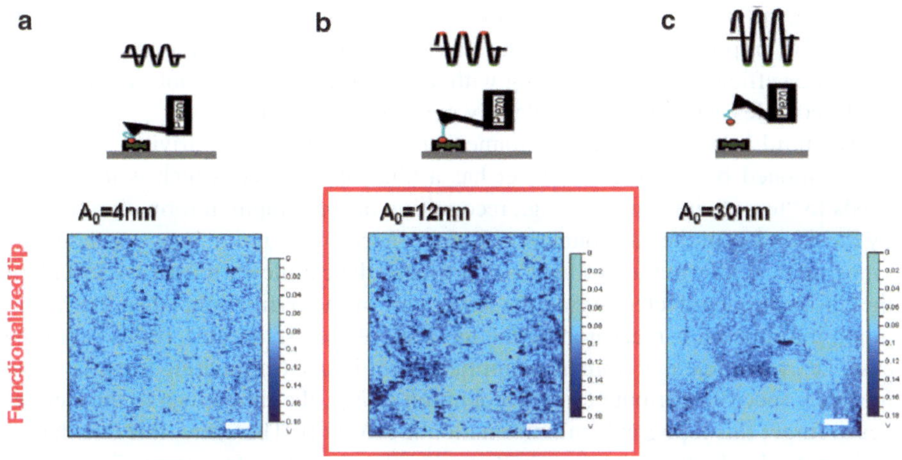

Fig. 2.6 Three different regimes of oscillation amplitudes; No recognition signal at (**a**) lowest (4 nm) and (**c**) highest (30 nm) amplitude. Pronounced recognition signal (**b**) was detected at efficient amplitude (12 nm)

image since the feedback itself cannot discriminate the different contributions to the amplitude reduction. The stretching of the polymer linker exhibits a strong nonlinear behavior so that for small linker extensions the force acting on the cantilever is negligible. It is, however, much higher at extensions approaching the linker's contour length. In other words, the value of the amplitude in the lower part of the oscillation is not affected by the linker stretching. This part is only affected by changes in topography, and is therefore used as feedback parameter in the so-called HA feedback loop, yielding the unbiased surface topography.

The adjustment of oscillation amplitude is the second important parameter of TREC imaging in which the proper recognition signal only occurs when the ligand–receptor complex survive until the tip has moved laterally away from the position of the receptor molecule. Therefore, the choosing of ideal amplitude, which is physically determined by the stretching behavior of the linker molecule, plays a critical role in TREC experiments. There are three different regimes of oscillation amplitudes as sketched in Fig. 2.6. For small amplitudes, the ligand (TCR) molecule has bound to the receptor (CD1D complex) on the cell surface. However, the linker does neither stretch nor exert an efficient force on the cantilever in the upward swing, generating no recognition signal (Fig. 2.6a). When the amplitude has been increased to the second regime, the linker is efficiently stretched without detaching the TCR from the CD1d complex in the upper part of the oscillation, resulting in a pronounced recognition signal and shown as dark spots in recognition image (Fig. 2.6b). When the amplitude is further increased in the last regime (III), the peak-to-peak value of the amplitude is already higher than the contour length of the linker molecule. Therefore, the TCR molecule unbinds from the CD1d complex as soon as the every top peak of the first oscillation cycle. Because continuous binding while scanning over the CD1d complex is prevented in this case, no recognition signals are generated (Fig. 2.6c). Consequently, the oscillation amplitude of the AFM tip must properly be adjusted as in regime (II) to ensure sufficient linker stretching without rupturing the TCR molecule from the CD1d complex when scanning laterally over the cell surface.

The third important imaging parameter is a properly chosen driving frequency. As mentioned before, the cantilever has a kind of memory, which is the time it needs to "forget" information (e.g., recognition or topography information) caused by the damping of the environment. To obtain a true recognition image without any topographical information or feedback artifacts, the cantilever must have lost all of the information collected during the lower part of the oscillation, when the recognition information is measured (at the top peak of the oscillation). If this is not the case, i.e., when features from the amplitude-error image (originating from the finite feedback speed) contribute to the recognition image, the time given to the cantilever to forget this topographical information is too short. Therefore, this time has to be increased, which can be done by lowering the excitation frequency. The contribution of the amplitude error to the contrast in the recognition image and its dependence on the excitation frequency have been demonstrated experimentally and explained in detail (Preiner et al. 2009).

2.4.3 Nanomapping of CD1d-Glycolipids Complexes on THP1 Cells by Using TREC

The localization and distribution of glycolipids pulsed CD1d complexes were examined by using TREC imaging. Before identifying the binding site, magnetically coated AFM tip was modified with the soluble biotinylated iNKT TCR via a heterobifunctional PEG cross-linker (Fig. 2.3). The affinity of iNKT TCR to glycolipid loaded CD1d complexes is already shown in the literature (Bozna et al. 2011; McCarthy et al. 2007). In this study, in order to locally identify the CD1d glycolipid complexes on THP1 cell surfaces, expressed CD1d molecules, the cells were incubated with three different glycolipids (α-GalCer, C20:2, and OCH12) for 16 and 4 h.

The oscillation of functionalized AFM tips was adjusted (~8 nm) slightly smaller than the extended PEG linker, which has length of ~10 nm to allow binding iNKT-TCR molecule on the tip to the glycolipid loaded CD1d molecule complex on the cell surface during scanning.

The recognized CD1d glycolipids microdomains were acquired by scanning THP1 cells which were loaded with α-GalCer, C20:2, and OCH12, respectively, and showed in Fig. 2.7 as representative simultaneous topography and recognition images. Figure 2.7a, b is the image of the control cell group which was not loaded with any glycolipids. The statistical analysis of the area distribution of microdomains was achieved by measuring recognition sites in four different areas (~1 × 1 μm^2) of the same THP1 cell surface and is shown in the right panels of Fig. 2.7.

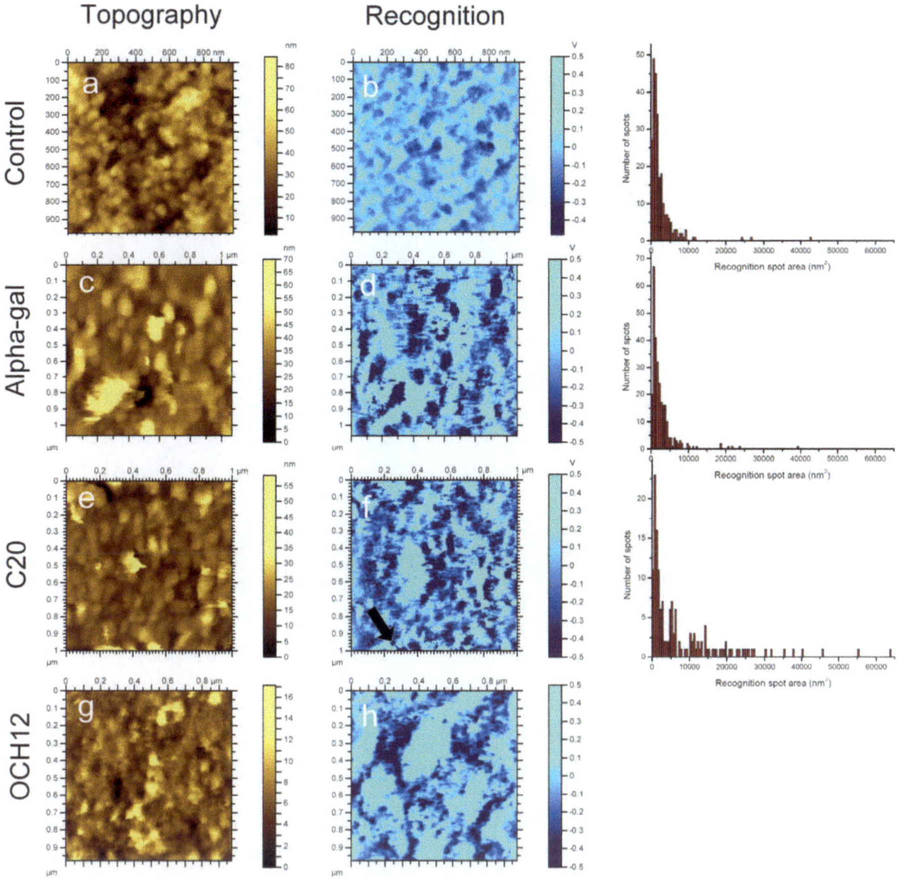

Fig. 2.7 Topography and recognition images of control (**a**, **b**), α-GalCer (**c**, **d**), C20 (**e**, **f**), and OCH12 (**g**, **h**) loaded THP1 cells. *Right panels* represent the area distribution of microdomains corresponding recognition images detected in four different areas of the same cell

In the case of the cells that were pulsed with α-GalCer and C20:2 for 16 h, the dimension (area) of the recognition spots (microdomains of the CD1d glycolipid complexes) were detected from the amplitude reduction, arising from interactions between iNKT-TCR and CD1d-glycolipid complex, and were revealed in recognition images (Fig. 2.7d, f). According to topographical images (Fig. 2.7c, e), the cellular membrane features were organized into typical spherical form with ~100–150 nm in length and heights varying from ~20 to ~70 nm. Analyzed recognition images (Fig. 2.7, right panels) showed that the α-GalCer and C20:2 loaded CD1d proteins formed microdomains with the dimension (area) from ~250 to ~10,000 nm^2 (mean ± SD, 2,219 ± 989, $n = 523$) and distributed nonuniformly. On the other hand, low-affinity profile of iNKT TCR to OCH12 affected the distribution and area size of the CD1d-OCH12 complexes. A closer look at recognition spots (Fig. 2.7h) reveals that they consist of larger connected microdomains.

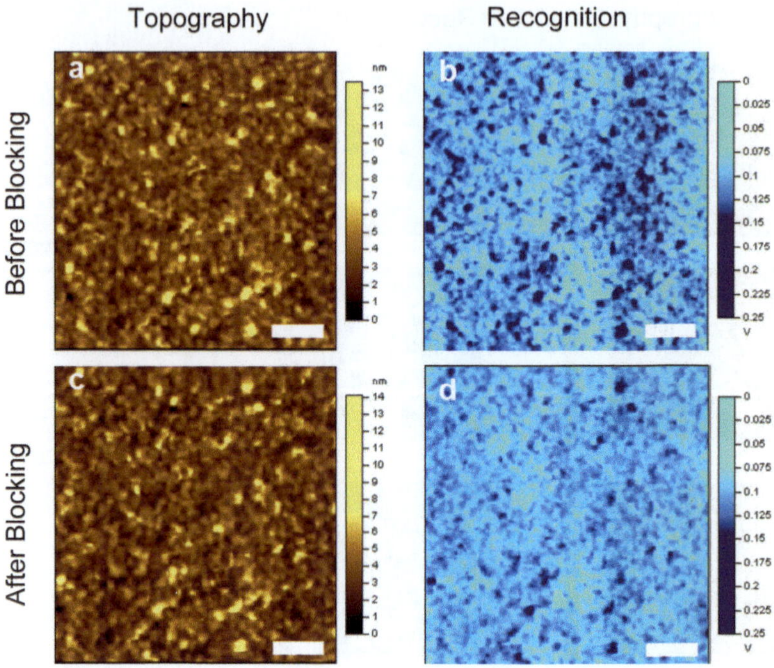

Fig. 2.8 Specificity of iNKT TCR functionalized AFM tip to CD1d-α-GalCer complexes. Topography (**a**, **c**) and recognition (**b**, **d**) images before and after addition of free anti-CD1d antibody solution. *Scale bar* is 100 nm.

The analyzed dimension area of the recognition spots was increased up to 30,000 nm^2 (mean ± SD, 8,197 ± 6,925, $n = 155$). Specificity of iNKT-TCR functionalized AFM tips to glycolipid pulsed CD1d complex was proven by using both control cell group, which was not loaded with any glycolipids, and blocking experiment. As expected and shown in Fig. 2.7b, no recognition event was detected when scanning the control cell group. Furthermore, specificity of recognition signal was also proven by adding anti-CD1d antibody to block CD1d molecules on the cell surface, while scanning the sample at the same position. The recognition clusters (Fig. 2.8b) partly disappeared (Fig. 2.8d) 1 h after addition of anti-CD1d antibody, whereas no changes in topography images (Fig. 2.8a, c) have been observed.

THP1 cells also pulsed with same glycolipids (α-GalCer, C20:2 and OCH12) for 4 h to detect the effect of short incubation time of glycolipids to distribution of CD1d-glycolipid complexes. When the cells were loaded with α-GalCer for 4 h, we could not identify pronounced recognition spots of CD1d-α-GalCer complexes (Fig. 2.9b). However, CD1d-C20:2 complexes were successfully detected even after 4 h incubation time (Fig. 2.9d). It is recently reported that while α-GalCer underwent intracellular loading and was presented on CD1d more slowly (longer internalization), C20:2 showed rapid kinetics of direct loading (without internalization) (Im et al. 2009). These findings are in agreement with our observations. In contrast to THP1 cell, incubated with OCH12 for 16 h, the cells that were pulsed

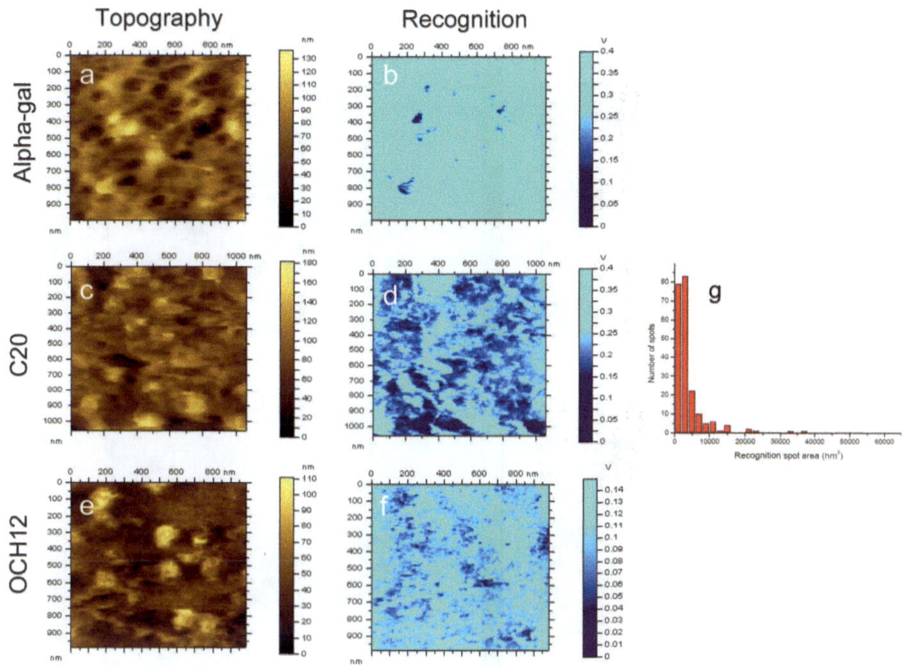

Fig. 2.9 Distribution and localization of CD1d molecules when they were loaded with α-GalCer (**a**, **b**), C20 (**c**, **d**), and OCH12 (**e**, **f**) for 4 h. The area distribution (*k*) of CD1d-C20 complex corresponding recognition images detected in two different areas of the same cell

with same glycolipids for 4 h did not show any pronounced recognition event (Fig. 2.9f). This is most likely caused by improper presentation of OCH12 on CD1d molecule during selected time period. Thereby the binding strength between iNKT-TCR functionalized AFM tip and CD1d-OCH12 complex was not sufficient to stretch PEG linker enough to create proper amplitude reduction for selected incubation time. After analyzing the dimension area of recognition images for 4 h C20:2 loaded cells, CD1d-C20:2 complexes again formed similar sizes (with a dimension area from ~250 to ~10,000 nm^2) microdomains with the 16 h pulsed ones (Fig. 2.9g).

Blocking experiments were also applied to show specificity of modified AFM tip by adding anti-CD1d antibody to liquid cell while acquiring the topography (Fig. 2.10a, c) and recognition (Fig. 2.10b, d) images on the same area of 4 h C20:2 pulsed THP1 cell surfaces. Almost all dark spots (recognition spots) in Fig. 2.10b were abolished in 1 h after injection of anti-CD1d antibody (Fig. 2.10d).

2.5 Concluding Remarks

"Simultaneous Topography and RECognition" (TREC) imaging is a combination of high-resolution AFM topography imaging with single-molecule force microscopy. This powerful AFM technique not only yields fine structural details about

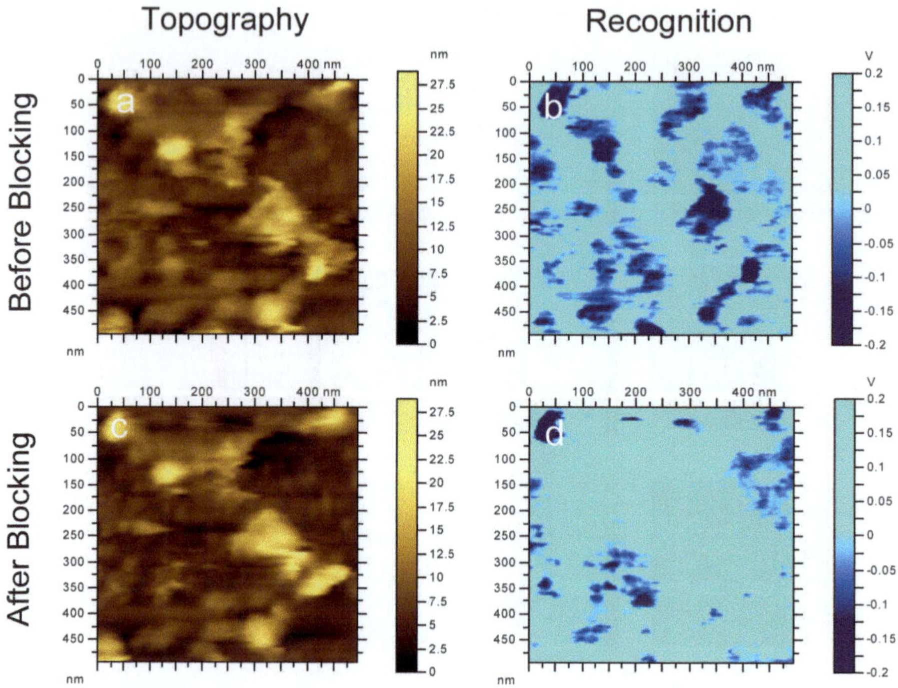

Fig. 2.10 Specificity of iNKT TCR functionalized AFM tip to 4 h pulsed C20-CD1d complex. (**a–c**) topographical images simultaneously recorded with recognition maps (**b–d**). After addition of free anti-CD1d antibody solution into the liquid, most of the recognition spots disappeared (**d**), while topographical images remained unchanged (**c**)

topography but also senses biochemical composition of native biological samples under physiological conditions. The present work shows a major advantage of TREC over optical approaches to cells with a spatial topographical and recognition resolution of ~5 nm. In the presented work, TREC has successfully been exploited to identify CD1d glycolipid complex sites on THP1 cells and to colocalize their position with membrane topographical features. The recognition events which were shown as dark spots in recognition image were revealed with the diameter ranging between 25 and 160 nm (Fig. 2.8f). Since the diameter of the iNKT-TCR/CD1d complex is ~3.5 nm (Borg et al. 2007) and the free orientation of the PEG linker allows binding 10 nm before and 10 nm after the binding sites, the expected diameter of a single-receptor recognition spot is 23.5 nm, which is the minimal patch size of recognition spots we observed in the recognition images (see the arrow in Fig. 2.7f).

Overall, TREC imaging allows to detect single-molecular interactions, and thus to visualize, identify, and quantify local receptor binding sites and assign their locations to the topographical features of cell surfaces. This study illustrates the great potential of TREC for the investigation and localization of membrane proteins on cell surfaces with several piconewton force resolution and a positional accuracy of a few nanometers. For these reasons, TREC is a promising tool for the identification and location of receptor binding sites on cells, organelles, and other subcellular structures.

References

Abramowitz M, Stegun IA (1964) Handbook of mathematical functions with formulas, graphs, and mathematical tables. Dover, New York, NY

Ahmad FS, Chtcheglova LA, Mayer B, Kuznetsov SA, Hinterdofer P (2011) Nanosensing of Fcγ receptors on macrophages. Anal Bioanal Chem 399(7):2359–2367

Ashkin A (1997) Optical trapping and manipulation of neutral particles using lasers. Proc Natl Acad Sci USA 94:4853–4860

Atkins PW (1998) Physical chemistry. Oxford University Press, Oxford

Baumgartner W, Hinterdorfer P, Schindler H (2000a) Data analysis of interaction forces measured with the atomic force microscope. Ultramicroscopy 82:85–95

Baumgartner W, Gruber HJ, Hinterdorfer P, Drenckhahn D (2000b) Affinity of trans-interacting VE-cadherin determined by atomic force microscopy. Single Mol 1:119–122

Bell GI (1978) Models for the specific adhesion of cells to cells. Science 200:618–627

Borg NA, Wun KS, Kjer-Nielsen L, Wilce MCJ, Pellicci DG, Koh R, Besra GS, Bharadwaj M, Godfrey DI, McCluskey J, Rossjohn J (2007) CD1d-lipid-antigen recognition by the semi-invariant NKT T-cell receptor. Nature 448:44–49

Bozna BL, Polzella P, Rankl C, Zhu R, Salio M, Shepherd D, Duman M, Cerundolo V, Hinterdorfer P (2011) Binding strength and dynamics of iNKT TCR/CD1d glycosphingolipids interaction on living cells by single molecule force spectroscopy. J Biol Chem 286:15973–15979

Butt HJ, Jaschke M (1995) Calculation of thermal noise in atomic force microscopy. Nanotechnology 6:1–7

Cerundolo V, Silk JD, Hajar MS, Salio M (2009) Harnessing invariant NKT cells in vaccination strategies. Nat Rev Immunol 9:28–38

Chtcheglova LA, Waschke J, Wildling L, Drenckhahn D, Hinterdorfer P (2007) Nano-scale dynamic recognition imaging on vascular endothelial cells. Biophys J 93:L11–L13

Dudko OK, Filippov AE, Klafter J, Urbakh M (2003) Beyond the conventional description of dynamic force spectroscopy of adhesion bonds. Proc Natl Acad Sci USA 100:11378–11381

Dudko OK, Hummer G, Szabo A (2006) Intrinsic rates and activation free energies from single-molecule pulling experiments. Phys Rev Lett 96:108101–108104

Ebner A, Kienberger F, Kada G, Stroh CM, Geretschläger M, Kamruzzahan ASM, Wildling L, Johnson WT, Ashcroft B, Nelson J, Lindsay SM, Gruber HJ, Hinterdorfer P (2005) Localization of single avidin-biotin interactions using simultaneous topography and molecular recognition imaging. Chemphyschem 6:897–900

Evans E, Ritchie K (1997) Dynamic strength of molecular adhesion bonds. Biophys J 72:1541–1555

Evans E, Ritchie K (1999) Strength of a weak bond connecting flexible polymer chains. Biophys J 76:439–2447

Evans E, Ritchie K, Merkel R (1995) Sensitive force technique to probe molecular adhesion and structural linkages at biological interfaces. Biophys J 68:2580–2587

Florin EL, Moy VT, Gaub HE (1994) Adhesion forces between individual ligand–receptor pairs. Science 264:415–417

Friedsam C, Wehle AK, Kühner F, Gaub HE (2003) Dynamic single-molecule force spectroscopy: bond rupture analysis with variable spacer length. J Phys Condens Matter 15:S1709–S1723

Fritz J, Katopidis AG, Kolbinger F, Anselmetti D (1998) Force-mediated kinetics of single P-selectin/ligand complexes observed by atomic force microscopy. Proc Nat Acad Sci USA 95:12283–12288

Fujii S, Shimizu K, Smith C, Bonifaz L, Steinman RM (2003) Activation of natural killer T cells by a-galactosylceramide rapidly induces the full maturation of dendritic cells *in vivo* and thereby acts as an adjuvant for combined CD4 and CD8 T cell immunity to a coadministered protein. J Exp Med 198:267–279

Haselgrübler T, Amerstorfer A, Schindler H, Gruber HJ (1995) Synthesis and applications of a new poly(ethylene glycol) derivative for the crosslinking of amines with thiols. Bioconjug Chem 6:242–248

Hermans IF, Silk JD, Gileadi U, Salio M, Mathew B, Ritter G, Schmidt R, Harris AL, Old L, Cerundolo V (2003) NKT cells enhance CD4+ and CD8+ T cell responses to soluble antigen in vivo through direct interaction with dendritic cells. J Immunol 171:5140–5147

Hinterdorfer P, Dufrene YF (2006) Detection and localization of single molecular recognition events using atomic force microscopy. Nat Methods 3:347–355

Hinterdorfer P, Baumgartner W, Gruber HJ, Schilcher K, Schindler H (1996) Detection and localization of individual antibody–antigen recognition events by atomic force microscopy. Proc Natl Acad Sci USA 93:3477–3481

Horber JKH, Miles MJ (2003) Scanning probe evolution in biology. Science 302:1002–1005

Hummer G, Szabo A (2003) Kinetics from nonequilibrium single-molecule pulling experiments. Biophys J 85:5–15

Hutter JL, Bechhoefer J (1993) Calibration of atomic-force microscope tips. Rev Sci Instrum 64:1868–1873

Im JS, Arora P, Bricard G, Molano A, Venkataswamy MM, Baine I, Jerud ES, Goldberg MF, Baena A, Yu KOA, Ndonye RM, Howell AR, Yuan W, Cresswell P, Chang YT, Illarionov PA, Besra GS, Porcelli SA (2009) Kinetics and cellular site of glycolipid loading control the outcome of natural killer T cell activation. Immunity 30:888–898

Kawano T, Cui J, Koezuka Y, Toura I, Kaneko Y, Motoki K, Ueno H, Nakagawa R, Sato H, Kondo E, Koseki H, Taniguchi M (1997) Science 278:1626–1629

Kellermayer MSZ (1997) Delayed dissociation of in vitro moving actin filaments from heavy meromyosin induced by low concentrations of Triton X-100. Biophys Chem 67:199–210

Kinjo Yuki et al (2006) Natural killer T cells recognize diacylglycerol antigens from pathogenic bacteria. Nat Immunol 7:978–986

Koch M, Stronge VS, Shepherd D, Gadola SD, Mathew B, Ritter G, Fersht AR, Besra GS, Schmidt RR, Jones EY, Cerundolo V (2005) The crystal structure of human CD1d with and without α-galactosylceramide. Nat Immunol 6:819–826

Kronenberg M (2005) Toward an understanding of NKT cell biology: progress and paradoxes. Annu Rev Immunol 23:877–900

Leckband D, Israelachvili J (1993) Molecular basis of protein function as determined by direct force measurements. Enzyme Microb Technol 15:450–459

Leckband DE, Israelachvili JN, Schmitt FJ, Knoll W (1992) Long-range attraction and molecular rearrangements in receptor-ligand interactions. Science 255:1419–1421

Lee GU, Kidwell DA, Colton RJ (1994) Sensing discrete streptavidin biotin interactions with atomic-force microscopy. Langmuir 10:354–357

Mattner J et al (2005) Exogenous and endogenous glycolipid antigens activate NKT cells during microbial infections. Nature 434:525–529

McCarthy C, Shepherd D, Fleire S, Stronge VS, Koch M, Illarionov PA, Bossi G, Salio M, Denkberg G, Reddington F et al (2007) The length of lipids bound to human CD1d molecules modulates the affinity of NKT cell TCR and the threshold of NKT cell activation. J Exp Med 204:1131–1144

Merkel R, Nassoy P, Leung A, Ritchie K, Evans E (1999) Energy landscapes of receptor–ligand bonds explored with dynamic force spectroscopy. Nature 397:50–53

Preiner J, Ebner A, Chtcheglova L, Zhu R, Hinterdorfer P (2009) Simultaneous topography and recognition imaging: physical aspects and optimal imaging conditions. Nanotechnology 20:215103

Rankl C, Kienberger F, Wildling L, Wruss J, Gruber HJ, Blaas D, Hinterdorfer P (2008) Multiple receptors involved in human rhinovirus attachment to live cells. Proc Natl Acad Sci USA 105 (46):17778–17783

Spada FM, Koezuka Y, Porcelli SA (1998) CD1d-restricted recognition of synthetic glycolipid antigens by human natural killer T cells. J Exp Med 188:1529–1534

Stroh C, Wang H, Bash R, Ashcroft B, Nelson J, Gruber H, Lohr D, Lindsay SM, Hinterdorfer P (2004a) Single-molecule recognition imaging microscopy. Proc Natl Acad Sci USA 101 (34):12503–12507

Stroh CM, Ebner A, Geretschlager M, Freudenthaler G, Kienberger F, Kamruzzahan AS, Smith-Gill SJ, Gruber HJ, Hinterdorfer P (2004b) Simultaneous topography and recognition imaging using force microscopy. Biophys J 87:1981–1990

Strunz T, Oroszlan K, Schumakovitch GH-G, Hegner M (2000) Model energy landscapes and the force-induced dissociation of ligand-receptor bonds. Biophys J 79:1206–1212

Svoboda K, Schmidt CF, Schnapp BJ, Block SM (1993) Direct observation of kinesin stepping by optical trapping interferometry. Nature 365:721–727

Zajonc DM, Cantu C III, Mattner J, Zhou D, Savage PB, Bendelac A, Wilson IA, Teyton L (2005) Structure and function of a potent a-galactosylceramide agonist for the semi-invariant NKT cell receptor. Nat Immunol 6:810–818

Chapter 3
Catch Bonds of Integrin/Ligand Interactions

Cheng Zhu and Wei Chen

3.1 Introduction

Molecular interactions are ubiquitous in biology. For example, interactions between specific adhesive receptors and ligands mediate adhesion of a cell to another cell or to the extracellular matrix (ECM). Cell adhesions are important to a wide variety of cellular functions, such as leukocyte trafficking and homing, and tissue cell migration, morphogenesis, and remolding (Alberts et al. 2002; Isacke and Horton 2000).

Because molecular interactions provide physical linkages between intracellular structures or between a cell to its extracellular environment, they are likely regulated by mechanical force, which can be either externally applied to the cell or internally generated by the cell. In the case of circulating leukocyte adhesion to the vascular surface under flow conditions, the adhesive forces, which anchor the white blood cell to the blood vessel wall against hemodynamic forces, are directly exerted on the receptor/ligand bonds between the leukocyte and the endothelial cell (McEver and Zhu 2010).

Bell proposed the first model on how force may regulate dissociation kinetics of a molecular bond based on experimental data of fracture of solids, assuming that force exponentially accelerates unbinding of molecular bonds (Bell 1978). Ten years later, Dembo et al. suggested that force could also decelerate unbinding of molecular bonds (Dembo et al. 1988). Bonds can be classified according to how they are regulated by force: slip bonds if force accelerates dissociation by shortening their lifetimes and catch bonds if force decelerates dissociation by prolonging their lifetimes (Dembo et al. 1988). Fifteen years later, catch bonds were demonstrated experimentally using interactions between P-selectin and

C. Zhu (✉) • W. Chen
The Coulter Department of Biomedical Engineering, Georgia Institute of Technology, Atlanta, GA, USA
e-mail: cheng.zhu@bme.gatech.edu

A.F. Oberhauser (ed.), *Single-molecule Studies of Proteins*, Biophysics for the Life Sciences 2, DOI 10.1007/978-1-4614-4921-8_3,

P-selectin glycoprotein ligand 1 (PSGL-1) (Marshall et al. 2003). Since then, catch bonds have been observed in other selectin/ligand (Lou et al. 2006; Sarangapani et al. 2004; Wayman et al.2010), actin/myosin (Guo and Guilford 2006), FimH/manose ligand (Yakovenko et al. 2008), platelet glycoprotein Ibα (GPIbα)/von Willebrand factor (VWF) (Yago et al. 2008), and integrin/ligand (Chen et al. 2010; Kong et al. 2009) interactions. In addition to intermolecular interactions, catch bond behavior has also been observed in intramolecular interactions such as protein unfolding (Wu et al. 2010).

Single-molecule force techniques have been used as the principal experimental approach to study catch bonds, including force-clamp assays with atomic force microscopy (AFM) (Kong et al. 2009; Marshall et al. 2003; Sarangapani et al. 2004; Wu et al. 2010; Yago et al. 2008; Yakovenko et al. 2008), laser tweezers (Guo and Guilford 2006), and a biomembrane force probe (BFP) (Chen et al. 2010; Lou et al. 2006; Wayman et al. 2010), jump-and-ramp assay with a BFP (Evans et al. 2004), and single-tether lifetime assay with a flow chamber (Lou et al. 2006; Marshall et al. 2003; Sarangapani et al. 2004; Wayman et al. 2010; Yago et al. 2004, 2008), although other flow chamber assays have also provided strong supporting evidence (Friedland et al. 2009; Thomas et al. 2002).

Structural analyses and molecular dynamics (MD) simulations have been used to elucidate the structural bases for catch bonds, including the sliding/rebinding (Lou et al. 2006; Lou and Zhu 2007; Yago et al. 2008) and allosteric (Le Trong et al. 2010; Xiang et al. 2011) mechanisms.

In this chapter, we will describe single molecular interaction experiments that demonstrate catch bonds between integrin $\alpha_L\beta_2$ or $\alpha_5\beta_1$ with their respective ligands, intercellular adhesion molecule 1 (ICAM-1) or fibronectin (FN), by using BFP and AFM, respectively. We will also discuss experimental analyses and models that relate integrin catch bonds to their structures and conformations.

3.2 Integrins

Integrins are a family of heterodimeric transmembrane adhesive receptors consisting of an α and a β subunit broadly expressed on a wide variety of cells (Hynes 2002; Isacke and Horton 2000). Their ligands include members of the immunoglobulin (Ig) superfamily on another cell (e.g., ICAM-1), ECM proteins (e.g., FN), and plasma proteins (e.g., VWF). There are 18 α and 8 β subunits known to form at least 24 integrins in humans. Ten α subunits have an additional A or inserted (I) domain and the other eight do not, forming 11 αA (or αI) domain-containing integrins, e.g., $\alpha_L\beta_2$ (Fig. 3.1a–c), and 13 αA domain-lacking integrins, e.g., $\alpha_5\beta_1$ (Fig. 3.1d–f). Integrins consist of a large ectodomain, a transmembrane segment, and a short cytoplasmic tail. The overall shape of an integrin ectodomain is that of a large head supported by two long legs (Arnaout et al. 2005; Luo and Springer 2006). The head of αA domain-lacking integrins consists of the β-propeller domain of the α subunit and the βA (or βI) domain of the β subunit (Fig. 3.1d–f). The two legs contain the thigh domain and the calf-1 and -2 domains connecting to the

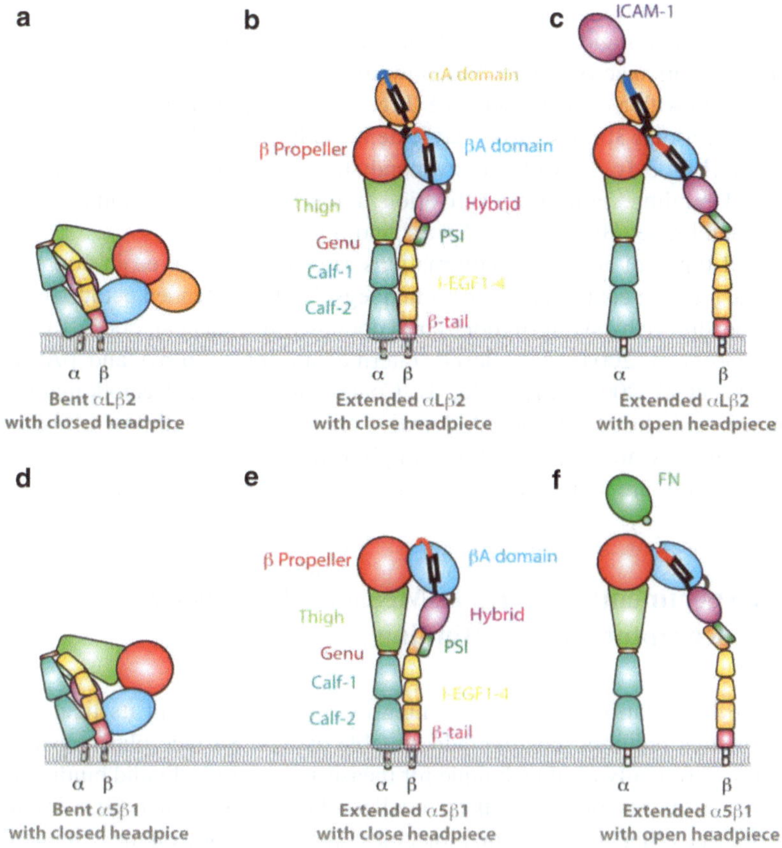

Fig. 3.1 Integrin domains at different conformations. αA domain-containing (**a–c**) or domain-lacking (**d–f**), integrin with bent (**a** and **d**) or extended (**b**, **c**, **e**, **f**) ectodomains, and closed (**a**, **b**, **d**, **e**) or open (**c** and **f**) headpieces. Only a simplified single domain schematic is depicted for the ligand in panels (**c**) and (**d**)

transmembrane segment of the α subunit and the hybrid, plexin-semaphorin-integrin (PSI), epidermal growth factor (EGF) 1–4 domains, and the β tail domain (βTD) connecting to the transmembrane segment of the β subunit (Fig. 3.1). For the domain-containing integrins, ligands bind to the metal ion-dependent adhesion site (MIDAS) on the top of the αA domain (Fig. 3.1c). For the domain-lacking integrins, ligands bind to the top of the headpiece between the βA and β-propeller domains (Fig. 3.1f). Integrin/ligand interactions are capable of transmitting force and transducing signals bidirectionally across the cell membrane, thereby playing a key role in mechanosensing and mechanotransduction (Schwartz and DeSimone 2008).

The ligand-binding affinities of integrins are closely related to their structures, which may undergo conformational changes in response to intracellular signals to the cytoplasmic tail (inside-out signaling), ligand binding to the integrin head (outside-in signaling), and/or force applied to the integrin tail or head (mechanical activation) (Takagi and Springer 2002). Under physiological conditions, integrins may assume an inactive state with low affinities for ligands and adopt a bent conformation with a closed headpiece (Fig. 3.1a, d). Upon activation by extracellular or intracellular stimuli, they may extend with a closed (Fig. 3.1b, e) or open headpiece (Fig. 3.1c, f), which corresponds to intermediate or high-affinity states for ligand binding. Following activation, integrins transduce signals across the plasma membrane (Hynes 2002; Takagi and Springer 2002).

Several types of integrin conformations have been observed in structural and functional studies: coalescent or separated legs, bent or extended ectodomain, close or open headpiece by swing-in/out hybrid domain (Nishida et al. 2006; Takagi et al. 2002; Xiao et al. 2004), and closed, intermediate, or open αA and βA domains (Shimaoka et al. 2003) (Fig. 3.1a, d). Coalescent legs, bent ectodomain, closed headpiece, and closed αA and βA domains are associated with low-affinity states. Leg separation, ectodomain extension, hybrid domain swing-out, and αA and βA domain opening are associated with high-affinity states (Fig. 3.1c, f).

3.3 Experimental Assays to Measure 2D Kinetics of Receptor/Ligand Binding

To define the relationship between integrin conformation and its ligand-binding characteristics requires sensitive and suitable assays. Surface plasmon resonance (SPR) is a commonly used technique for measuring receptor/ligand binding affinity and kinetics in solution, i.e., the so-called three-dimensional (3D) interaction. However, like other cell adhesion molecules, integrins reside on the cell membrane, the organization and diffusion of which is restricted by their interactions with membrane microstructures and/or the cytoskeleton. Since these cellular microenvironmental factors have been shown to impact receptor/ligand binding kinetics, SPR may not be a suitable assay to accurately quantify kinetics of membrane molecules. To overcome this limitation, two-dimensional (2D) assays, e.g., the adhesion frequency assay (Chesla et al. 1998) and the thermal fluctuation assay (Chen et al. 2008a), have been used to measure 2D association and force-free dissociation kinetics of surface-anchored receptor/ligand interaction. As alluded to earlier, integrin/ligand bonds anchor cells within tissues and provide traction for their migration. Consequently, these bonds are often subjected to mechanical force, which can regulate their dissociation kinetics. Single-bond force-ramp and force-clamp assays have been used to measure the force-dependent off-rates of integrin/ligand dissociation.

The adhesion frequency assay determines receptor/ligand binding kinetics mechanically (Chen et al. 2008b; Chesla et al. 1998; Zarnitsyna and Zhu 2011). Any sensitive force technique can be used as an adhesion sensor, e.g., a human red blood cell (RBC), which is also used as well as a presenting cell of the ligands for the integrins expressed on the target cell. Biotin-streptavidin or chromium chloride coupling can be used to coat ligands on RBCs. By motorized micromanipulation,

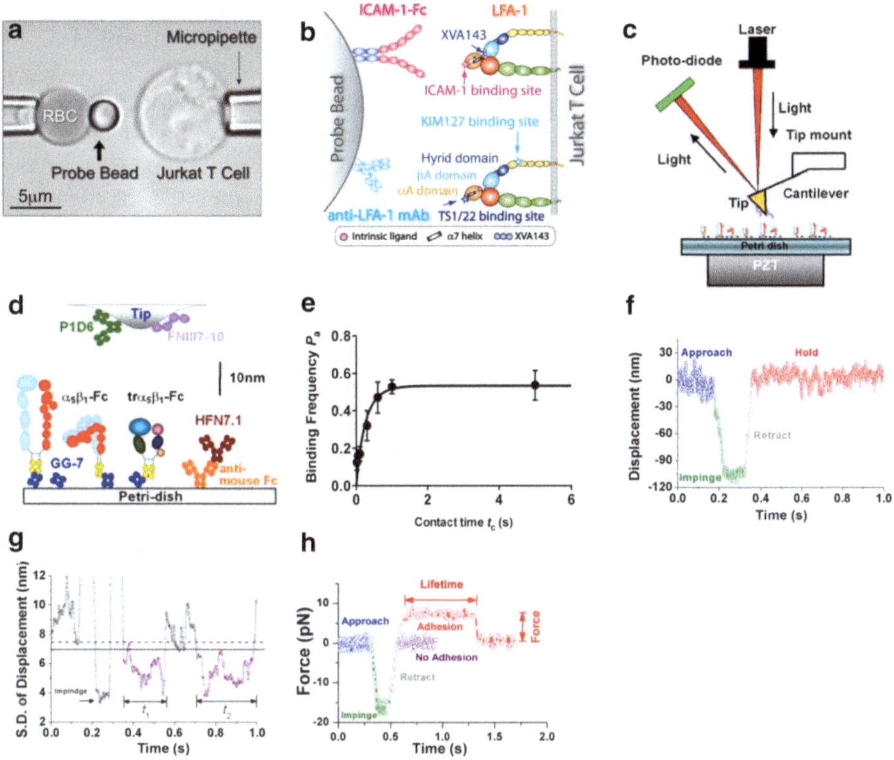

Fig. 3.2 Experimental setups and 2D kinetics assays (a–d). Setups (a, c) and composites of all interacting molecules (b, d) of BFP (a, b) and AFM (c, d) experiments for measuring 2D integrin/ligand interaction kinetics. (e) Example plot of binding frequency vs. contact duration from an adhesion frequency assay fitted by Eq. (3.1). (f, g) Example plots of displacements vs. time and standard deviation of displacement vs. time from a thermal fluctuation assay. (h) Example plots of force vs. time of a lifetime event and a no adhesion event from a force-clamp assay. Reproduced from (Chen et al. 2010; Kong et al. 2009) with permission

the target cell is brought into contact with the RBC to enable integrin/ligand interaction within a preset contact duration and retracted away from the RBC to allow for visualization of whether integrin/ligand interaction(s) is (are) present at the end of the contact from elongation of (or the lack thereof) the RBC. The binary outcome of this contact cycle is collected repeatedly many times to obtain the frequency of adhesion observed for that contact duration t_c. The adhesion frequency P_a is measured over a range of contact durations, which is fitted by a mathematical model (Eq. 3.1) to extract the 2D affinity K_a and off-rate k_{off} (Fig. 3.2e) (Chesla et al. 1998):

$$P_a = 1 - \exp\{-m_r m_l A_c K_a [1 - \exp(-k_{off} t_c)\}, \qquad (3.1)$$

where m_r and m_l are respective site densities of integrin and ligand on the target cell and RBC independently measured by flow-cytometry, and A_c is the contact area kept constant (a few percent of 3 μm^2 depending on the smoothness of the target cell surface) for all experiments. By multiplying the effective 2D affinity A_cK_a by k_{off}, the effective 2D on-rate A_ck_{on} can be obtained.

The thermal fluctuation assay also is a mechanical-based assay (Fig. 3.2f,g) (Chen et al. 2008a). It uses a BFP (Fig. 3.2a), which represents a high-tech version of the micropipette system by attaching a glass bead to the RBC apex to allow its tracking by high-precision, high-speed image analysis with a 0.7-ms temporal, 3-nm spatial, and 1-pN force resolution (Chen et al. 2008b). In contrast to the adhesion frequency assay that only detects the presence of interaction at the end of a contact duration, the thermal fluctuation assay pinpoints the association and dissociation events during the entire period of close contact without retracting the target cell to reveal RBC elongation. This greatly enhances the quantity, quality, and reliability of the information obtained, which makes kinetic measurements much simpler and more robust. The idea of the assay is to take advantage of the susceptibility of an ultrasensitive force probe to thermal fluctuations. Formation of a molecular bond adds the molecular spring (with a constant k_m) to the force probe spring (with a constant k_p), thereby giving rise to a stiffer system spring (with a constant $k_s = k_m + k_p$) that is equivalent to two springs in parallel, which reduces the thermal fluctuations (Chen et al. 2008a, b). Thus, the bond formation and dissociation events are signified by reductions and resumptions of the BFP thermal fluctuations, as measured by abrupt decreases and increases of standard deviation of the BFP bead position (Fig. 3.2g). Lifetime of a bond is measured from the moment of its formation to that of its dissociation. The moment of the last bond dissociation to that of the bond formation is also measured and called waiting time. The 2D off-rates and on-rates can be derived from analyses of the distributions of bond lifetimes and waiting times, respectively. For example, off-rate can be estimated from the negative slope of the linear fit to the natural log of the survival frequency vs. bond lifetime plot (cf. Fig. 3.3c) or the reciprocal of average bond lifetimes (Marshall et al. 2003).

The aforementioned two assays are used to characterize force-free 2D kinetic rates. By comparison, force-clamp lifetime assay (Fig. 3.2h) is commonly used for measuring force-dependent off-rate using BFP (Fig. 3.2a) or AFM (Fig. 3.2c). In this assay, the receptor-expressing surface (target cell, target bead, or polystyrene dish) is driven to approach and contact the ligand-coated force sensor (BFP probe bead or AFM cantilever tip) with a small (<20 pN) compressive force for a short duration (e.g., 0.1 s) to allow bond formation, retract at a certain speed for adhesion detection, hold at desired force to wait for bond dissociation, and return to the original position to complete the test cycle, which is then repeated. Adhesion (if present) is detected by tensile force on retraction. Lifetime is measured from the instant when force reaches a desired level to the instant of bond dissociation (Chen et al. 2010; Kong et al. 2009). Lifetimes of integrin/ligand interactions measured under force are often not distributed as a single exponential, suggesting coexistence of multiple bond states that dissociate at different off-rates (Chen et al. 2010; Kong et al. 2009). A commonly used simple metric is the average lifetime, which equals the reciprocal of apparent off-rate (cf. Figs. 3.5, 3.6, 3.7, and 3.8).

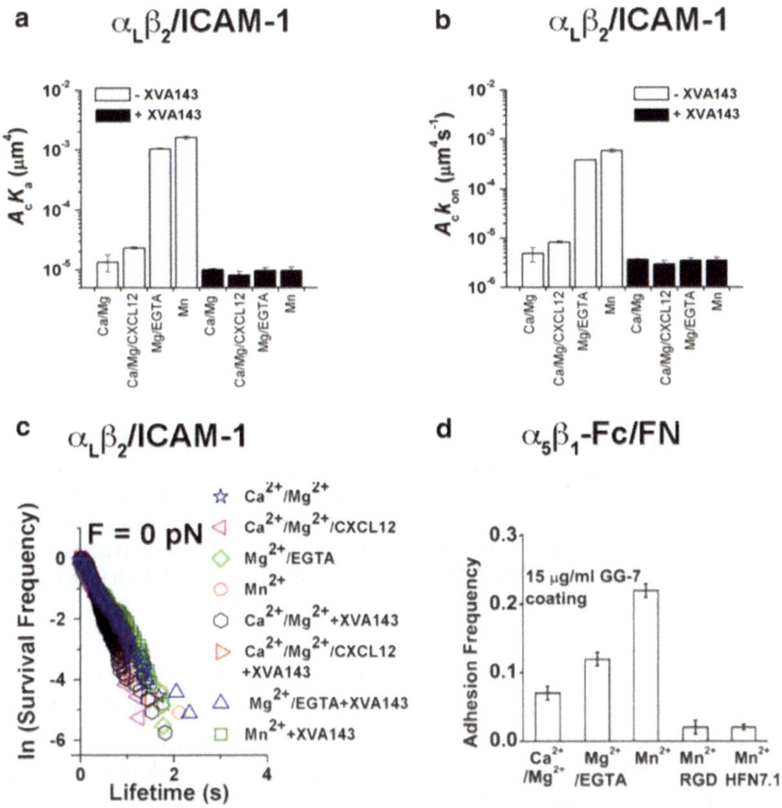

Fig. 3.3 Regulation of integrin-binding frequency and kinetics (**a–c**). Two-dimensional affinity (**a**), on-rates (**b**), and zero-force off- rates (**c**) of $\alpha_L\beta_2$/ICAM-1 interactions in $Ca^{2+}Mg^{2+}$, $Ca^{2+}Mg^{2+}$/CXCL12, Mg^{2+}/EGTA or Mn^{2+} without or with 1 μM XVA143. (**d**) Binding frequencies of $\alpha_5\beta_1$ interacting with FN in $Ca^{2+}Mg^{2+}$, Mg^{2+}/EGTA or Mn^{2+} in the absence or presence of solution RGD (to block integrin) or antibody HFN7.1 (to block FN). Reproduced from (Chen et al. 2010; Kong et al. 2009) with permission

3.4 Regulation of Integrin/Ligand Binding by Conformational Changes

Integrin conformation and ligand-binding affinity can be regulated by many factors, including divalent cations, small molecule antagonists, and inside-out signaling. For example, compared to $Ca^{2+}Mg^{2+}$, Mg^{2+}/EGTA or Mn^{2+} (Chen et al. 2010; Salas et al. 2004; Shamri et al. 2005) induces $\alpha_L\beta_2$ (Fig. 3.3a–c) and $\alpha_5\beta_1$ (Fig. 3.3d) integrins to bind their respective ligands, ICAM-1 and FN, with much higher 2D affinities (Fig. 3.3a) by greatly enhancing 2D on-rates (Fig. 3.3b) (Chen et al. 2010), as reflected by much higher adhesion frequencies (Fig. 3.3d) (Kong et al. 2009). Both integrins assume a bent conformation with the hybrid domain in a closed position in $Ca^{2+}Mg^{2+}$. Mg^{2+}/EGTA and Mn^{2+} increase the fraction of integrins in

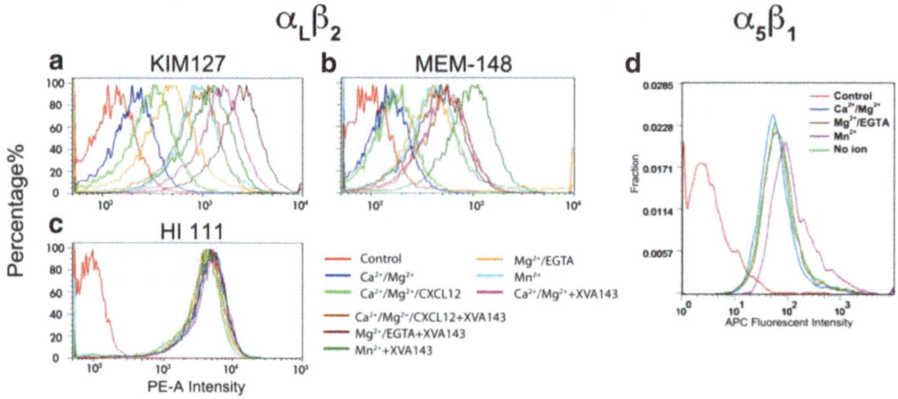

Fig. 3.4 Characterization of integrin conformations. Fluorescence intensity histograms of Jurkat cells stained by isotype-matched irrelevant IgG (control), KIM127 plus a PE-conjugated secondary goat–anti-mouse antibody (**a**), PE-conjugated MEM-148 (**b**), PE-conjugated HI111 (**c**) in $Ca^{2+}Mg^{2+}$, $Ca^{2+}Mg^{2+}$/CXCL12, Mg^{2+}/EGTA or Mn^{2+} without or with XVA143 or $\alpha_5\beta_1$-Fc coated beads by APC-conjugated extension reporting antibody 9EG7 or isotype-matched irrelevant IgG (control) in $Ca^{2+}Mg^{2+}$, Mg^{2+}/EGTA or Mn^{2+} (**d**). Reproduced from (Chen et al. 2010; Kong et al. 2009) with permission

the extended conformation and with a swing-out hybrid domain, as indicated by increased binding of respective reporter antibodies, KIM127 and MEM148 for the case of $\alpha_L\beta_2$ (Fig. 3.4a, b). A small molecule antagonist XVA143 induces $\alpha_L\beta_2$ extension and hybrid domain swing-out while locking its αA domain MIDAS in a closed conformation. Correspondingly, the affinity and kinetic rates of $\alpha_L\beta_2$/ICAM-1 interaction are similar to those in $Ca^{2+}Mg^{2+}$ (Fig. 3.3a–c). Binding of a soluble chemokine CXCL12 to the CXC chemokine receptor 4 (CXCR4) on Jurkat cells only slightly increases the fraction of extended $\alpha_L\beta_2$ (Fig. 3.4a) without inducing the hybrid domain swing-out (Fig. 3.4b). Correspondingly, CXCL12 only slightly increase $\alpha_L\beta_2$/ICAM-1 2D affinity and on-rate (Fig. 3.3a, b). None of these agents induce open conformation of α_L MIDAS as indicated by the lack of change in the binding of HI111, an antibody reporting the closed α_L MIDAS (Fig. 3.4c). Correspondingly, the zero-force off-rate of the $\alpha_L\beta_2$/ICAM-1 interaction is indifferent to all these treatment, as measured using the thermal fluctuation assay by BFP (Fig. 3.3c).

3.5 Force Prolongs Integrin/Ligand Mean Bond Lifetime: Catch Bonds

Integrin catch bonds have been observed in both αA domain-lacking ($\alpha_5\beta_1$) and αA domain- containing ($\alpha_L\beta_2$) integrins interacting with their physiological ligands FN and ICAM-1, respectively, using either purified ($\alpha_5\beta_1$) or cell surface ($\alpha_L\beta_2$)

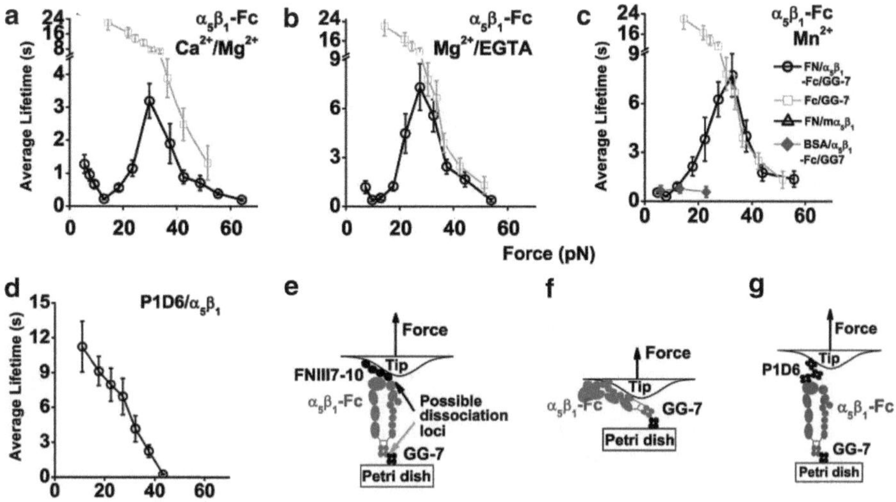

Fig. 3.5 Force-dependent lifetimes of $\alpha5\beta1$ bonds with FN in $Ca^{2+}Mg^{2+}$ (**a**), Mg^{2+}/EGTA (**b**), or Mn^{2+} (**c**) or with anti- $\alpha_5\beta_1$ P1D6 (**d**). The *black curves* denote data measured using molecules depicted in E (for **a–c**) and G (for **d**) with two possible dissociation loci: between $\alpha_5\beta_1$ and its ligand (fragment of type III FN containing domains 7–10, for **a–c**) or P1D6 (for **d**) and between Fc and anti-Fc antibody GG7. The *gray curves* denote data measured using molecules depicted in (**f**) with only one dissociation locus: between Fc and anti-Fc antibody GG7. The overlaps between the *black* and *gray curves* in the 30–60-pN force regime in (**b**) and (**c**) suggest that the slip bond is due to dissociation of $\alpha_5\beta_1$-Fc from GG7. Reproduced from (Kong et al. 2009) with permission

integrins measured by either AFM (Kong et al. 2009) or BFP (Chen et al. 2010) in force-clamp lifetime experiments.

In all three divalent cation conditions ($Ca^{2+}Mg^{2+}$, Mg^{2+}/EGTA and Mn^{2+}) that activate integrin to different conformations, increasing force initially prolongs average lifetimes of bonds of $\alpha_5\beta_1$ ectodomain fused with IgG Fc (Fig. 3.2d) interacting with FN, i.e., catch bonds (Fig. 3.5a–c). Beyond ~30 pN, further increase of force shortens average bond lifetimes of $\alpha_5\beta_1$/FN interaction, i.e., slip bond. Lifetimes at low forces (<10 pN) were similarly short (<2 s), but were prolonged less by force in $Ca^{2+}Mg^{2+}$ (Fig. 3.5a) than Mg^{2+}/EGTA (Fig. 3.5b), which were similar to Mn^{2+} (Fig. 3.5c). Catch bonds of integrin are specific for its physiological ligand, as $\alpha_5\beta_1$ only forms slip bond with its antibody PD16 (Fig. 3.5d).

Similar to $\alpha_5\beta_1$/FN catch bonds, in $Ca^{2+}Mg^{2+}$, Mg^{2+}/EGTA and Mn^{2+} and at low forces (from 0 to 15–20 pN), increasing force prolongs $\alpha_L\beta_2$/ICAM-1 bond lifetimes, as measured by the average of lifetimes <10 s (Fig. 3.6a) as well as the fraction of lifetimes >10 s (Fig. 3.6b). Increasing force further shortens the short (<10 s) lifetime average (Fig. 3.6a) and reduces the long (>10 s) lifetime fraction (Fig. 3.6b). Priming by inside-out signaling with CXCL12 strengthens the catch bond by shifting the lifetime vs. force curve rightward to a higher peak force and upward to a longer peak short lifetime average and a greater peak long lifetime

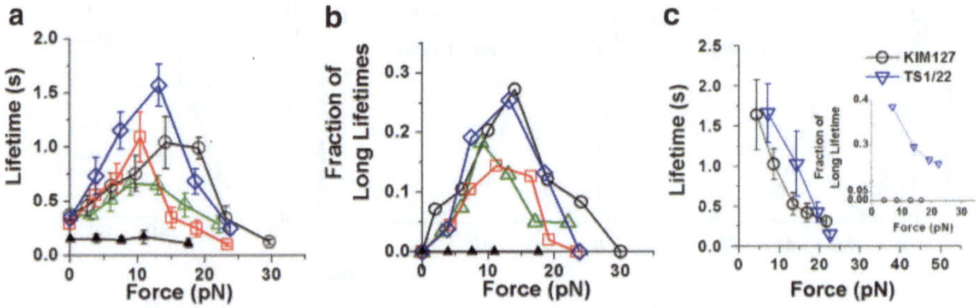

Fig. 3.6 Force-dependent lifetimes of $\alpha L\beta 2$bonds with ICAM-1 or antibodies (**a**) and (**b**). Plots of short (<10 s) lifetime average (± s.e.m.) (**a**) and long (>10 s) lifetime fraction (**b**) vs. force of $\alpha_L\beta_2$/ICAM-1 bonds in $Ca^{2+}Mg^{2+}$ (*open square*), $Ca^{2+}Mg^{2+}$/CXCL12 (*open diamond*), Mg^{2+}/ EGTA (*open triangle*), or Mn^{2+} (*open circle*), or of nonspecific adhesions with 10 μg/ml TS1/18 blocking antibody in Ca^{2+}/Mg^{2+} (*filled triangle*). Lifetimes at zero and at peak forces are significantly different ($p < 0.01$, student *t*-test). (**c**) Plot of short (<10 s) lifetime average (± s.e.m.) and long (>10 s) lifetime fraction (insert) vs. force of bonds of $\alpha_L\beta_2$with TS1/22 in Ca^{2+}/Mg^{2+} (*inverted open triangle*) or with KIM127 in Mn^{2+} plus 100 μg/ml soluble ICAM-1 (to expose the KIM127 epitope) (*open circle*). Reproduced from (Chen et al. 2010) with permission

fraction (Fig. 3.6a, b). Again, interactions of $\alpha_L\beta_2$ with mAbs KIM127 and TS1/22 exhibit only slip bonds in the force region tested (Fig. 3.6c), indicating the specificity of the $\alpha_L\beta_2$/ICAM-1 catch bond.

3.6 Structural Analysis of Integrin Catch Bonds

Integrin/ligand catch bonds are regulated by integrin conformational changes. A switchblade model has been proposed for global changes of integrin conformations from bent to extended during activation (Beglova et al. 2002; Takagi et al. 2002). Intuitively, pulling via a ligand engaged to the top of an integrin head would extend the bent integrin, as observed in MD simulations (Chen et al. 2011; Matsumoto et al. 2008). This argues that force may convert a low-affinity, short-lived integrin/ligand bond to one with high-affinity and long bond lifetime, thereby giving rise to a catch bond behavior (Zhu et al. 2005). MD simulated pulling of an integrin headpiece induces swing-out of the hybrid domain that corresponds to a high-affinity conformation (Provasi et al. 2009; Puklin-Faucher et al. 2006; Puklin-Faucher and Vogel, 2009). However, the direction of pulling may impact the outcome, as vertical pulling did not induce hybrid domain swing-out (Chen et al. 2011; Zhu et al. 2008), rather, simulated lateral pulling that forces the closed legs to open induces hybrid domain swing-out (Zhu et al. 2008). A "traction" model proposes that extension "primes" integrins to states with intermediate affinity for ligand. Transition to the high-affinity state may require an energy-dependent cytoskeleton that exerts lateral force (traction) to fully separate the α and β tails (Schurpf and Springer 2011; Zhu et al. 2008).

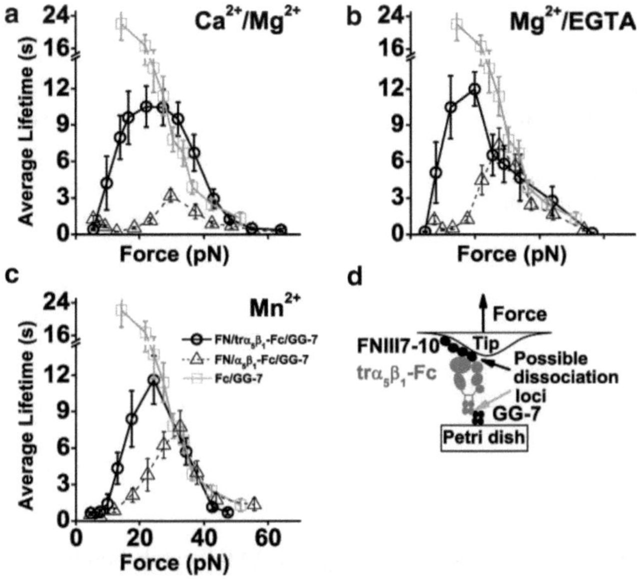

Fig. 3.7 Force-dependent lifetimes of bonds between a leg-truncated $\alpha_5\beta_1$ and FN measured in $Ca^{2+}Mg^{2+}$ (**a**), Mg^{2+}/EGTA (**b**), or Mn^{2+} (**c**). (**d**) Schematic showing different dissociation loci. Reproduced from (Kong et al. 2009) with permission

Experimentally, changing the cation compositions from $Ca^{2+}Mg^{2+}$ to Mg^{2+}/EDTA or Mn^{2+} progressively increases the fractions of integrins in extended conformations (Fig. 3.4a, d) and with a swing-out hybrid domain (Fig. 3.4b). Yet the biphasic force-dependent lifetime curves indicative of catch–slip bonds are observed in all three divalent cation conditions for both $\alpha_5\beta_1$ (Fig. 3.5a–c) and $\alpha_L\beta_2$ (Fig. 3.6a, b) integrins interacting with their respective ligands. Also, catch bonds have been observed in all three divalent cation conditions between FN and a truncated construct of $\alpha_5\beta_1$ that deletes the integrin legs (Fig. 3.7). Thus, integrins are not required to assume a particular conformation prior to force application to form catch bonds with physiological ligands.

A bell-rope model has been suggested to relate different ligand-binding affinities to distinct conformations of the αA domain, such that pulling down the α_7-helix at the bottom opens the αA domain by rearranging the MIDAS on the top to change from the low to intermediate and high-affinity states (Shimaoka et al. 2003). The downward movement of the αA domain α_7-helix may result from binding of an intrinsic ligand on the α_7-helix to the βA domain MIDAS, thereby connecting the conformational changes of the αA domain to those of the βA and other downstream domains of the integrin (Fig. 3.1c) (Jin et al. 2004; Xiang et al. 2011). Similar α_7-helix downward movements in the βA domain are believed to regulate binding not only of intrinsic ligand at the C-terminal of αA domain α_7-helix in αA domain-containing integrins but also of extrinsic ligand in αA domain-lacking integrins.

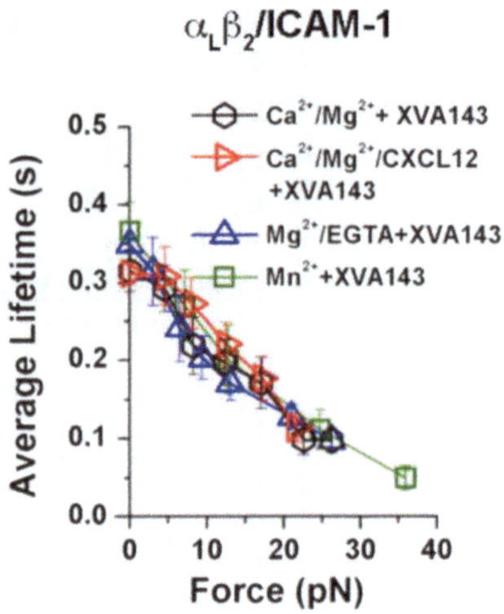

Fig. 3.8 Force-dependent lifetimes of $\alpha_L\beta_2$/ICAM-1 bonds in the presence of a small molecule antagonist XVA143 in $Ca^{2+}Mg^{2+}$, Mg^{2+}/EGTA, or Mn^{2+}, or $Ca^{2+}Mg^{2+}$ plus CXCL12. Reproduced from (Chen et al. 2010) with permission

Experimentally, the formation of $\alpha_L\beta_2$/ICAM-1 catch bonds requires force to transmit through the internal linkage between the αA and βA domains. Blocking this linkage by a small molecule antagonist XVA143, which inhibits ligation of the αA domain α_7-helix to the βA domain MIDAS, impairs the movement of the αA domain α_7-helix from the "up" to "down" position. Consequently, the $\alpha_L\beta_2$/ICAM-1 catch bonds are completely abolished and the biphasic force-dependent lifetime curves are converted to monotonically decreasing curves typical of slip bonds that are indistinguishable among different cation conditions (Fig. 3.8).

3.7 A Physical Model for $\alpha_L\beta_2$/ICAM-1 Catch Bonds

The $\alpha_L\beta_2$/ICAM-1 catch bonds (Fig. 3.6a, b) and their elimination by XVA143 (Fig. 3.8) have been explained by a three-state model (Chen et al. 2010; Xiang et al. 2011). It assumes that an $\alpha_L\beta_2$ integrin can bind ICAM-1 with different on-rates depending on the global conformation, but always dissociates with the same off-rate characteristic of short-lived state before force is applied. External force switches the bond to an intermediate-lived state and then a long-lived state, with slower and slower off-rates. The off-rates of the three states, the respective probabilities for an $\alpha_L\beta_2$/ICAM-1 bond to reside in these states, and the transition probabilities between neighboring states depend on the force level. Lifetimes pooled from many single-bond measurements represent a mixed population. The force-dependent lifetime curves (Fig. 3.6a, b) reflect the weighted average over three fractions of bonds associated with the three states. Dissociation from each of the three states is governed by the Bell model (Bell 1978) but each has a unique stress-free off-rate (k_0) and reactive compliance (a), as shown in Fig. 3.9a–e and summarized in Table 3.1. The short-lived state has a k_0 value an order of

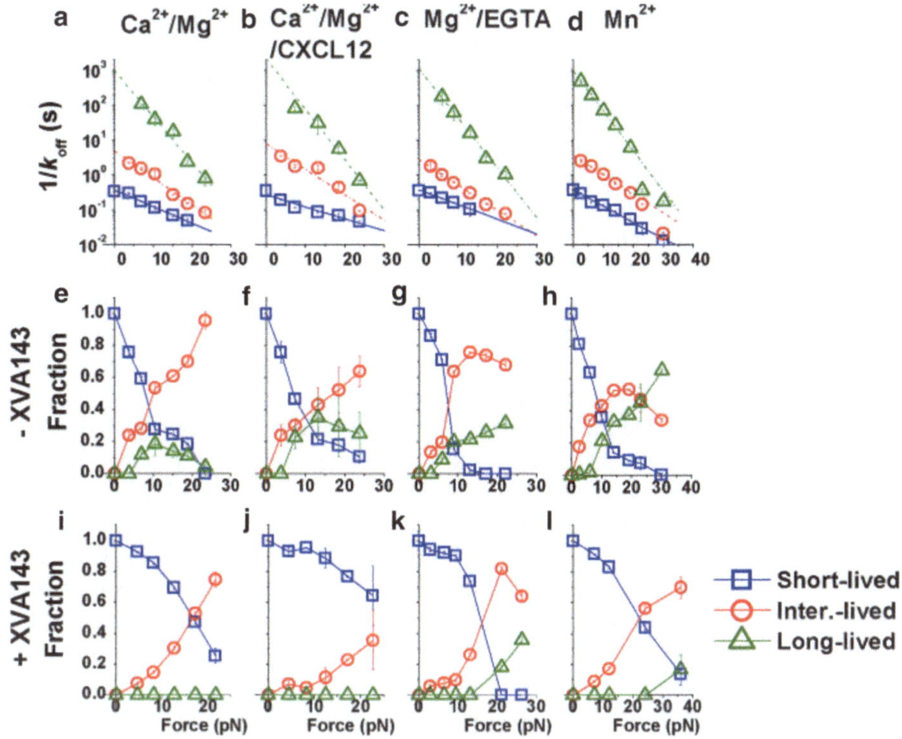

Fig. 3.9 Analysis of lifetimes by a three-state model. Plots of reciprocal off-rates (**a–d**) or fraction of bonds (**e–l**) of the short- (*open square*), intermediate- (*open circle*), and long (*open triangle*)-lived states vs. lifetime data measured in $Ca^{2+}Mg^{2+}$ (A, E, I), $Ca^{2+}Mg^{2+}$/CXCL12 (**b**, **f**, **j**), Mg^{2+}/EGTA (C, G, K), or Mn^{2+} (**d**, **h**, **l**) in the absence (**e–h**) and presence (**i–l**) of XVA143. The reciprocal off-rates in (**a–d**) were required to fit both data in the absence and presence of XVA143, which were fitted by the Bell model, $k_i = k_{0i}\exp(a_i f/k_B T)$, $i = 1 - 3$ with parameters listed in Table 3.1. Reproduced from (Chen et al. 2010) with permission

Table 3.1 Bell model parameters of the three-state model under different conditions

	Ca^{2+}/Mg^{2+}		Ca^{2+}/Mg^{2+}/CXCL12		Mg^{2+}EGTA		Mn^{2+}		All	
Off-rate states	k_0 (s^{-1})	a (nm)	k_0 (s^{-1})	a (nm)	k_0 (s^{-1})	a (nm)	k_0 (s^{-1})	a (nm)	k_0 (s^{-1})	a (nm)
Short-lived	2.66	0.47	3.7	0.33	2.6	0.41	2.65	0.45	2.79	0.43
Intermediate-lived	0.2	0.75	0.13	0.71	0.36	0.7	0.28	0.53	0.22	0.69
Long-lived	0.0001	1.28	0.00097	1.24	0.0011	1.33	0.0011	1.1	0.00091	1.29

Reproduced from (Chen et al. 2010) with permission

magnitude faster (smaller $1/k_0$ in Fig. 3.9a–d) than that of the intermediate-lived state, which is ~240-fold faster than that of the long-lived state. However, force-induced decline in the reciprocal off-rate is faster for the long-lived state than the intermediate-lived state, which is in turn faster than the short-lived state (steeper slopes of the linear fits in Fig. 3.9e and larger *a* values in Table 3.1).

The fractions of bonds in the three states are evaluated from data measured in the absence of XVA143 (Fig. 3.9e–h) and in the presence of XVA143 (Fig. 3.9i–l), in different cation and chemokine conditions. Without XVA143, increasing force causes the fractions of intermediate- and long-lived bonds to grow at the expense of the short-lived bond fraction (Fig. 3.9e–h), giving rise to the catch bonds. This is countered by the rapid declines of the reciprocal off-rates of all three states with increasing force at high levels, giving rise to the eventual slip bonds beyond an "optimal" force where lifetimes reach maximal.

With XVA143, the decline of the short-lived state fractions and the growth of the intermediate-lived state fractions occur at higher forces and slower paces (Fig. 3.9i–l) compared to the corresponding conditions without XVA143 (Fig. 3.9e–h). The growth of the long-lived state fractions is eliminated in $Ca^{2+}Mg^{2+}$ and $Ca^{2+}Mg^{2+}$/CXCL12 and further right-shifted in Mg^{2+}/EGTA and Mn^{2+} to force so high that even the timescale of the long-lived state becomes very short. These results indicate that the external force applied by a bound ICAM-1 elicits $\alpha_L\beta_2$ catch bonds via an internal force that pulls on the αA domain α_7-helix. Blocking the internal ligation between the αA and βA domains reduces the portion of the force shared by this α_7-helix, which are internally distributed within the stretched $\alpha_L\beta_2$. This reduction has to be compensated by a higher external force to generate comparable pulling of the α_7-helix via its C-terminal connection to the β-propeller domain.

Steered molecular dynamics (SMD) simulations on the αA domain further support the three-state model and provide structural mechanism for catch bond. In the SMD simulations, pulling the αA domain α_7-helix induced its conformational changes and revealed a stable intermediate state along the pathway of α_7-helix downward movement. Downward movement of the α_7-helix was quantified by the root mean square distance (RMSD) between the simulated structure and its initial "up" position, which is plotted vs. simulation time in Fig. 3.10a. The "up" position was stable, as applying a 250-pN force to pull at the α_7-helix C terminus in the first 3.6 ns did not increase the RMSD significantly. Then state transitions happened when the RMSD rapidly increased from 3 to 6 Å in 0.2 ns. A stable intermediate state was revealed with a 4.5-Å RMSD held for 0.2 ns when the short period of sudden jumps was zoomed with a magnified timescale (Fig. 3.10a, inset). After two abrupt changes, the RMSD was gradually stabilized around 8 Å for the next 10 ns under constant pulling. At 15 ns, the force was released to allow the system to relax (Fig. 3.10a, dashed vertical line). The α_7-helix stayed at the down position for 2 ns but shifted back to the up position at 17 ns, which remained in that conformation for the next 18 ns of free dynamics simulation. Figure 3.10b–e shows snapshots of the simulated structure at 0, 3.7, 13, and 35 ns with the α_7-helix shown in green superimposed with its structures in the "down" (red) and "up" (blue) positions.

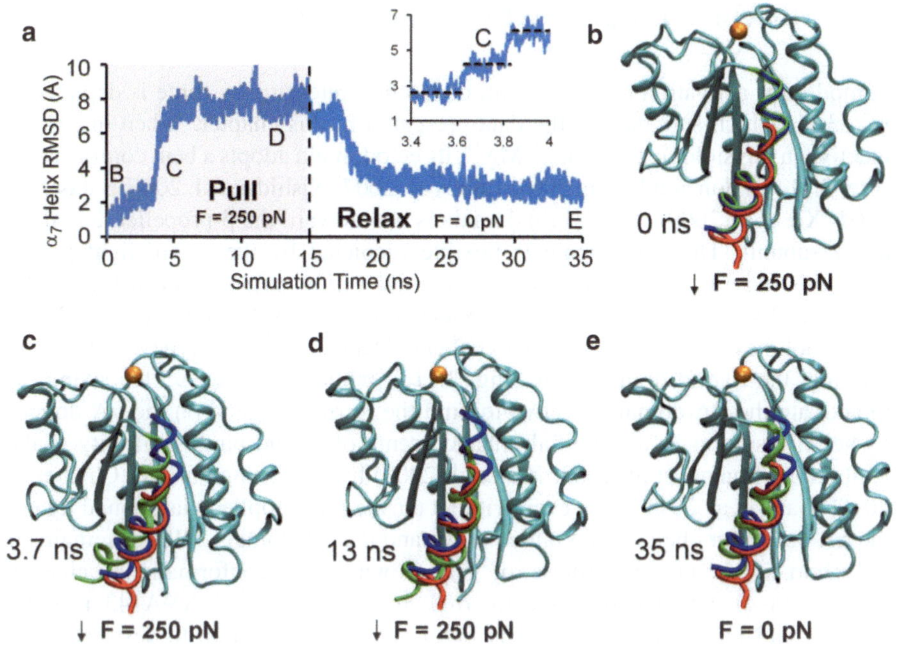

Fig. 3.10 SMD simulation of pulling the α_7-helix of the $\alpha_L\beta_2$ αA domain. (**a**) Time course of RMSD between the simulated α7-helix structure and the equilibrated structure at the up position. Letters indicate time points where the snapshots shown in (**b–e**) were taken. *Dashed vertical line* marks the time when force was released to allow free dynamics simulation. The *inset* shows the conformational changes at 3.4–4.0 ns. *Dashed lines* mark the three stable conformations. (**b–e**) Snapshots of the simulated structures (*cyan*) at the indicated times with the α_7-helix shown in *green*. The equilibrated α7-helix at the up (*blue*) and down (*red*) positions is also superimposed for comparison. Reproduced from (Xiang et al. 2011) with permission

The $\alpha_L\beta_2$/ICAM-1 catch–slip bonds and their elimination by XVA143 can be explained by a structural model for force to allosterically regulate bond states. The model includes two competing force-induced mechanisms: switching $\alpha_L\beta_2$ from the short-lived to intermediate- and long-lived states and accelerating dissociation. At low forces, the former mechanism dominates, resulting in catch bonds. At high forces, the latter mechanism takes over, resulting in slip bonds. XVA143 greatly impairs the former mechanism, converting the catch–slip bond behavior to that of slip-only. Without force, ICAM-1 engagement alone does not induce state transitions, suggesting that the α_7-helix remains in the "up" position despite that the βA domain may have already been activated by the integrin extension (Fig. 3.4a) and by the hybrid domain swing-out (Fig. 3.4b) via inside-out signaling or treatment with Mg^{2+}/EGTA or Mn^{2+}, while the MIDAS keeps in the close conformation (Fig. 3.4c).

3.8 Model for Integrin Activation by Force

A model for activation of αA domain-containing integrins by force is depicted in Fig. 3.11, which summarizes the data presented in this chapter. When integrin is inactive in physiological ions $Ca^{2+}Mg^{2+}$, its ectodomain adopts a bent conformation (Fig. 3.11a) (Alon and Dustin 2007; Luo et al. 2007; Nishida et al. 2006; Salas et al. 2004; Xie et al. 2009). The hybrid domain is bent toward the β-propeller domain of the α subunit. The αA domain is flexible (depicted by the dotted light yellow ellipsoids moving from both sides of the solid yellow αA domain in Fig. 3.11a) (Xie et al. 2009) with a "closed" MIDAS and "up" α_7-helix. Priming by Mg^{2+}, Mn^{2+}, or chemokines induces the integrin to unbend. Depending on the type and degree of priming, the hybrid domain may (Fig. 3.11c) or may not (Fig. 3.11b) swing out to activate the βA domain MIDAS to bind the intrinsic ligand on the αA domain α_7-helix, resulting in differential enhancements of ligand-binding affinity (Salas et al. 2004; Shamri et al. 2005; Xie et al. 2009) by increasing on-rate (Chen et al. 2010; Xiang et al. 2011). However, without tensile force applied via a bound ligand, the αA domain α_7-helix is not pulled down and the αA domain MIDAS remains in the short-lived state regardless of any downstream conformational changes (Fig. 3.11a–c). Compared to the internal ligand, binding of XVA143 is much stronger and longer lasting, as it can block the αA/βA ligation to destabilize the αA domain orientation and can induce conformational changes that propagate from the βA domain further downstream (Fig. 3.11e).

Force applied via ICAM-1 would pull the αA domain α_7-helix up unless it is anchored to the βA domain. The ligation between the αA and βA domains allows the applied force to be transmitted along the α_7-helix (indicated by an arrow), which pulls it down (relative to the rest of the αA domain that is being pulled up) to induce transition to the intermediate- and long-lived states in the αA domain MIDAS to form a catch bond with the ligand, regardless of the conformations of the downstream domains (Fig. 3.11a–d). The prolonged lifetimes of the $\alpha_L\beta_2$/ICAM-1 bond at increased force require continued down pulling of the α_7-helix, implying prolonged lifetimes of its bond with the βA domain MIDAS at increased force, i.e., an internal catch bond between the αA and βA domains (Fig. 3.11d).

XVA143 blocks the αA/βA internal ligation and prevents force transmission via the internal ligation from the αA to the βA domains (Fig. 3.11f). This force component is now shifted to other connections between the αA and β-propeller domains, which reduces the total force borne by the α_7-helix (indicated by a smaller arrow). To generate a similar internal force to pull down the α_7-helix to induce the intermediate- and long-lived states, a much larger force is required because the majority of it is now borne by its C-terminal linker to the β-propeller domain (Fig. 3.11f).

The aforementioned model highlights the important role of force in outside-in signaling. Without force, the interactions between the αA and βA domains are weak and transient. Transient ligand engagement per se cannot change the αA domain MIDAS conformation (as assessed by the unchanged off-rate for ligand dissociation), cannot move its α_7-helix downward, and cannot induce conformational changes

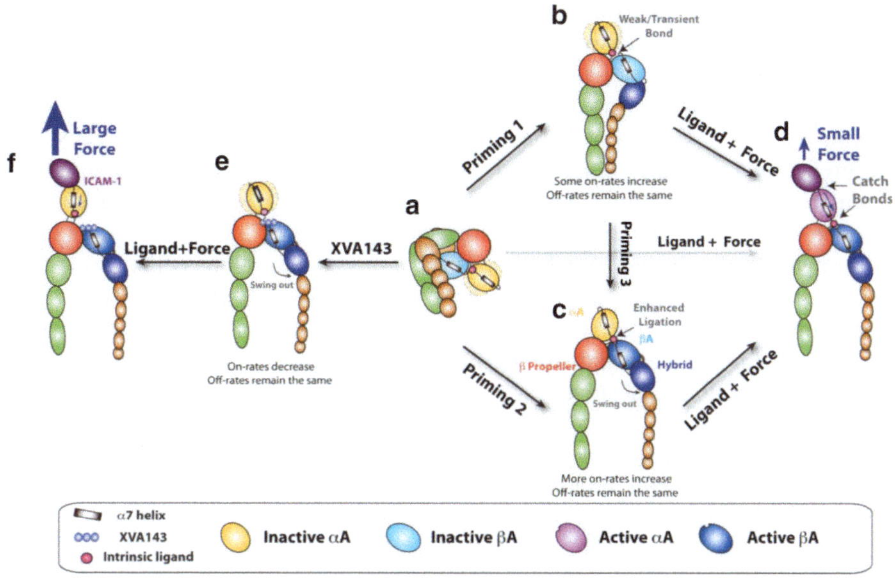

Fig. 3.11 Model for integrin activation. (**a**) Inactive, bent αLβ2with a closed headpiece. The hybrid domain is closed. The αA domain is in the inactive conformation with the α7-helix in the "up" position and the MIDAS in the closed conformation. (**b, c**) Priming by inside-out signaling or metal ions extends αLβ2with legs that may or may not separate, a hybrid domain that may or may not swing out, and the ligation between the αA and βA domain that may or may not be stable. The αA domain flexibility is depicted by the *dotted light yellow* ellipsoids moving about the *solid yellow* αA domain in (**b**), which disappears in (**c**) when the ligation between the αA and βA domains is enhanced by the hybrid domain swing-out. The on-rate for ICAM-1 is increased by different amounts depending on this ligation. Without force, however, the αA domain remains in a confirmation that yields the same fast off- rate for ICAM-1 dissociation. (**d**) Force applied by ICAM-1 is transmitted from the αA domain MIDAS to the α7-helix (indicated by a *curved arrow* inside the αA domain) to elicit catch bonds between the intrinsic ligand and the βA domain, which pulls the α7-helix down (indicated) to induce the intermediate- and long-lived states, resulting in αLβ2/ICAM-1 catch bonds. (**e**) Regardless of priming, binding of XVA143 blocks the αA/βA ligation, makes the αA domain flexible, and pushes the βA domain α7-helix further downward, inducing conformational changes that propagate further downstream, including hybrid domain swing-out and integrin extension. On-rate for ICAM-1 is reduced but zero-force off-rate is the same. (**f**) Blocking the internal ligation shifts a portion of the force originally transmitted via the αA/βA ligation to other contacts between the αA and β-propeller domains, which reduces the force borne by the α_7-helix (indicated by a *smaller arrow*). A much larger external force applied by ICAM-1 is needed to generate a similar internal force to pull down the αA domain α7-helix to induce the intermediate- and long-lived states; but the mechanism of forcing acceleration of dissociation dominates at such high forces, converting the αLβ2/ICAM-1 catch–slip bonds to that of slip-only. Reproduced from (Chen et al. 2010) with permission

downstream. In other words, force applied via a bound ligand is required to induce conformational changes in the αA domain MIDAS and α_7-helix, to strengthen the engagement of the βA domain MIDAS with the internal ligand, and to propagate the conformational changes further downstream to induce signaling. Catch bonds are key elements of this allosteric model of the integrin mechanochemistry.

As a family of cell adhesion molecules, integrins function as mechanical connectors. They also transduce signals bidirectionally from inside out and from outside in across the cell membrane. Such a dual role makes integrins prime candidates for force-sensing molecules in mechanotransduction (Schwartz and DeSimone 2008). Catch bonds provide a physical mechanism for force sensing if different bond lifetimes correspond to different activation states that transduce distinct signals. The ability of integrin/ligand bonds to be strengthened by force may be of great importance not only for leukocyte trafficking, which occurs under hydrodynamic force, but also for the migration of many other cell types, which involves cyclic adhesion and detachment between the cell and the ECM. Catch bonds may provide a mechanical mechanism for the cell to regulate adhesion by applying different forces at different times and locations when and where different adhesion strengths are desired.

Acknowledgments We thank the coauthors of our original papers, the results of which are summarized here, especially Fang Kong and Jizhong Lou. This work was supported by NIH grant R01AI044902.

References

Alberts B, Johnson A, Lewis J, Raff M, Roberts K, Walter P (2002) Molecular biology of the cell. Garland Science, New York, NY

Alon R, Dustin ML (2007) Force as a facilitator of integrin conformational changes during leukocyte arrest on blood vessels and antigen-presenting cells. Immunity 26:17–27

Arnaout MA, Mahalingam B, Xiong JP (2005) Integrin structure, allostery, and bidirectional signaling. Annu Rev Cell Dev Biol 21:381–410

Beglova N, Blacklow SC, Takagi J, Springer TA (2002) Cysteine-rich module structure reveals a fulcrum for integrin rearrangement upon activation. Nat Struct Biol 9:282–287

Bell GI (1978) Models for the specific adhesion of cells to cells. Science 200:618–627

Chen W, Evans EA, McEver RP, Zhu C (2008a) Monitoring receptor-ligand interactions between surfaces by thermal fluctuations. Biophys J 94:694–701

Chen W, Zarnitsyna V, Sarangapani K, Huang J, Zhu C (2008b) Measuring receptor ligand binding kinetics on cell surfaces: from adhesion frequency to thermal fluctuation methods. Cellu Mol Bioeng 1:276–288

Chen W, Lou J, Zhu C (2010) Forcing switch from short- to intermediate- and long-lived states of the αA domain generates LFA-1/ICAM-1 catch bonds. J Biol Chem 285:35967–35978

Chen W, Lou J, Hsin J, Schulten K, Harvey SC, Zhu C (2011) Molecular dynamics simulations of forced unbending of integrin $\alpha_V\beta_3$. PLoS Comput Biol 7:e1001086

Chesla SE, Selvaraj P, Zhu C (1998) Measuring two-dimensional receptor-ligand binding kinetics by micropipette. Biophys J 75:1553–1572

Dembo M, Torney DC, Saxman K, Hammer D (1988) The reaction-limited kinetics of membrane-to- surface adhesion and detachment. Proc R Soc Lond B Biol Sci 234:55–83

Evans E, Leung A, Heinrich V, Zhu C (2004) Mechanical switching and coupling between two dissociation pathways in a P-selectin adhesion bond. Proc Natl Acad Sci USA 101:11281–11286

Friedland JC, Lee MH, Boettiger D (2009) Mechanically activated integrin switch controls $\alpha_5\beta_1$ function. Science 323:642–644

Guo B, Guilford WH (2006) Mechanics of actomyosin bonds in different nucleotide states are tuned to muscle contraction. Proc Natl Acad Sci. USA 103:9844–9849

Hynes RO (2002) Integrins: bidirectional, allosteric signaling machines. Cell 110:673–687

Isacke CM, Horton MA (2000) The adhesion molecule factsbook. Elsevier, Amsterdam

Jin M, Andricioaei I, Springer TA (2004) Conversion between three conformational states of integrin I domains with a C-terminal pull spring studied with molecular dynamics. Structure 12:2137–2147

Kong F, Garcia A, Mould A, Humphries M, Zhu C (2009) Demonstration of catch bonds between an integrin and its ligand. J Cell Biol 185:1275–1284

Le Trong I, Aprikian P, Kidd BA, Forero-Shelton M, Tchesnokova V, Rajagopal P, Rodriguez V, Interlandi G, Klevit R, Vogel V, Stenkamp RE, Sokurenko EV, Thomas WE (2010) Structural basis for mechanical force regulation of the adhesin FimH via finger trap-like beta sheet twisting. Cell 141:645–655

Lou J, Zhu C (2007) A structure-based sliding-rebinding mechanism for catch bonds. Biophys J 92:1471–1485

Lou J, Yago T, Klopochi AG, Metha P, Chen W, Zarnitsyna VI, Bovin NV, Zhu C, McEver RP (2006) Flow-enhanced adhesion regulated by a selectin interdomain hinge. J Cell Biol 174:1107–1117

Luo BH, Springer TA (2006) Integrin structures and conformational signaling. Curr Opin Cell Biol 18:579–586

Luo BH, Carman CV, Springer TA (2007) Structural basis of integrin regulation and signaling. Annu Rev Immunol 25:619–647

Marshall BT, Long M, Piper JW, Yago T, McEver RP, Zhu C (2003) Direct observation of catch bonds involving cell-adhesion molecules. Nature 423:190–193

Matsumoto A, Kamata T, Takagi J, Iwasaki K, Yura K (2008) Key interactions in integrin ectodomain responsible for global conformational change detected by elastic network normal-mode analysis. Biophys J 95:2895–2908

McEver RP, Zhu C (2010) Rolling cell adhesion. Annu Rev Cell Dev Biol 26:363–396

Nishida N, Xie C, Shimaoka M, Cheng Y, Walz T, Springer TA (2006) Activation of leukocyte β_2 integrins by conversion from bent to extended conformations. Immunity 25:583–594

Provasi D, Murcia M, Coller BS, Filizola M (2009) Targeted molecular dynamics reveals overall common conformational changes upon hybrid domain swing-out in beta3 integrins. Proteins 77:477–489

Puklin-Faucher E, Vogel V (2009) Integrin activation dynamics between the RGD-binding site and the headpiece hinge. J Biol Chem 284:36557–36568

Puklin-Faucher E, Gao M, Schulten K, Vogel V (2006) How the headpiece hinge angle is opened: new insights into the dynamics of integrin activation. J Cell Biol 175:349–360

Salas A, Shimaoka M, Kogan AN, Harwood C, von Andrian UH, Springer TA (2004) Rolling adhesion through an extended conformation of integrin $\alpha_L\beta_2$ and relation to α I and β I-like domain interaction. Immunity 20:393–406

Sarangapani KK, Yago T, Klopochi AG, Lawrence MB, Fieger CB, Rosen SD, McEver RP, Zhu C (2004) Low force decelerates L-selectin dissociation from P-selectin glycoprotein ligand-1 and endoglycan. J Biol Chem 279:2291–2298

Schurpf T, Springer TA (2011) Regulation of integrin affinity on cell surfaces. EMBO J 30:4712–4727

Schwartz MA, DeSimone DW (2008) Cell adhesion receptors in mechanotransduction. Curr Opin Cell Biol 20:551–556

Shamri R, Grabovsky V, Gauguet JM, Feigelson S, Manevich E, Kolanus W, Robinson MK, Staunton DE, von Andrian UH, Alon R (2005) Lymphocyte arrest requires instantaneous induction of an extended LFA-1 conformation mediated by endothelium-bound chemokines. Nat Immunol 6:497–506

Shimaoka M, Xiao T, Liu JH, Yang Y, Dong Y, Jun CD, McCormack A, Zhang R, Joachimiak A, Takagi J, Wang JH, Springer TA (2003) Structures of the α_L I domain and its complex with ICAM-1 reveal a shape-shifting pathway for integrin regulation. Cell 112:99–111

Takagi J, Springer TA (2002) Integrin activation and structural rearrangement. Immunol Rev 186:141–163

Takagi J, Petre BM, Walz T, Springer TA (2002) Global conformational rearrangements in integrin extracellular domains in outside-in and inside-out signaling. Cell 110:599–611

Thomas WE, Trintchina E, Forero M, Vogel V, Sokurenko EV (2002) Bacterial adhesion to target cells enhanced by shear force. Cell 109:913–923

Wayman AM, Chen W, McEver RP, Zhu C (2010) Triphasic force dependence of E-selectin/ligand dissociation governs cell rolling under flow. Biophys J 99:1166–1174

Wu T, Lin J, Cruz MA, Dong JF, Zhu C (2010) Force-induced cleavage of single VWF-A1A2A3 tridomains by ADAMTS-13. Blood 115:370–378

Xiang X, Lee C-, Li T, Chen W, Lou J, Zhu C (2011) Structural basis and kinetics of force-induced conformational changes of an αA domain-containing integrin. PLoS One 6(11):e27946

Xiao T, Takagi J, Coller BS, Wang JH, Springer TA (2004) Structural basis for allostery in integrins and binding to fibrinogen-mimetic therapeutics. Nature 432:59–67

Xie C, Zhu J, Chen X, Mi L, Nishida N, Springer TA (2009) Structure of an integrin with an αI domain, complement receptor type 4. EMBO J. 29:666–679

Yago T, Wu J, Wey CD, Klopocki AG, Zhu C, McEver RP (2004) Catch bonds govern adhesion through L-selectin at threshold shear. J Cell Biol 166:913–923

Yago T, Lou J, Wu T, Yang J, Miner JJ, Coburn L, Lopez JA, Cruz MA, Dong JF, McIntire LV, McEver RP, Zhu C (2008) Platelet glycoprotein Ibα forms catch bonds with human WT vWF but not with type 2B von Willebrand disease vWF. J Clin Invest 118:3195–3207

Yakovenko O, Sharma S, Forero M, Tchesnokova V, Aprikian P, Kidd B, Mach A, Vogel V, Sokurenko E, Thomas WE (2008) FimH forms catch bonds that are enhanced by mechanical force due to allosteric regulation. J Biol Chem 283:11596–11605

Zarnitsyna VI, Zhu C (2011) Adhesion frequency assay for in situ kinetics analysis of cross-junctional molecular interactions at the cell-cell interface. J Vis Exp e3519

Zhu C, Lou J, McEver RP (2005) Catch bonds: physical models, structural bases, biological function and rheological relevance. Biorheology 42:443–462

Zhu J, Luo BH, Xiao T, Zhang C, Nishida N, Springer TA (2008) Structure of a complete integrin ectodomain in a physiologic resting state and activation and deactivation by applied forces. Mol Cell 32:849–861

Chapter 4
AFM Visualization of Protein–DNA Interactions

Yuri L. Lyubchenko

4.1 Introduction

Proteins are the principal mediators of DNA processes and organization. Enzymes and regulatory proteins perform cellular processes, such as replication, transcription, and recombination; while architectural proteins are involved in the organization of DNA within the cell into structures, such as chromatin. DNA is a long polymeric molecule, ranging in size of several thousand monomeric units (bacteriophage genome) to a million base pairs (eukaryotic genome); therefore, nanoscale imaging tools are instrumental in understanding the mechanisms of genetic processes and intracellular DNA organization. Historically, electron microscopy (EM) was widely employed to study the various properties of DNA. Of particular note, EM imaging was critical in proving the nucleosomal organization of DNA within the cell by graphically illustrating the beads-on-a-string model of unfolded chromatin and elucidating the higher order structure of chromatin [see Watson (2008) and numerous references therein]. The advent of atomic force microscopy (AFM), and the application of this nanoimaging technique to molecular biology, opens new prospects for studies of protein–DNA complexes. AFM technology allows for a gentle sample preparation, easing concerns about sample preservation. Moreover, AFM is capable of imaging samples in fully hydrated states by scanning in aqueous solutions. Therefore, in addition to nanoscale static structural data of protein–DNA complexes, AFM visualizes the dynamics of conformational transitions of DNA and various nucleoprotein complexes.

Y.L. Lyubchenko (✉)
Department of Pharmaceutical Sciences, University of Nebraska Medical Center, Omaha, NE, USA
e-mail: ylyubchenko@unmc.edu

A.F. Oberhauser (ed.), *Single-molecule Studies of Proteins*,
Biophysics for the Life Sciences 2, DOI 10.1007/978-1-4614-4921-8_4,

4.2 AFM Methodology

4.2.1 AFM Instrumentation

In AFM, a sharp stylus reads the profile of the sample by scanning over the sample in a raster pattern. The tip is attached to a cantilever that works as a spring, pressing the tip against the sample to reproduce the surface profile. The vertical position of the tip and corresponding *x–y* coordinate of the tip are translated by the electronics to the sample topography. The *x–y* position of the tip is controlled by the scanner with accuracy greater than 1 nm, with the vertical displacement of the tip determined with subnanometer accuracy. Therefore, the sample topography can be obtained with subnanometer resolution, which has been confirmed by numerous high-resolution AFM studies. A recent study reported superatomic resolution for isolated organic molecules imaged with AFM (Gross et al. 2010). Imaging was performed under low temperature (5 K) and high vacuum conditions.

The intermediate or intermittent contact mode (IC mode), has proven to be efficient in numerous biological applications. Initially, this mode was termed Tapping mode (Zhong et al. 1993), but the name has been trademarked by the AFM manufacturer, Veeco/Bruker-Nano. Therefore, other manufacturers utilize this methodology under different names, such as alternating contact (AC). In the contact mode, the tip–sample distance is maintained via measuring the deflection of the tip cantilever determined by the van der Waals repulsion forces. In AC/IC/TM-AFM, a cantilever is deliberately vibrated at a frequency close to the cantilever resonant frequency by a piezoelectric modulator with a small amplitude. As the tip approaches a surface, the van der Waals attractive force between the tip and the sample changes both the amplitude and the phase of the cantilever vibration. These changes are monitored by a Z-servo system feedback loop to control the tip–sample distance. The ability of the AC/IC/TM mode to gently image biological samples is the major attractive feature of this operating mode. Figure 4.1 shows images of two samples to illustrate the power of AC/IC/TM mode for biological applications. An image of plasmid DNA with an extruded cruciform structure (indicated with an arrow) with RuvA protein bound specifically to the cruciform is shown in Fig. 4.1a (Shlyakhtenko et al. 2000). Figure 4.1b shows the AFM images of a large DNA assembly termed kinetoplast DNA (kDNA) (Lyubchenko et al. 1995). This natively existing kDNA assembly consists of several thousands of DNA circles of approximately 2 kb (minicircles) and a few dozen of much larger DNA molecules, macrocircles interlocked into a stable network. Individual or a few interlocked minicircles are seen in the background. The gentle operation of AFM, combined with appropriate sample preparation procedures, were two critical issues enabling the observation of intact large and fragile biological systems such as kDNA (Lyubchenko et al. 1995).

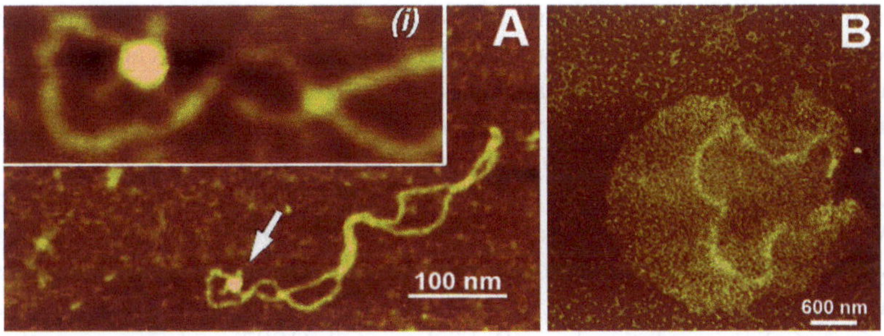

Fig. 4.1 AFM images of two different samples. (**a**) Supercoiled pUC8F14C plasmid DNA with the cruciform protrusion (indicated with an *arrow*) with the RuvA tetramer bound to the cruciform. A large rescanned image is shown in the *inset* (i). See Shlyakhtenko et al. (2000) for specific details. (**b**) Kinetoplast DNA. Kinetoplast DNA is a giant network composed of several thousand DNA minicircles and a few dozen microcircles, which are topologically interlocked. See Lyubchenko et al. (1995) for more information. The images in (**a**) were acquired in air and in propanol (**b**) with a NanoScope III microscope operated in the Tapping Mode. Z-range is 5 nm for (**a**) and 20 nm for (**b**). Reproduced from Shlyakhtenko, L.S., Hsieh, P., Grigoriev, M., Potaman, V.N., Sinden, R.R. and Lyubchenko, Y.L. (2000). A cruciform structural transition provides a molecular switch for chromosome structure and dynamics. *Journal of Molecular Biology* 296: 1169–1173, with permission provided by Elsevier

4.2.2 Sample Preparation for AFM Protein–DNA Complexes

Sample preparation is a key step for any imaging technique and inevitable sweeping effects of the AFM probe highlights the importance of the sample preparation issue. Progress in reliable imaging of DNA was made in the early 1990s when two methods for DNA sample preparation for AFM were proposed (Bustamante et al. 1992; Lyubchenko et al. 1992a, b; Vesenka et al. 1992). In both methods, mica was used as the AFM substrate, primarily due to its very smooth surface topography. The mica surface is almost atomically flat over microns of area. However, the surface is negatively charged and cannot bind negatively charged DNA molecules. The methods described below overcome this problem.

In the first method (Bustamante and Rivetti 1996; Bustamante et al. 1992; Lyubchenko et al. 1992a, b; Vesenka et al. 1992), Mg^{2+} is used to increase the affinity of the negatively charged mica surface for DNA. As a result, the DNA molecules are held in place strongly enough to permit reliable imaging by AFM. Instead of Mg^{2+} cations, other divalent cations can be used (Thundat et al. 1992), with Ni^{2+} cations being the second-most widely used cation in AFM studies of DNA and protein–DNA complexes [reviewed in Murphy et al. (2011)].

An alternative method [reviewed in Lyubchenko et al. (2001)] utilizes the vapor deposition of aminopropyltriethoxy silane (APTES) on mica and other silica surfaces to functionalize the surface with amino groups. The methodology termed AP-mica allows one to prepare smooth functionalized mica, capable of binding negatively charged DNA in wide ranges of pH, ionic strengths, and temperatures.

In this method, divalent cations are not required, and the ability of AP-mica to retain DNA-binding activity for a few weeks after preparation, provides another attractive feature. The use of AP-mica was critical in studies of local and global conformational transitions of DNA and properties of different protein–DNA complexes, including chromatin, which do not need divalent cations [reviewed in (Lyubchenko et al. 2001; Lyubchenko 2004)]. The images of complexes of RuvA with plasmid DNA and large DNA complexes, such as kinetoplast DNA, are shown in Fig. 4.1, and were obtained using the AP-mica methodology. Note a clear plectonemic shape of the supercoiled DNA in Fig. 4.1a, which is in agreement with EM data and numerous studies in solution [e.g., Rybenkov et al. (1997) and references therein]. The cation-assisted methods failed to obtain this morphology of supercoiled DNA, making these methods problematic for studies of supercoiled DNA (Lyubchenko 2004). Another mica functionalization technique utilizes aminopropyl silatrane to yield APS-mica (Shlyakhtenko et al. 2003). This substrate has properties very similar to AP-mica, but is better suited for imaging DNA and protein–DNA complexes in aqueous solutions (Lyubchenko and Shlyakhtenko 2009; Lyubchenko et al. 2009) including high-speed AFM imaging (Miyagi et al. 2011; Shlyakhtenko et al. 2009). Polylysine has also been used as a mica coating for DNA imaging (Bussiek et al. 2003) and was later applied for imaging of reconstituted chromatin (Bussiek et al. 2005) including the time-lapse AFM in aqueous solutions (Suzuki et al. 2010).

4.3 Complexes of DNA with Functional DNA-Binding Proteins

In this section we describe AFM applications to study protein–DNA complexes involved in a number of DNA functions. We focus on a few DNA recombination and repair systems as well as DNA restriction accomplished by restriction enzymes. These proteins bind either specific DNA lesions (DNA repair) or specific DNA sequences (DNA recombination and restriction). The AFM analysis of these systems utilizes three major capabilities of AFM (1) the specific binding localization of the protein to a target (a lesion or specific DNA sequence); (2) the characterization of protein stoichiometry within the complex; and (3) the direct visualization of the dynamics and interactions within the complex by using the capability of AFM to perform scanning in aqueous solutions. The subsections below illustrate applications of these AFM approaches to a number of systems.

4.3.1 Site-Specific Protein–DNA Complexes

DNA restriction is a protective mechanism by which bacterial cells degrade foreign DNA within the cell. They recognize a specific sequence in the DNA duplex and cleave the DNA if the sequence is not methylated. The mechanism by which

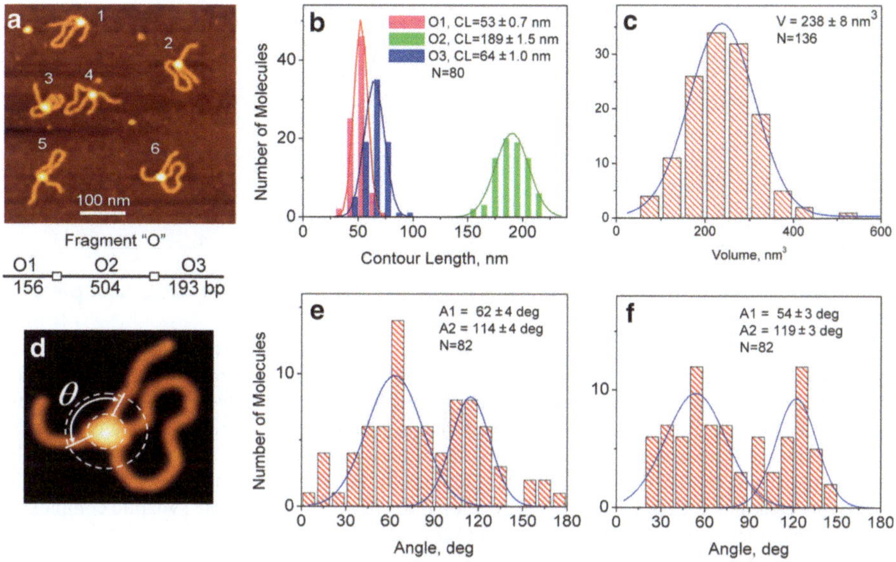

Fig. 4.2 AFM images and the results of AFM analysis for SfiI complexes in *cis* with fragment "O". (**a**) AFM image of the 500 × 500 nm^2 area with six molecules in the field. A model of the DNA fragment is shown below the image. All molecules on the image have a loop. (**b**) The histograms for the contour length (short arm O1—*red*, long arm O3—*green*, and a circumference of the loop O2—*blue*). (**c**) The histogram for SfiI volume measurements. (**d**) The schematic illustrating the procedure for measuring the angle between DNA strands. (**e**) The histogram for the angle between DNA arms outside the loop. (**f**) The histogram for the angle between DNA strands inside the loop. Reproduced from Lushnikov, A.Y., Potaman, V.N., and Lyubchenko, Y.L., (2006a). Site-specific labeling of supercoiled DNA. *Nucleic Acids Res* 34: e111, with permission provided by ACS publications

restriction enzymes recognize the DNA-binding sites, form a complex, and cleave the site, are the primary questions addressed by AFM. The SfiI restriction enzyme was studied in Lushnikov et al. (2006a, b). SfiI cleaves DNA only if it finds two specific sites and brings them together to form a synaptic precleavage complex. The formation of synaptic complexes is directly evidenced from AFM studies. AFM images of SfiI complexes with DNA fragments containing two binding sites for SfiI are shown in Fig. 4.2a. The formation of the synaptic complex leads to the looped complexes that are seen on the images. All DNA molecules on this image have a clear-cut looped morphology with a bright spot at the crossover. The looped DNA structure is the predominant morphology for this type of complex, comprising 85 % of the DNA molecules observed. The specificity of protein binding is verified by the length measurements and the data shown in the histogram in Fig. 4.2b are in agreement with the position of the specific region on the DNA substrate shown schematically in Fig. 4.2a. The volume measurements (Fig. 4.2c) suggest that SfiI in the synaptic complex is a tetramer. The measurements of the angle between the DNA arms within the complex (Fig. 4.2d–f) demonstrate that the two DNA duplexes are crossed at an angle of approximately 60°. Moreover, additional

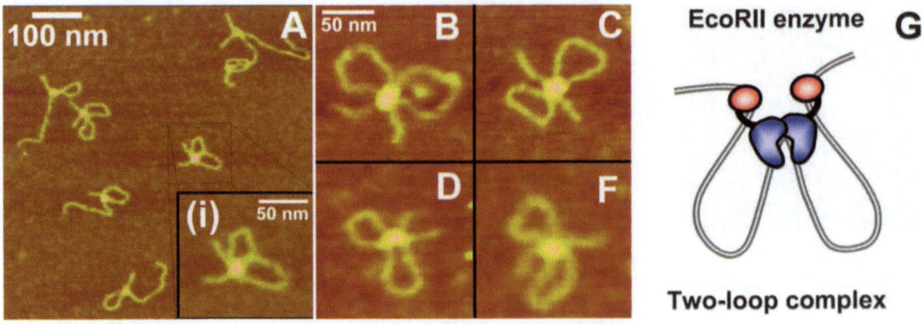

Fig. 4.3 AFM images of the EcoRII–DNA complexes. (**a**) Large scan with various complexes. A two-looped complex in the *middle* of the scan is boxed. A zoomed image of this two-looped complex is shown as *inset* (i). (**b–f**) Gallery of AFM images for two-looped DNA protein complexes, bar size = 50 nm, is applied to all images. (**g**) Model for the two-looped complex of EcoRII with the DNA fragment bound at all three sites. Reproduced with permission provided by ACS publications from Shlyakhtenko, L.S., Gilmore, J., Portillo, A., Tamulaitis, G., Siksnys, V. and Lyubchenko, Y.L., 2007. Direct visualization of the EcoRII–DNA triple synaptic complex by atomic force microscopy. *Biochemistry* 46: 11128–11136

analysis on specially designed DNA substrates led to the conclusion that two different orientations of the DNA recognition sites are possible in the synaptosome and the central part of the DNA sequence should have direct contact with the protein. Importantly, these properties of the synaptosome are in concordance with an independently published crystallographic study (Vanamee et al. 2005).

AFM was instrumental in characterizing rather unusual DNA synaptosomes in which the protein holds three specific sites (Shlyakhtenko et al. 2007). In the majority of synaptic protein–DNA complexes, two DNA duplexes are involved in the synaptosome assembly. However, a three-site synaptic complex is formed during Mu DNA transposition (Kobryn et al. 2002; Watson and Chaconas 1996). In the His-mediated inversion reaction, a tripartite complex is formed when the His tetramers, holding hixL, and hixR sites, interact with the Fis dimer, which brings an enhancer site to the complex (Merickel and Johnson 2004). These are complex multiprotein systems in which the participation of multimers of the same protein (three MuA transposases for Mu system) or more than two proteins (His-mediated invertasome) are involved. It was proposed, based on biochemical data, that efficient cleavage mediated by the EcoRII restriction endonuclease also requires the formation of a synaptic complex involving three DNA recognition regions (Tamulaitis et al. 2006). This finding was surprising, as EcoRII is a rather simple system and exists in solution as a dimer and, according to early biochemical studies, EcoRII is capable of forming synaptic complexes by cooperative binding of two DNA regions (Reuter et al. 1998). To test this novel model for the synaptic complex formation, DNA substrates with a well-defined protein-binding-site position were complexed with EcoRII and used for AFM studies. The AFM image in Fig. 4.3a shows that in addition to one-looped structures corresponding to the coupling of two sites on DNA, three-site binding was also identified by the appearance of two loops. A gallery of two-loop complexes is shown in Fig. 4.3b–f. The volume measurements led to the conclusion that the protein exists in dimeric stoichiometry within the three-site synaptic complexes. The results further suggested a model in which the N-domain of EcoRII binds two sites, whereas the third site is bound by two C-domains of the protein monomer (Fig. 4.3g).

4.3.2 DNA Repair Systems

Another biologically important problem to which AFM was applied is DNA mismatch repair. Although the DNA replication machinery utilizes a proofreading mechanism to achieve a high degree of accuracy, some misincorporated nucleotides remain undetected and become mismatches between the template and the newly synthesized DNA strands. If mismatches are not repaired, the second round of DNA replication leads to mutation. The mission of the DNA mismatch repair system (MMR) is to detect misincorporated nucleotides and repair the defect.

In *Escherichia coli*, mismatches are recognized by a dimer of MutS protein that in complex with MutL, activates MutH endonuclease and cleaves one DNA strand near the mismatch. In the pioneering work of Wang et al. (2003), the authors demonstrate that during nonspecific binding, MutS bends the DNA duplex and undergoes a conformational change upon approaching the mismatch and the formation of the initial recognition complex (IRC), in which the DNA is kinked. Upon the formation of the ultimate recognition complex (URC), the protein undergoes an additional conformational change accompanied by straightening of the DNA (Yang et al. 2005). In this work, the authors also show that MutS is capable of binding to the DNA ends, and this finding is explained by earlier results demonstrating the low binding constant for MutS obtained by traditional bulk measurements. Insight into the mechanism of the search for mismatches and the role of the conserved Phe and Glu residues of MutS in the formation of complexes with the mismatch was made in another publication by the same team, in which complexes of wild type MutS and two mutants were analyzed by AFM (Tessmer et al. 2008). These studies show that Glu is critical in DNA bending in the nonspecific complex formation of MutS, but it also promotes DNA straightening during the formation of specific complexes. However, Phe is not required for these conformational DNA changes. The authors combined their findings and crystallographic data to propose the model for mismatch search, which involves the conformational changes of MutS and DNA between the IRC and URC complexes.

The mechanism by which MutS initiates the mismatch repair process was studied in the recent AFM paper by Jiang and Marszalek (2011). The *E. coli* MMR system is initiated by MutS binding to the mismatch, followed by the cleavage of one DNA strand by MutH endonuclease at the GATC site (in the complex with MutS and MutL) as far as 1,000 bp or more from the mismatch. The interaction of the distant sites via the formation of DNA loops is one of the plausible mechanisms that could explain how the MMR system functions and a

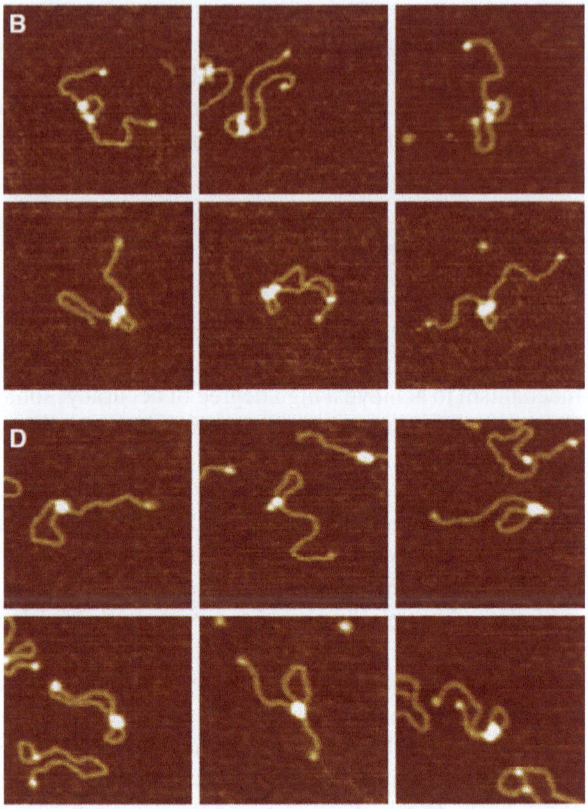

Fig. 4.4 AFM images showing single-site binding and loop binding of MutS to heteroduplex DNA. (**b**) Images of MutS deposited in the absence of nucleotides show MutS loop binding. (**d**) Images of MutS deposited in the presence of 0.5 mM ATP show loop binding. In both cases, the ends of the DNA were previously bound with streptavidin. The scale bar indicates 200 nm. Reproduced from Jiang, Y. and Marszalek, P.E. (2011). Atomic force microscopy captures MutS tetramers initiating DNA mismatch repair. *Embo J* 30: 2881–2893, with permission provided by NPG

paper by Jiang and Marszalek (2011) addressed a number of specific questions related to the DNA looping process. First, it was demonstrated that DNA looping can be performed by the MutS protein only. This finding is illustrated by Fig. 4.4b, in which the AFM images of complexes of MutS with 1,100 bp DNA containing a mismatch in the middle are shown. An interesting feature of these images is that the two protein blobs are clearly resolved on these images. The volume measurements show that each blob is the MutS dimer. However, adding ATP dramatically changes the morphology of the loops. As seen in Fig. 4.4d, the protein appears as a bright, large, one-blob structure. The volume measurements suggest that each blob is the MutS tetramer. Similar studies of one-site bound protein showed that the addition of ATP changes the protein dimer size, suggesting that ATP induces conformational changes of MutS. Based on these observations, as well as experiments with the roadblock placed on the DNS duplex, the authors proposed the model in which one MutS dimer resides at the mismatch, whereas another dimer binds a nonspecific site and accomplishes the search for the cleavage site by translocating the nonspecific DNA dusplex. ATP is involved in this process; however, whether its hydrolysis is required for translocation remains an unanswered question.

4.4 AFM of Complexes of DNA with Architectural Proteins

Architectural proteins are involved in all DNA functions, including recombination, replication, and transcription. The subsections below briefly outline the AFM studies of single-stranded DNA-binding proteins (SSB) of *E. coli*. SSB have a high affinity for single-stranded DNA (ssDNA) and participate in all genetic processes involving ssDNA as an accessory protein [reviewed in Bochkarev and Bochkareva (2004), Kur et al. (2005), Shereda et al. (2008)]. Similar proteins are found in other organisms, including eukaryotes, and their primary function is to stabilize unwound single strands and prevent ssDNA from reforming a double helix during all genetic processes. Another type of architectural protein is the histones, which are involved in the formation of nucleosomes, the fundamental unit of chromatin. AFM was instrumental in elucidating the structure and dynamics of nucleosomal proteins.

4.4.1 Single-Stranded DNA-Binding Proteins

SSB proteins play a role in DNA replication, where they stabilize unwound single strands and prevent ssDNA from reforming a double helix. During the homologous recombination process, SSB proteins are involved in the formation of RecA–DNA filaments, regulating the process of loading RecA onto ssDNA (Shereda et al. 2008). The interaction of SSB with other recombination mediator proteins, such as RecO, facilitates the annealing step via the SSB-induced conformational change of the SSB–ssDNA filament (Ryzhikov et al. 2011). According to Lohman and Ferrari (1994), SSB has two distinct binding modes, depending on the salt concentration. At low monovalent salt concentrations (<10 mM NaCl), SSB protein binds ssDNA cooperatively with only two subunits of the SSB tetramer interacting with ssDNA. This mode is termed $(SSB)_{35}$ mode. At high ionic strength (>200 mM NaCl), the $(SSB)_{65}$- binding mode, all four subunits of SSB interact with ssDNA in a noncooperative manner.

AFM and electron microscopy (EM) are widely used for the nanoscale characterization of SSB–DNA complexes. Pioneering work (Chrysogelos and Griffith 1982), in which EM was applied to SSB complexes with fd phage ssDNA as a substrate, demonstrated that images of the SSB–DNA complex appeared as beaded loops, with approximately 170 nucleotides per loop. Cross-linking of the protein to DNA with glutaraldehyde was used during EM to stabilize the SSB–DNA complexes. The advantage of AFM compared with EM is that a gentle sample preparation procedure is used, allowing the creation of protein–DNA complexes without the cross-linking step. AFM images of SSB–DNA complexes were obtained in Hamon et al. (2007). A circular DNA substrate (M13 ssDNA) was used and images at high and low ionic strengths were obtained. The complex, consisting of a beads-on-a-string morphology, with bright globular SSB blobs

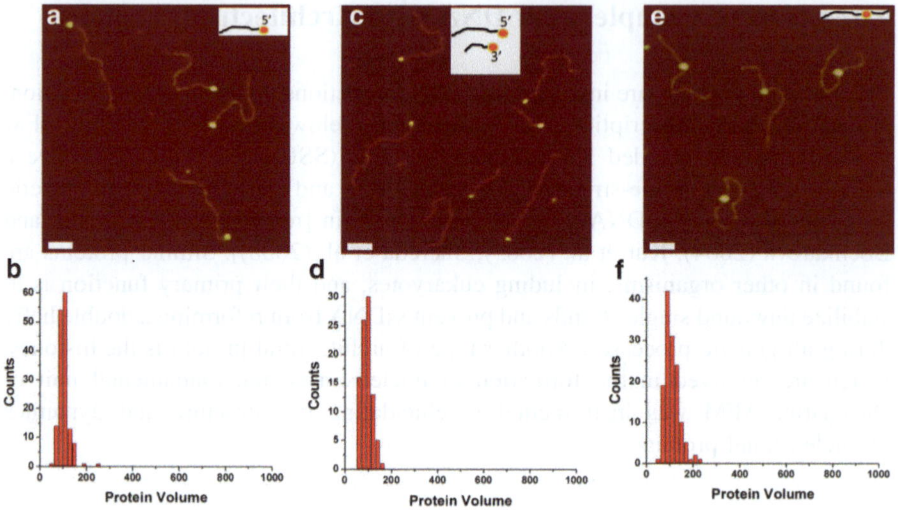

Fig. 4.5 AFM data for complexes of SSB with hybrid DNA of different types. (**a**), (**c**), and (**e**) are AFM images of complexes with 5′-tail-DNA, mixture of 5′-tail-DNA, and 3′-tail-DNA and gap-DNA, respectively. The *insets* show the schematic presentations of all complexes. Bar sizes are 50 nm. The histogram for the protein volumes are shown in (**b**), (**d**), and (**e**) for 5′-tail-DNA, mixture of 5′-tail-DNA, and 3′-tail-DNA, and gap-DNA, respectively. Reproduced from Shlyakhtenko, L.S., Lushnikov, A.Y., Miyagi, A. and Lyubchenko, Y.L. (2012). Specificity of binding of the single-stranded DNA binding protein to the target. *Biochemistry* (in press), with permission provided by ACS publications

assembled on the DNA substrate is seen for high-salt and low-salt conditions. The bead sizes increase as the ionic strength decreases, and the results are in agreement with the different modes of SSB binding to DNA under different ionic conditions (Lohman and Ferrari 1994; Shereda et al. 2008).

The effect of ionic strength on the conformation of SSB and on the specificity of the complex formation with ssDNA was studied in Shlyakhtenko et al. (2012). The mica functionalization methodology was utilized, allowing the preparation of samples in a wide range of ionic strengths (Lyubchenko and Shlyakhtenko 2009; Lyubchenko et al. 2009; Shlyakhtenko et al. 2003), leading to the generation of a novel approach to study the interaction of specific and nonspecific complexes with SSB. In this approach, ssDNA is attached either to the end of the DNA duplex (tail-DNA) or between the two DNA duplexes (gap-DNA). Therefore, the SSB complex with ssDNA (specific complex) can be distinguished from nonspecific complexes (SSB with DNA duplex) by the protein location on the DNA substrate, as seen in the AFM images (a, c, and e) and the results of the volume measurements (b, d, and f) in Fig. 4.5. The protein, appearing as a blob, binds very specifically to the ssDNA segment of the DNA substrates, either at the ends for tail-DNAs (a and c), or internal sites (e) for gap-DNA. The protein did not bind to the DNA duplex. The volume measurements are in agreement with the tetrameric stoichiometry of SSB in the complexes. The images were obtained in a buffer containing 10 mM Mg^{2+} cations.

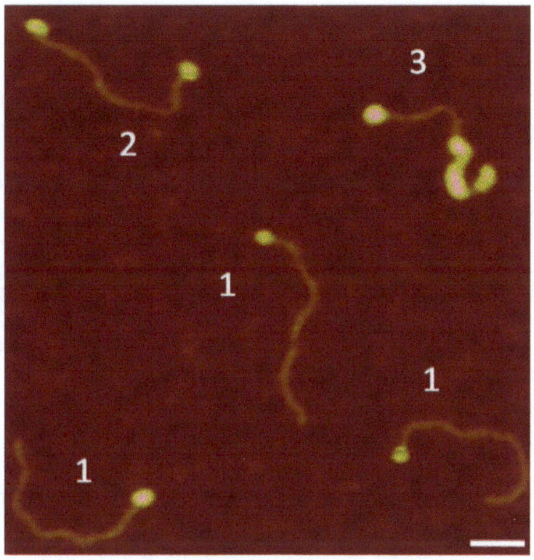

Fig. 4.6 AFM images of complexes 69-tail-DNA with SSB protein assembled in low-salt conditions. Different types of complexes are indicated with 1, 2, and 3. Bar size is 50 nm. Reproduced from Shlyakhtenko, L.S., Lushnikov, A.Y., Miyagi, A. and Lyubchenko, Y.L. (2012). Specificity of binding of the single-stranded DNA binding protein to the target. *Biochemistry* (in press), with permission provided by ACS publications

The pattern of the protein binding changes dramatically if Mg^{2+} cations are removed from the reaction mixture. A large-scale image containing a set of complexes is shown in Fig. 4.6. Under these conditions, SSB binds not only to the ssDNA region of the substrate (marked with 1), but it also binds to the dsDNA end (indicated with 2) and to internal sites of the dsDNA region, as clusters of blobs. Interestingly, large end-bound individual blobs, in addition to the one observed in the presence of Mg^{2+} cations, are observed on the low-salt images. These data suggest that the specificity of SSB binding to ssDNA is strongly dependent on ionic conditions. The high-salt conditions, which are close to physiological conditions, provides the specificity for SSB binding, but the stringency of the interaction is provided by the Mg^{2+} cations. It has also been proposed that the SSB conformation depends on ionic strength, and under low-salt conditions, the protein adopts a conformation facilitating interprotein interactions and the formation of protein assemblies as large as octamers.

4.4.2 Chromatin Structure

DNA inside cells is condensed by interacting with specialized architectural proteins, which assemble DNA into chromosomes in eukaryotes. Figure 4.7 [figure was taken from (Sputnam. http://isite.lps.org/sputnam/Biology/U3Cell/chromosome_1.png)] schematically shows several levels of chromatin compaction to form chromosomes. Five major types of histones—H1, H2A, H2B, H3, and H4—are the major architectural proteins enabling this DNA assembly. The basic structural unit of chromatin is the nucleosome, which was described by Roger Kornberg in 1974 (Watson 2008). In this structure, a DNA segment of 147 bp makes 1 and 3/4

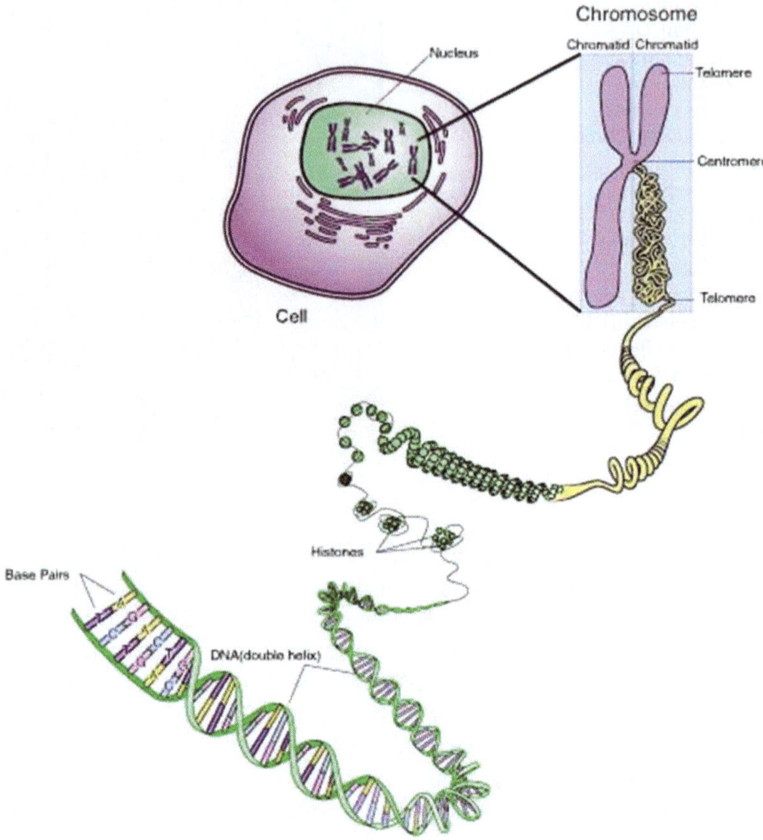

Fig. 4.7 Schematic of the DNA assembly process into chromatin and chromosomes. Reproduced from: http://isite.lps.org/sputnam/Biology/U3Cell/chromosome_1.png

turns to make a compact unit. This level of DNA compaction into nucleosomes yields a bead-on-a-string chromatin fiber of approximately 10 nm in diameter. The chromatin fiber is condensed by coiling into a 30-nm fiber, followed by further condensation of chromatin, resulting in the formation of the chromosome. Electron microscopy (EM) and later AFM were the most important nanoimaging tools enabling the visualization of chromatin at different levels of the assembly.

The nucleosomal beads-on-a-string morphology was confirmed by imaging of chromatin with EM; however, it required fixation of the chromatin with glutaraldehyde. Fixation was not necessary for AFM due to the allowance of a gentle sample preparation. AFM images of the nucleosomal array assembled on the alpha-satellite DNA repeat is shown in Fig. 4.8 [taken from Bussiek et al. (2007)]. In addition to a large scan (a), zoomed images are shown in (b–f), prepared at different ionic conditions. The sample prepared under low-salt conditions yielded the expected

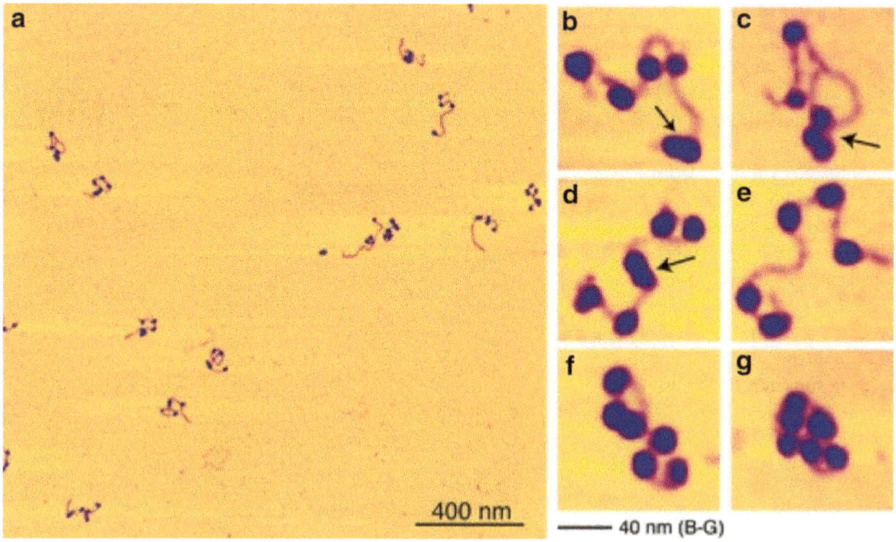

Fig. 4.8 Atomic force microscopy imaging in air of nucleosome arrays. The chromatin was reconstituted at a nominal octamer to DNA weight ratio of 0.5 and adsorbed onto the surface in Hepes–NaOH buffer (10 mM, pH 8.0). (**a**) Overview scan of 172-α7 arrays adsorbed in the buffer additionally containing 10 mM NaCl. (**b–g**) Zooms of individual 172-α7 arrays [adsorption buffer + 10 mM NaCl (**b**, **c**), no additional salt (**d**, **e**), and 10 mM NaCl, 2 mM $MgCl_2$ (**f**, **g**)]. Reproduced from Bussiek, M., Muller, G., Waldeck, W., Diekmann, S. and Langowski, J. (2007). Organisation of nucleosomal arrays reconstituted with repetitive African green monkey alpha-satellite DNA as analyzed by atomic force microscopy. *Eur Biophys* J 37: 81–93, with permission provided by *European Biophysics Journal*

extended conformations (Fig. 4.8b–e, h). However, in the presence of $MgCl_2$, more compact conformations were observed (Fig. 4.8f, g). AFM measurements of internucleosomal distances indicated that a regular nucleosomal arrangement is defined by sequence-specific DNA–histone interactions. Moreover, the analysis of the occupancy of different sites within the DNA template revealed a preferential occupancy of the major site. The alpha-satellite DNA studied in this work is a segment of the centromere region of the chromatin, containing multiple repeats, and the preferential occupancy of the major site is in agreement with results obtained from in vivo studies. Altogether, these data led the authors to speculate that satellite DNA sequences define the sequence-specific structure of centromeric chromatin.

The ability of AFM to image nonfixed chromatin samples was critical in studies of the structure and dynamics of mononucleosomes, nucleosome core particles (NCP) (Shlyakhtenko et al. 2009). A typical AFM image is shown in Fig. 4.9. The nucleosomes are clearly seen on these images as bright blobs with the DNA arms at both sides of the particle; however, the morphology of the particles is different, suggesting that the number of DNA turns vary between the nucleosomes. AFM provides the height of the topographic images, in addition to the lateral sizes,

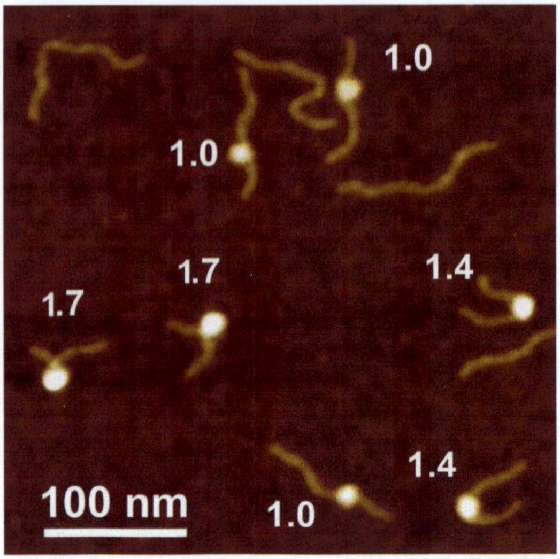

Fig. 4.9 AFM image taken in air of the nucleosome samples. Nucleosomes marked 1, 2, and 3 have about 1.7, 1.4, and 1.0 turns of DNA wrapped around the core particle. Reproduced from Shlyakhtenko, L.S., Lushnikov, A.Y. and Lyubchenko, Y.L. (2009). Dynamics of nucleosomes revealed by time-lapse atomic force microscopy. *Biochemistry* 48: 7842–7848, with permission provided by ACS publications

allowing one to calculate the volume values of the particles. Both of the parameters were used for interpreting the images of NCP (Fig. 4.9) in terms of the number of DNA turns around the histone core, and the measurements showed that these particles contain 1.7, 1.4, and 1.0 DNA turns, respectively. The figures for each individual NCP particle correspond to the numbers of DNA turns calculated for each particle. The values are distributed in a wide range, suggesting that nucleosomes in solution are dynamic and capable of large-scale unwrapping. Similar results were obtained in Filenko et al. (2012), in which the dinucleosome design was studied.

The mononucleosome system was instrumental in understanding the role of NCP concentration and solvent properties on nucleosome stability. It was found, in various single molecule studies of nucleosomes, that their stability decreases at low concentrations of particles, and the use of detergents stabilizes the NCP. In Menshikova et al. (2011), the effect of 3-[(3-Cholamidopropyl)dimethylammonio]-L-propanesulfonate (CHAPS) on the stability of nucleosomes was analyzed. These studies showed that in the presence of millimolar concentrations, CHAPS, nucleosomes, even at subnanomolar concentrations, remain intact over days, compared to a complete dissociation of the same nucleosome sample over 10 min in the absence of CHAPS. Importantly, CHAPS does not change the conformation of nucleosomes, as confirmed by AFM analysis. Moreover, 16 μM CHAPS stabilizes nucleosomes up to and exceeding 1 h of incubation, in a solution containing as low as 0.4 nM nucleosomes. Interestingly, the specificity of the nucleosome formation in regards to the DNA sequence in the presence of CHAPS decreased. This finding suggests that the sequence positioning of nucleosomes can be modulated by environmental conditions.

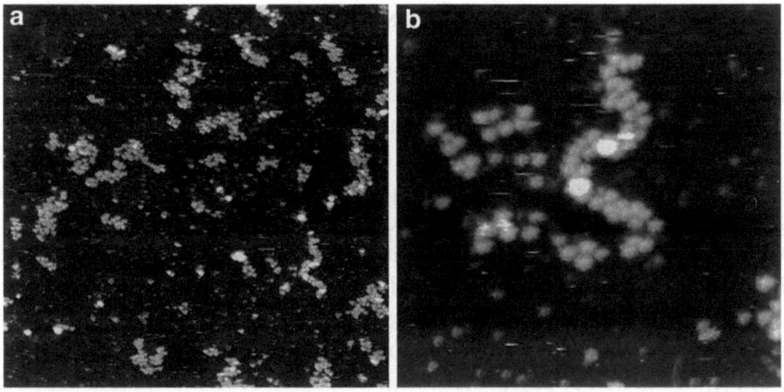

Fig. 4.10 AFM images of native unfixed chicken erythrocyte chromatin fibers (containing linker histones) deposited on glass and imaged in a low ionic strength buffer (5 mM TEA, 1 mM sodium butyrate, 0.1 mM EDTA, pH 7.0). Images are 2,000 × 2,000 (**a**) and 500 nm × 500 nm (**b**); height scale from 0 to 15 nm. Reproduced from Bustamante, C., Zuccheri, G., Leuba, S.H., Yang, G. and Samori, B. (1997). Visualization and analysis of chromatin by scanning force microscopy. *Methods* 12: 73–83, with permission provided by Elsevier

Structural characterization of the higher order structures of chromatin is a challenge due to the low stability of chromatin at this state and the high dependence of the structure on ionic conditions. Early EM observations of microsections of metaphase HeLa chromosomes revealed fibrils of approximately 30 nm (Marsden and Laemmli 1979). Later, X-ray diffraction studies revealed side-to-side packing of nucleosomes in the direction of the 300A filament, and radial packing around it. These results are consistent with the "solenoid" model of Finch and Klug (Widom and Klug 1985). The advent of AFM allowed chromatin to be visualized through alternative preparation procedures. Chromatin gently isolated from chicken erythrocytes was imaged in Bustamante et al. (1997). Importantly, in addition to glutaraldehyde fixed samples, nonfixed samples were studied as well. Figure 4.10 shows AFM images for the nonfixed samples. Long condensed chromatin fibrils appear on a large-scale image (a), along with short oligomers of nucleosomes. A small-scale image of one long fibril is shown in the right image (b). Computer modeling of the chromatin fibrils was performed to reveal the zigzag organization of nucleosomes within the fibril. These studies revealed that variability of the linker length contributed to the irregularity in the fibril morphology observed by AFM. Another conclusion of these analyses is that the length of DNA wrapped around nucleosomes varies in a broad range, between one and two turns. The recent AFM studies of mononucleosomes are consistent with this conclusion (Filenko et al. 2012; Menshikova et al. 2011; Miyagi et al. 2011; Shlyakhtenko et al. 2009). At the current time, the analysis of higher order chromatin structure has not been completed. A range of single molecule techniques is currently being applied to unravel the problem (Bednar and Dimitrov 2011), and development of AFM instrumentation and methodology will aid in the progress of these studies.

4.5 Time-Lapse AFM Analyses of Protein–DNA Complexes

The time-lapse AFM technique was successfully applied to studies of protein–DNA complexes of various kinds. An early study (Guthold et al. 1994) visualized the assembly process of RNA polymerase (RNAP)–DNA complexes in a series of time-lapse images. Later, the same group visualized the sliding of the polymerase along DNA (Kasas et al. 1997), concluding that diffusion of RNAP along DNA constitutes a mechanism for accelerated promoter location. Further studies (Guthold et al. 1999) showed that in addition to sliding, an enzyme could jump from one site to another. A similar immobilization approach enabled AFM observation of the interaction of DNA with photolyase (van Noort et al. 1998). A detailed study of the interaction of p53 protein with DNA was described in (Jiao et al. 2001). These studies revealed the protein jumping around the DNA duplex as well as protein sliding. The flipping process suggests that relatively strong, nonsequence-specific interactions between p53 and DNA take place. The observation is in line with an early theoretical model in which sliding and three-dimensional diffusions were proposed as modes for the site search mechanism by a protein (Guthold et al. 1999; Winter et al. 1981). In addition to the protein recognizing one site on the DNA, there is a large family of proteins that need two specifc sites for their functions. The analysis of the process for the search of two sites is a serious complication for a theoretical study, and two-site binding also narrows the assortment of experimental approaches. Recent studies showed that AFM is a method of choice for characterizing protein–DNA systems of this type (Lushnikov et al. 2006a, b; Lyubchenko and Shlyakhtenko 2009).

Time-lapse high-speed AFM [HS AFM, (Ando et al. 2007)] in liquid has been recently applied to image interactions of the ATP-dependent type III restriction enzyme, EcoP15I, with DNA (Crampton et al. 2007). This restriction enzyme needs to interact with two recognition sites, separated by up to 3,500 bp, before it can cleave DNA. This system was imaged in real time, at scan rates of 1–3 frames/s. The study showed that EcoP15I translocated over the DNA for long distances, at a rate of 79 ± 33 bp/s, resulting in accumulated supercoiled DNA in an ATP-dependent manner. It was concluded from this study that EcoP15I used two distinct mechanisms to communicate between two recognition sites: diffusive DNA loop formation and ATPase-driven translocation of the intervening DNA contour.

Another multisite system, the EcoRII–DNA complex, has been studied using the same HS-AFM approach (Gilmore et al. 2009). Similar to the one-site DNA-binding proteins, EcoRII shows multiple binding–unbinding events and translocations along the DNA. Measurements of the diffusion coefficients revealed 50 times faster two-dimensional mobility compared with the linear diffusion. However, most striking was the observation of the dynamics of the system, in which the protein occupies two specific sites on the DNA. Figure 4.11 displays 8 frames of the event, spanning 20 s, imaged at a rate of 2 frames/s, in which the initial specific complex with a short loop (2.5 s) undergoes sliding dynamics (5 s–12.5 s), resulting in the formation of the specific complex (15 s), involving

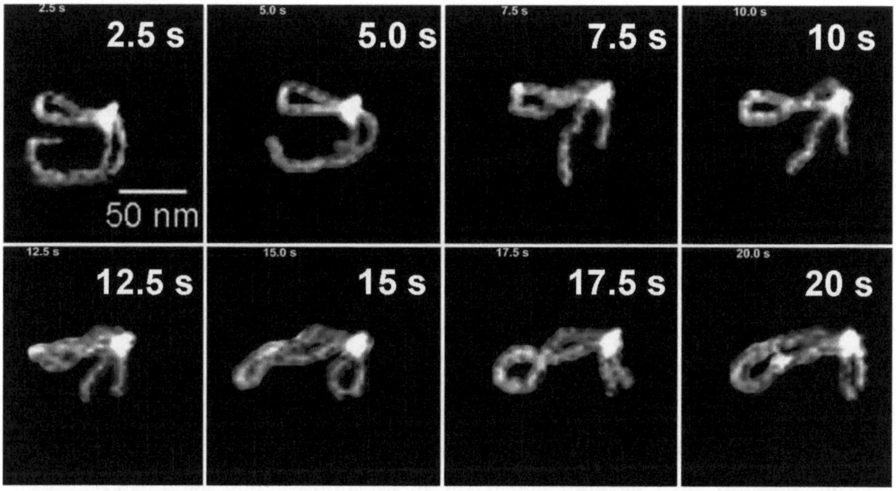

Fig. 4.11 High-speed AFM analysis of EcoRII translocation. Individual frames are shown every 2.5 s. As the long arm gradually gets shorter, the loop length gradually increases. The sliding process over the length of 300 bp occurs within 10 s. Reproduced with permission from Gilmore, J.L., Suzuki, Y., Tamulaitis, G., Siksnys, V., Takeyasu, K. and Lyubchenko, Y.L. (2009). Single-molecule dynamics of the DNA–EcoRII protein complexes revealed with high-speed atomic force microscopy. *Biochemistry* 48: 10492–10498

the third specific site on DNA, identified by the formation of a stable large loop. The contour length of the entire DNA throughout the imaging period remains constant, whereas other parameters change. The short arm also does not change in length, staying at a value close to the expected 34 nm, for specific binding of 100 bp from the end, which shows that the protein remains specifically bound at this recognition site. On the contrary, at about 6 s, another arm decreases in size in parallel with the increase of the loop size. After 10 s, the protein stops translocating and binds to another specific DNA site. This is validated by the length measurements, which show that arm 2 stops at the length of 34 nm, expected for binding 100 bp from the end where the third specific binding site is located. The change in the loop length occurs over a period of about 10 s, covering a distance of about 300 bp (102 nm) until it stops at another recognition site. This means that the protein in this complex moves at a rate of about 30 bp/s (10.2 nm/s).

Two important conclusions emerge from this chapter. First, multisite binding proteins, in addition to the facilitated diffusion mechanisms, utilize a novel loop mechanism in which the protein bound to one DNA site searches for another site by threading the DNA filament through the complex. This threading process continues until the enzyme finds the second recognition site, upon which it forms a stable synaptic complex. Second, threading of DNA is not accompanied by DNA rotating, as it was anticipated based on the helical geometry of DNA, suggesting that the sliding process occurs by a linear movement of the protein along the DNA surface with no rotation. These are entirely novel findings never reported before. Importantly, only time-lapse AFM imaging is capable of detecting such phenomena. Moreover, the process is fast, taking less than 10 s, which corresponds to a rate of about 30 bp/s (10.2 nm/s); therefore, the unique capabilities of HS AFM are required to detect these events, thereby allowing direct visualization of this process.

4.6 Conclusions

The ability of AFM to image biological samples in a fully hydrated state is coming of age. The tip geometry is the major resolution-limiting factor for AFM. Recent advances in the tip manufacturing process provided AFM practitioners with probes as sharp as a few nanometers. At the same time, there is room for improvement in instrumentation, such as high speed for large samples and sample preparation techniques. The ability to use time-lapse AFM to image DNA in liquid in complex with various proteins, including ligands, enables direct imaging of critical genetic processes, such as transcription, replication, and recombination. Recent advances in high-speed AFM (HS AFM) have enabled the technique to become available to a broader biomedical community. The unique capability of time-lapse AFM to provide the dynamics at the nanoscale range opens novel prospects for nanoimaging. Moreover, the current HS AFM design operates with the oscillation amplitude in the range of a nanometer, thereby dramatically minimizing the potential deformation effect of the tip. This technological advance is an important step for further improvement of the instrument, enabling AFM to operate in the least invasive mode (attractive tip–sample interaction regime). The latter will almost eliminate the tip-induced modification of the sample. Finally, the technological advances of AFM analysis of DNA and protein–DNA complexes can be extended to other biomolecular systems and thus will provide molecular biologists with a powerful nanoimaging tool.

Acknowledgments I am grateful to Luda Shlyakhtenko for valuable comments and critical reading of the manuscript, and current and former members of the group for their contribution to works incorporated into the manuscript. The work was supported by grants from National Institutes of Health Grants (1P01GM091743-01A1 and 1 R01 GM096039-01A1), US Department of Energy (DE-FG02-08ER64579), National Science Foundation (EPS – 1004094), and Nebraska Research Initiative grant to Y.L.L.

References

Ando T, Uchihashi T, Kodera N, Yamamoto D, Taniguchi M, Miyagi A, Yamashita H (2007) High-speed atomic force microscopy for observing dynamic biomolecular processes. J Mol Recognit 20:448–458

Bednar J, Dimitrov S (2011) Chromatin under mechanical stress: from single 30 nm fibers to single nucleosomes. FEBS J 278:2231–2243

Bochkarev A, Bochkareva E (2004) From RPA to BRCA2: lessons from single-stranded DNA binding by the OB-fold. Curr Opin Struct Biol 14:36–42

Bussiek M, Mucke N, Langowski J (2003) Polylysine-coated mica can be used to observe systematic changes in the supercoiled DNA conformation by scanning force microscopy in solution. Nucleic Acids Res 31:e137

Bussiek M, Toth K, Brun N, Langowski J (2005) DNA-loop formation on nucleosomes shown by in situ scanning force microscopy of supercoiled DNA. J Mol Biol 345:695–706

Bussiek M, Muller G, Waldeck W, Diekmann S, Langowski J (2007) Organisation of nucleosomal arrays reconstituted with repetitive African green monkey alpha-satellite DNA as analysed by atomic force microscopy. Eur Biophys J 37:81–93

Bustamante C, Vesenka J, Tang CL, Rees W, Guthold M, Keller R (1992) Circular DNA molecules imaged in air by scanning force microscopy. Biochemistry 31:22–26

Bustamante C, Rivetti C (1996) Visualizing protein-nucleic acid interactions on a large scale with the scanning force microscope. Annu Rev Biophys Biomol Struct 25:395–429

Bustamante C, Zuccheri G, Leuba SH, Yang G, Samori B (1997) Visualization and analysis of chromatin by scanning force microscopy. Methods 12:73–83

Chrysogelos S, Griffith J (1982) *Escherichia coli* single-strand binding protein organizes single-stranded DNA in nucleosome-like units. Proc Natl Acad Sci USA 79:5803–5807

Crampton N, Yokokawa M et al (2007) Fast-scan atomic force microscopy reveals that the type III restriction enzyme EcoP15I is capable of DNA translocation and looping. Proc Natl Acad Sci USA 104:12755–12760

Filenko NA, Palets DB, Lyubchenko YL (2012) Structure and dynamics of dinucleosomes assessed by atomic force microscopy. J Amino Acids 2012:650840

Gilmore JL, Suzuki Y, Tamulaitis G, Siksnys V, Takeyasu K, Lyubchenko YL (2009) Single-molecule dynamics of the DNA-EcoRII protein complexes revealed with high-speed atomic force microscopy. Biochemistry 48:10492–10498

Gross L, Mohn F, Moll N, Meyer G, Ebel R, Abdel-Mageed WM, Jaspars M (2010) Organic structure determination using atomic resolution scanning probe microscopy. Nat Chem 2:821–825

Guthold M, Bezanilla M, Erie DA, Jenkins B, Hansma HG, Bustamante C (1994) Following the assembly of RNA polymerase-DNA complexes in aqueous solutions with the scanning force microscope. Proc Natl Acad Sci USA 91:12927–12931

Guthold M, Zhu X et al (1999) Direct observation of one-dimensional diffusion and transcription by *Escherichia coli* RNA polymerase. Biophys J 77:2284–2294

Hamon L, Pastre D, Dupaigne P, Le Breton C, Le Cam E, Pietrement O (2007) High-resolution AFM imaging of single-stranded DNA-binding (SSB) protein–DNA complexes. Nucleic Acids Res 35:e58

Jiang Y, Marszalek PE (2011) Atomic force microscopy captures MutS tetramers initiating DNA mismatch repair. EMBO J 30:2881–2893

Jiao Y, Cherny DI, Heim G, Jovin TM, Schaffer TE (2001) Dynamic interactions of p53 with DNA in solution by time-lapse atomic force microscopy. J Mol Biol 314:233–243

Kasas S, Thomson NH et al (1997) *Escherichia coli* RNA polymerase activity observed using atomic force microscopy. Biochemistry 36:461–468

Kobryn K, Watson MA, Allison RG, Chaconas G (2002) The Mu three-site synapse: a strained assembly platform in which delivery of the L1 transposase binding site triggers catalytic commitment. Mol Cell 10:659–669

Kur J, Olszewski M, Dlugolecka A, Filipkowski P (2005) Single-stranded DNA-binding proteins (SSBs) – sources and applications in molecular biology. Acta Biochim Pol 52:569–574

Lohman TM, Ferrari ME (1994) *Escherichia coli* single-stranded DNA-binding protein: multiple DNA-binding modes and cooperativities. Annu Rev Biochem 63:527–570

Lushnikov AY, Potaman VN, Lyubchenko YL (2006a) Site-specific labeling of supercoiled DNA. Nucleic Acids Res 34:e111
Lushnikov AY, Potaman VN, Oussatcheva EA, Sinden RR, Lyubchenko YL (2006b) DNA strand arrangement within the SfiI-DNA complex: atomic force microscopy analysis. Biochemistry 45:152–158
Lyubchenko YL, Gall AA, Shlyakhtenko LS, Harrington RE, Jacobs BL, Oden PI, Lindsay SM (1992a) Atomic force microscopy imaging of double stranded DNA and RNA. J Biomol Struct Dyn 10:589–606
Lyubchenko YL, Jacobs BL, Lindsay SM (1992b) Atomic force microscopy of reovirus dsRNA: a routine technique for length measurements. Nucleic Acids Res 20:3983–3986
Lyubchenko YL, Jacobs BL, Lindsay SM, Stasiak A (1995) Atomic force microscopy of nucleoprotein complexes. Scanning Microsc 9:705–724, discussion 724–707
Lyubchenko YL (2004) DNA structure and dynamics: an atomic force microscopy study. Cell Biochem Biophys 41:75–98
Lyubchenko YL, Shlyakhtenko LS (2009) AFM for analysis of structure and dynamics of DNA and protein-DNA complexes. Methods 47:206–213
Lyubchenko YL, Shlyakhtenko LS, Gall AA (2009) Atomic force microscopy imaging and probing of DNA, proteins, and protein DNA complexes: silatrane surface chemistry. Methods Mol Biol 543:337–351
Lyubchenko YL, Gall AA, Shlyakhtenko LS (2001) Atomic force microscopy of DNA and protein-DNA complexes using functionalized mica substrates. In: Moss T (ed) DNA-protein interactions; principles and protocols, vol 148, Methods in molecular biology. Humana, Totowa, NJ, pp 569–578
Marsden MP, Laemmli UK (1979) Metaphase chromosome structure: evidence for a radial loop model. Cell 17:849–858
Menshikova I, Menshikov E, Filenko N, Lyubchenko YL (2011) Nucleosomes structure and dynamics: effect of CHAPS. Int J Biochem Mol Biol 2:129–137
Merickel SK, Johnson RC (2004) Topological analysis of Hin-catalysed DNA recombination in vivo and in vitro. Mol Microbiol 51:1143–1154
Miyagi A, Ando T, Lyubchenko YL (2011) Dynamics of nucleosomes assessed with time-lapse high-speed atomic force microscopy. Biochemistry 50:7901–7908
Murphy PJ, Shannon M, Goertz J (2011) Visualization of recombinant DNA and protein complexes using atomic force microscopy. J Vis Exp: JoVE 53: pil: 3061. doi:10.3791/3061
Reuter M, Kupper D, Meisel A, Schroeder C, Kruger DH (1998) Cooperative binding properties of restriction endonuclease EcoRII with DNA recognition sites. J Biol Chem 273:8294–8300
Rybenkov VV, Vologodskii AV, Cozzarelli NR (1997) The effect of ionic conditions on the conformations of supercoiled DNA I. Sedimentation analysis. J Mol Biol 267:299–311
Ryzhikov M, Koroleva O, Postnov D, Tran A, Korolev S (2011) Mechanism of RecO recruitment to DNA by single-stranded DNA binding protein. Nucleic Acids Res 39:6305–6314
Shereda RD, Kozlov AG, Lohman TM, Cox MM, Keck JL (2008) SSB as an organizer/mobilizer of genome maintenance complexes. Crit Rev Biochem Mol Biol 43:289–318
Shlyakhtenko LS, Hsieh P, Grigoriev M, Potaman VN, Sinden RR, Lyubchenko YL (2000) A cruciform structural transition provides a molecular switch for chromosome structure and dynamics. J Mol Biol 296:1169–1173
Shlyakhtenko LS, Gall AA, Filonov A, Cerovac Z, Lushnikov A, Lyubchenko YL (2003) Silatrane-based surface chemistry for immobilization of DNA, protein-DNA complexes and other biological materials. Ultramicroscopy 97:279–287
Shlyakhtenko LS, Gilmore J, Portillo A, Tamulaitis G, Siksnys V, Lyubchenko YL (2007) Direct visualization of the EcoRII-DNA triple synaptic complex by atomic force microscopy. Biochemistry 46:11128–11136
Shlyakhtenko LS, Lushnikov AY, Lyubchenko YL (2009) Dynamics of nucleosomes revealed by time-lapse atomic force microscopy. Biochemistry 48:7842–7848

Shlyakhtenko LS, Lushnikov AY, Miyagi A, Lyubchenko YL (2012) Specificity of binding of the single stranded DNA binding protein to the target. Biochemistry 51(7):1500–1509

Suzuki Y, Higuchi Y, Hizume K, Yokokawa M, Yoshimura SH, Yoshikawa K, Takeyasu K (2010) Molecular dynamics of DNA and nucleosomes in solution studied by fast-scanning atomic force microscopy. Ultramicroscopy 110(6):682–688

Tamulaitis G, Sasnauskas G, Mucke M, Siksnys V (2006) Simultaneous binding of three recognition sites is necessary for a concerted plasmid DNA cleavage by EcoRII restriction endonuclease. J Mol Biol 358:406–419

Tessmer I, Yang Y, Zhai J, Du C, Hsieh P, Hingorani MM, Erie DA (2008) Mechanism of MutS searching for DNA mismatches and signaling repair. J Biol Chem 283:36646–36654

Thundat T, Allison DP, Warmack RJ, Brown GM, Jacobson KB, Schrick JJ, Ferrell TL (1992) Atomic force microscopy of DNA on mica and chemically modified mica. Scanning Microsc 6:911–918

van Noort SJ, van der Werf KO, Eker AP, Wyman C, de Grooth BG, van Hulst NF, Greve J (1998) Direct visualization of dynamic protein-DNA interactions with a dedicated atomic force microscope. Biophys J 74:2840–2849

Vanamee ES, Viadiu H, Kucera R, Dorner L, Picone S, Schildkraut I, Aggarwal AK (2005) A view of consecutive binding events from structures of tetrameric endonuclease SfiI bound to DNA. EMBO J 24:4198–4208

Vesenka J, Guthold M, Tang CL, Keller D, Delaine E, Bustamante C (1992) Substrate preparation for reliable imaging of DNA molecules with the scanning force microscope. Ultramicroscopy 42–44:1243–1249

Wang H, Yang Y et al (2003) DNA bending and unbending by MutS govern mismatch recognition and specificity. Proc Natl Acad Sci USA 100:14822–14827

Watson JD (2008) Molecular biology of the gene. Pearson/Benjamin Cummings, San Francisco, CA

Watson MA, Chaconas G (1996) Three-site synapsis during Mu DNA transposition: a critical intermediate preceding engagement of the active site. Cell 85:435–445

Widom J, Klug A (1985) Structure of the 300Å chromatin filament: X-ray diffraction from oriented samples. Cell 43:207–213

Winter RB, Berg OG, von Hippel PH (1981) Diffusion-driven mechanisms of protein translocation on nucleic acids. 3. The *Escherichia coli* lac repressor–operator interaction: kinetic measurements and conclusions. Biochemistry 20:6961–6977

Yang Y, Sass LE, Du C, Hsieh P, Erie DA (2005) Determination of protein-DNA binding constants and specificities from statistical analyses of single molecules: MutS-DNA interactions. Nucleic Acids Res 33:4322–4334

Zhong Q, Inniss D, Kjoller K, Elings VB (1993) Fractured polymer/silica fiber surface studied by tapping mode atomic force microscopy. Surf Sci Lett 290:L688–L692

Chapter 5
Nanovisualization of Proteins in Action Using High-Speed AFM

Takayuki Uchihashi, Noriyuki Kodera, and Toshio Ando

5.1 Introduction

The dynamics of proteins, such as their conformational change and dynamic interplay with interaction partners, is crucial to their biological functions. To reveal the dynamic behavior of proteins, single-molecule approaches are indispensable because the molecules behave in an unsynchronized manner, which makes it difficult to monitor their dynamics by ensemble average experiments. Fluorescence microscopy has been widely used to study the functional behavior of single-protein molecules (Peterman et al. 2004; Joo et al. 2008; Roy et al. 2008; Yanagida and Ishii 2008). However, it only visualizes featureless fluorescent spots, meaning that protein molecules themselves are invisible. Moreover, what we can analyze using fluorescence techniques is limited to the behavior of a selected portion where a fluorophore is attached. Therefore, inferences have to be made to bridge the gap between the observed behavior of fluorescent spots and the actual behavior of labeled protein molecules. Consequently, it takes a considerably long time to reach a persuasive conclusion on how a protein dynamically behaves to function.

Atomic force microscopy (AFM) (Binnig et al. 1986) has made it possible for the first time to visualize the surface topography of samples at single-nanometer resolution even in liquid environments (Drake et al. 1989; Weisenhorn et al. 1989). In the early stage, pioneering studies explored this new capability by visualizing various biological samples under physiological conditions (Lindsay et al. 1989; Weisenhorn et al. 1989; Hansma et al. 1993; Schabert and Engel 1994; Müller et al. 1995; Shao

T. Uchihashi • T. Ando (✉)
Department of Physics and Bio-AFM Frontier Research Center, Kanazawa University, Kanawaza, Japan
e-mail: uchihast@staff.kanazawa-u.ac.jp; tando@staff.kanazawa-u.ac.jp

N. Kodera
Bio-AFM Frontier Research Center, Kanazawa University, Kanawaza, Japan
e-mail: nkodera@staff.kanazawa-u.ac.jp

A.F. Oberhauser (ed.), *Single-molecule Studies of Proteins*, Biophysics for the Life Sciences 2, DOI 10.1007/978-1-4614-4921-8_5,

and Yang 1995; Walz et al. 1996; Müller and Engel 1999). Moreover, several groups attempted to visualize dynamic biological processes (Bezanilla et al. 1994; Guthold et al. 1994; Kasas et al. 1997; Lin et al. 1999; Oberleithner et al. 2000). For example, the clotting process of fibrin (Drake et al. 1989), viral infection of isolated cells (Häberle et al. 1992), and antibody binding to an S-layer protein (Ohnesorge et al. 1992) were imaged at ~1 min or longer intervals. Therefore, in the early stage after the advent of AFM, researchers were already well aware that innovative biological studies would be opened by a new approach, i.e., visualization of dynamic events occurring with biological samples using AFM. In fact, one of the inventers of AFM, Gerd Binnig, mentioned as follows: "In biology, use of the force microscope will probably become quite common because of its ability to deliver films of processes" (Binnig 1992). However, we had to wait for this prediction to come true until 2008 when the instrumentation of high-speed AFM was completed.

Establishing both high-speed and low-invasive performances of AFM required lengthy efforts for solving various technical problems. In the early stage, small cantilevers (Schäffer et al., 1996; Ando et al. 2001), a fast scanner (Ando et al. 2001), a fast amplitude detector (Ando et al. 2001), and an optical beam deflection (OBD) sensor for detecting deflection of a small cantilever (Schäffer et al. 1996; Ando et al. 2001) were developed. Using HS-AFM equipped with some of these devices, fast imaging was attempted to visualize dynamic events of proteins (Vianni et al., 2000; Ando et al. 2001). However, it was evident that the bandwidth of feedback control to maintain the tip–sample interaction force constant was not high enough. Then, we further developed smaller cantilevers (Kitazawa et al. 2003; Ando et al. 2008), a faster *z*-scanner (Fukuma et al 2008), an active damping technique for the z-scanner (Kodera et al. 2005), a new proportional-integral-derivative (PID) feedback controller (Kodera et al. 2006), a compensator for drift in cantilever excitation efficiency (Kodera et al. 2006), and a fast amplitude detector with low noise performance (Ando et al. 2008).

Going through these developmental and proof-of-principle stages, HS-AFM has now moved to a new phase where we can use this microscopy to gain new insights into the functional mechanism of proteins. Various biomolecules and dynamic events have already been imaged by HS-AFM (Table 5.1). Although only imaged from the top direction, the whole topography of proteins and their minute temporal changes are visualized in some of these studies. Unlike fluorescence microscopy, various dynamic events of a molecule appear in the AFM movie without preselection, meaning that both expected and unexpected dynamic molecular behaviors can be visualized simultaneously. This gives a great chance of learning details of several facts of molecular dynamics in one imaging experiment. Moreover, the molecular dynamics that appears on video can often be interpreted quite easily without sophisticated analyses. These unique features of HS-AFM imaging significantly facilitate our understanding of the functional mechanism of proteins. In this chapter, we first briefly overview key technical developments in HS-AFM and then highlight our recent representative studies on dynamical properties of proteins. For other AFM imaging studies, see a recent review (Ando 2012).

Table 5.1 Summary of published high-speed AFM imaging studies

Subjects	Targets	References
Structure dynamics of proteins	Myosin V	Ando et al. (2001, 2002); Kodera et al. (2010)
	Dynein C	Ando et al. (2005)
	GroEL	Yokokawa et al. (2006a)
	Intrinsically disordered protein	Miyagi et al. (2008)
	$P2X_4$ receptor	Shinozaki et al. (2009)
	Bacteriorhodopsin	Shibata et al. (2010, 2011)
	Acid-sensing ion channel	Yokokawa et al. (2010)
	F_1-ATPase	Uchihashi et al. (2011)
Dynamic processes of proteins	Actin gliding	Ando et al. (2005)
	GroEL-ES	Ando et al. (2005); Yamamoto et al. (2009)
	Cellulase on cellulose	Igarashi et al. (2009, 2011)
	ClpX-FtzZ	Sugimoto et al. (2010)
	Amyloid-like fibrils	Milhiet et al. (2010)
	C-ring of F_0	Casuso et al. (2010)
Protein–DNA interactions	*ApaI* endnuclease-DNA	Yokokawa et al. (2006b)
	Type III restriction enzyme-DNA	Crampton et al. (2007)
	*Eco*RII restriction enzyme-DNA	Gilmore et al. (2009)
	Histone-nucleosome	Suzuki et al. (2010)
	Synthetic molecular transporter-DNA	Wickham et al. (2011)
Dynamics of 2D - protein crystals	Streptavidin	Yamamoto et al. (2008)
	Bacteriorhodopsin	Yamashita et al. (2009); Casuso et al (2009)
Dynamics of DNA	3D hollow structure of DNA origami	Endo and Sugiyama (2009); Endo et al. (2009)
	DNA origami jigsaw pieces	Endo et al. (2010a); Rajendran et al. (2011)
	DNA methylation	Endo et al. (2010b); Sannohe et al. (2010)
	G-quadruplex in DNA nanoscaffold	
	DNA cuboid	Endo et al. (2011)
Others	Process of cell death by antimicrobial peptide	Fantner et al. (2010)
	Lipid bilayer formation	Giocondi et al. (2010)
	Morphological transformation of detergent micelles	Inoue et al. (2011)

5.2 Overview of High-Speed-AFM Instrumentation

5.2.1 Overall Design of HS-AFM System

Among various operation modes of AFM, our HS-AFM system (Fig. 5.1) uses the tapping mode in which the cantilever tip is vertically oscillated at the first resonant frequency of the cantilever (Hansma et al. 1993; Zhong et al. 1993; Putman et al.

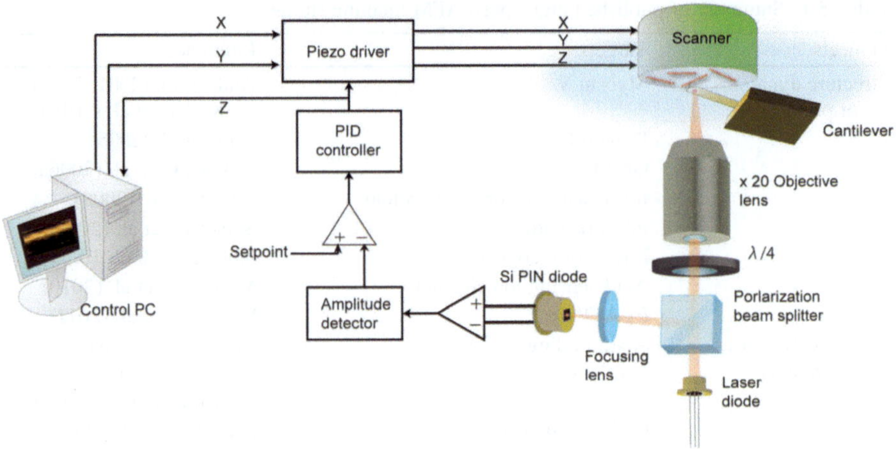

Fig. 5.1 Schematic of HS-AFM system

1994). The intermittent tip–sample contact eliminates the lateral force acting between tip and sample and thus prevents weakly adsorbed proteins from being dislodged from the substrate surface. Upon tip–sample contact, the cantilever oscillation amplitude decreases. During lateral scanning of the sample stage, the change in the amplitude is measured and then the sample stage is moved in the z-direction to recover the amplitude back to a given value (i.e., set point) through feedback control. For fast and nondestructive imaging of biological samples, the feedback control should be performed fast and precisely. Therefore, all devices contained in the feedback loop, i. e., the cantilever, the OBD sensor, the amplitude detector, the PID feedback circuit, the piezoactuator-based scanner, and the piezodriver, are optimized for their fast response.

The HS-AFM system includes an original inverted optical microscope. An objective lens with a long working distance used for the optical microscope is a part of the OBD sensor and is also used for viewing the cantilever and sample stage under the optical microscope, via a digital camera or a CCD camera. A glass slide, to which a cantilever holder and a liquid cell are attached, is placed on the optical microscope stage. A cantilever chip is held in the holder so that its tip points upwards (opposite to the way in conventional AFM). A sample stage attached to the z-scanner and facing downwards is placed over the cantilever. The instrument is now commercially available from the Research Institute of Biomolecule Metrology, Ltd. (Tsukuba, Japan).

5.2.2 *Small Cantilevers and OBD Sensor*

The cantilevers to be used in HS-AFM should have a high-resonant frequency and a small spring constant to meet both fast and low-invasive imaging capabilities, and thus, they should be fabricated to a small size. We use small rectangular cantilevers, so-called beak-like cantilevers, made of silicon nitride (Kitazawa et al. 2003). The dimensions are ~6 μm long, ~2 μm wide, and ~90 nm thick (BL-AC7DS-KU5, custom-made by Olympus). They have resonant frequencies ~3.5 MHz in air and ~1.2 MHz in water, a spring constant ~0.2 N/m, and a quality factor ~2 in water. Slightly larger cantilevers (9–10 μm long, 2 μm wide, and 130 nm thick; BL-AC10DS, Olympus, Tokyo) with resonant frequencies 1.5 MHz in air and 0.6 MHz in water, a spring constant 0.1 N/m, and a quality factor $Q \sim 2$ in water are available from Olympus, Atomic Force F&E GmbH (Mannheim, Germany), Asylum Research (Santa Barbara, USA), and Bruker (Calle Tecate, USA). However, the apex of the beak-like tip is not small enough (25–100 nm) to acquire high-resolution images. We usually use electron-beam-deposited (EBD) tip grown on the original beak-like tip using a field emission scanning electron microscope (FE-SEM) (Wendel et al. 1995). As a sublimate for the EBD tip, phenol powder is used. A spot-mode electron beam is focused onto each original tip to produce a ~1 μm long needle composed of amorphous carbon at a growth rate of ~17 nm/s. The EBD tip is sharpened by oxygen or nitrogen plasma etching to an apex radius of ~4 nm (~0.5 nm at the best case).

In our OBD sensor (Ando et al. 2001), an incident laser beam is focused onto a small cantilever through a ×20 objective lens with a long working distance of 24 mm (Fig. 5.1). The light reflected back from the rear side of the cantilever is collected and collimated using the same objective lens and guided to a quadrant-cell Si PIN photodiode. The focused spot is 3–4 μm in diameter. The incident and reflected beams are separated using a quarter-wavelength plate and a polarization beam splitter. The small cantilevers provide the OBD sensor with high deflection detection sensitivity (10–20 times higher than conventional large cantilevers) because the angle change at the cantilever's free end is 10–20 times larger for a given vertical displacement of the tip.

5.2.3 Amplitude Detector and Feedback Controller

The oscillation amplitude of the cantilever should be detected as fast as possible with a high signal-to-noise ratio. A conventional RMS-to-DC converter and a lock-in amplifier require at least several oscillation cycles to output an accurate amplitude signal. We developed two types of fast amplitude detectors, a peak-hold type (Ando et al. 2001) and a Fourier type (Kokavecz et al. 2006; Ando et al. 2008). In the peak-hold analogue circuit, the peak and bottom voltages of the output signal from the OBD senor are held and their difference is output as an amplitude signal. The amplitude signal is renewed every half cycle of cantilever oscillation. In the Fourier type (analogue/digital hybrid system), Fourier cosine and sine coefficients (a_1, b_1) are calculated every one cycle of cantilever oscillation and a signal proportional to $(a_1{}^2 + b_1{}^2)^{1/2}$ is output. The Fourier method is less susceptible to cantilever's thermal fluctuations in the deflection than the peak-hold method.

For the feedback control under a small tapping force condition, the amplitude set point should be set close to the free oscillation amplitude of the cantilever. Under this condition, tip "parachuting" often occurs, and thus the error signal is saturated at a small value for a while. "Parachuting" means that the cantilever tip completely detaches from the sample surface at a steep down-hill region of the sample and time elapses until the tip lands on the surface again. To reduce or eliminate the parachuting time, we developed a dynamic PID controller in which the gain parameters are dynamically changed depending on the cantilever oscillation amplitude (Kodera et al. 2006). This dynamic PID controller can maintain the feedback bandwidth at a high level even when the amplitude set point is close to the cantilever free oscillation amplitude, making high-speed imaging compatible with low-invasive imaging. In addition, an electronics developed to compensate for the drift in the cantilever excitation efficiency allows stable imaging under the small tapping force condition (Kodera et al. 2006).

5.2.4 Scanner and Active Damping

We developed three types of scanners with different maximum scan ranges. Type-1: $x = 1$ μm, $y = 4$ μm, $z = 1$ μm; Type-2: $x = 5$ μm, $y = 5$ μm, $z = 2$ μm; and Type-3: $x = 40$ μm, $y = 40$ μm, $z = 6$ μm. Type-1 is used for dynamic imaging of isolated proteins, while the latter two types are used for dynamic imaging of larger objects such as bacteria and eukaryotic cells. Each type of scanner is composed of three piezoactuators that displace the sample along the x-, y-, and z-axes. The slowest y-scanner displaces a block of the x- and z-scanners, the x-scanner displaces a block of the z-scanner, and the z-scanner displaces a glass-rod sample stage. Each block is connected with the surrounding base through a pair of flexures to achieve small crosstalk between the three displacement axes (Ando et al. 2008). Since the x-piezo is sandwiched between two flexures, the center of mass of the x-piezo is unchanged, minimizing the impulsive force exerting to the surrounding base and thus minimizing the induction of vibrations. To counteract the impulsive force produced by the z-scanner displacement, two identical piezoactuators are attached to a supporting base and are displaced simultaneously with the same length in the opposite direction (Ando et al. 2001). For the fastest z-scanner with a first resonant frequency 370 kHz, a piezoactuator is held at the four rims parallel to the displacement direction in order to counteract the impulsive force (Fukuma et al. 2008).

The first resonant frequencies of the z-scanners are those of the actuators used, meaning that resonance arising from the designed structure occurs at higher frequencies. We developed an active damping technique to eliminate the resonant vibrations of the z-scanners using a new Q-controller with corresponding mock z-scanners composed of LRC circuits (Kodera et al. 2005).

5.2.5 *Highest Possible Imaging Rate*

As a consequence of the whole development described above, we have achieved the highest feedback bandwidth $f_B \sim 110$ kHz. This is at least 10 times higher than other nominal high-speed AFM instruments. The highest possible imaging rate R_{max} is a function of f_B, the scan range in the x-direction W, the number of scan lines N, the smallest sample surface corrugation to be observed λ, and the sample fragility (Ando et al. 2008; Ando 2012). The overall relationship can be expressed as

$$R_{max} = \frac{2\lambda f_B \theta_m}{\pi W N}, \tag{5.1}$$

where θ_m represents an sample fragility-dependent allowable maximum phase delay in tracing the sample surface. Even for relatively fragile molecules ($\theta_m < \sim 20°$), we can capture images at 12 frames/s (fps) under the condition of $W = 200$ nm, $N = 100$, and $\lambda = 10$ nm without disturbing the physiological function of the molecules.

5.3 Imaging of Protein Molecules in Action

Some HS-AFM movies showing the dynamic behavior of proteins described below can be seen at http://www.s.kanazawa-u.ac.jp/phys/biophys/index.htm or at the corresponding publishers' homepages.

5.3.1 *Myosin V*

5.3.1.1 General Properties of Myosin V

Myosin V (M5) is a two-headed molecular motor that transports intracellular vesicles along actin filaments from the central regions of cells to the periphery (see a review; Sellers and Weisman 2008). Each N-terminal motor domain containing actin-binding and ATP hydrolysis sites is connected to an extended neck with six calmodulins (or five calmodulins and a light chain). The region containing the motor and neck domains is called "head." After the neck region, M5 is dimerized by a coiled-coil tail, which is followed by a C-terminal globular tail domain capable of binding to various cargos via adapter proteins (Fig. 5.2a). In order to elucidate the functional mechanism of M5, numerous studies have been done using biophysical and biochemical techniques. Using fluorescence microscopy and optical tweezers, it has been demonstrated that single molecules move processively along actin filaments towards the plus ends of the filaments (Mehta et al. 1999;

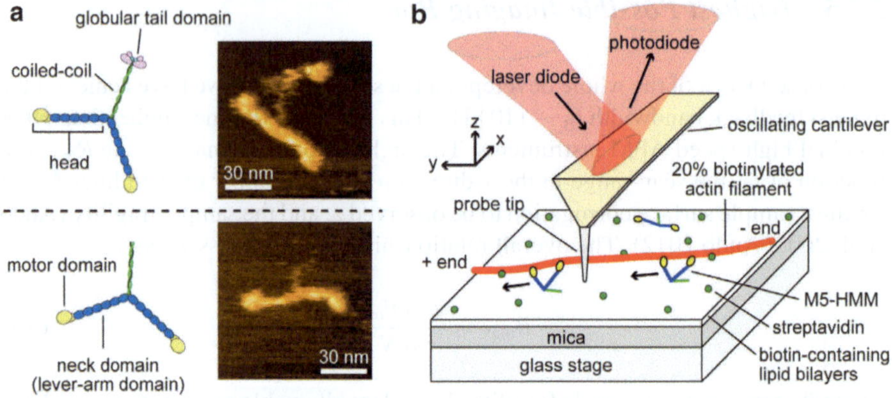

Fig. 5.2 (**a**) Schematic models and AFM images of M5 (*upper*) and M5-HMM (*lower*). (**b**) Schematic showing assay system for HS-AFM imaging of walking M5-HMM molecules

Sakamoto et al. 2000) in a hand-over-hand fashion (Forkey et al. 2003; Yildiz et al. 2003) with a 36-nm step for every ATP hydrolysis cycle (Sakamoto et al. 2008). Here, "hand-over-hand" means that the two heads alternate the leading and trailing positions at each step, very analogous to "walking." However, the protein molecules themselves are invisible in these studies and have therefore been visualized by electron microscopy only in the stationary states (Walker et al. 2000; Burgess et al. 2002; Oke et al. 2010). Thus, a comprehensive description of the molecule in action has not been accomplished until our recent HS-AFM study (Kodera et al. 2010). We briefly describe below some of our results.

5.3.1.2 Unidirectional Hand-Over-Hand Movement

As a substrate, we used a mica-supported lipid bilayer composed of an electrically natural phospholipid, a biotinylated phospholipid, and a positively charged phospholipid. The bilayer is in a gel phase at room temperature (Yamamoto et al. 2010). Partially biotinylated actin filaments were immobilized on the surface through streptavidin molecules with a low surface density (Fig. 5.2b). The positively charged lipid facilitated observing the characteristic sideways configuration of moving M5, allowing us to acquire high-resolution images. The successive AFM images captured at 7 fps in the presence of ATP (Fig. 5.3a) show M5-HMM (tail-truncated M5, bottom in Fig. 5.2a) moving processively along an actin filament with discrete ~36-nm steps. The two-headed bound M5-HMM showed unique structural features without exception, as illustrated in the bottom of Fig. 5.3a; the neck–motor domain junction appears smooth in the leading head whereas it is V-shaped in the trailing head, because the neck regions emerge from different parts of the motor domain. The short coiled-coil tail was mostly tilted towards the

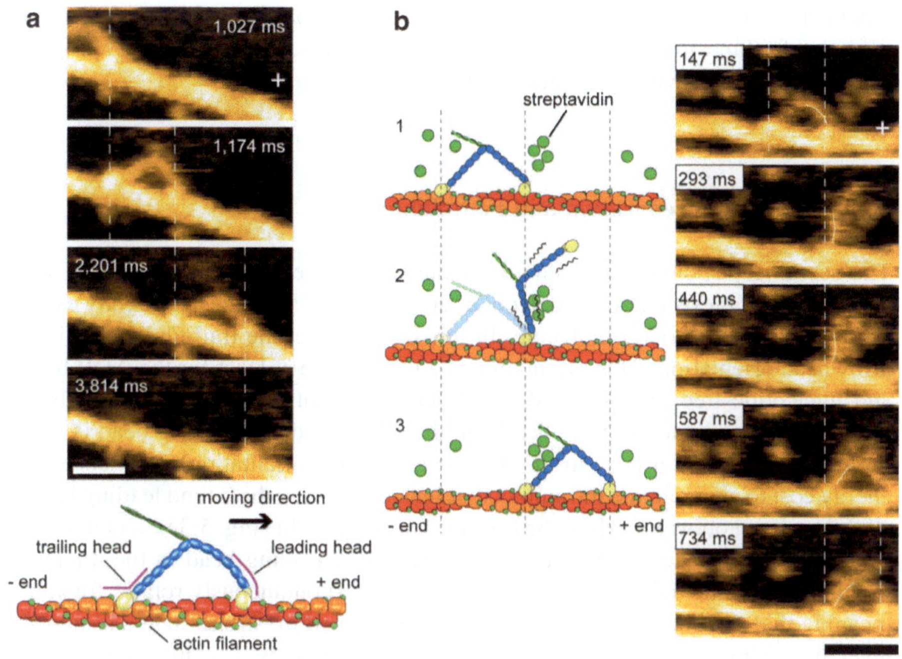

Fig. 5.3 (**a**) Successive AFM images showing unidirectional processive movement of M5-HMM observed in 1 μM ATP (*upper panels*; scale bar, 30 nm) and schematic showing two-headed bound M5-HMM (*bottom*). (**b**) Successive AFM images (*right panel*) showing hand-over-hand movement of M5-HMM in 1 μM ATP (scale bar, 50 nm) and schematic explaining the AFM images (*left panel*). The swinging lever is highlighted with the *thin white lines* (*right panel*). The *vertical dashed lines* in (**a**) and (**b**) show the *centers* of mass of the motor domains, and the *plus signs* indicate the plus ends of actin. All AFM images were taken at 7 fps

direction opposite to the moving direction. These features are totally consistent with those shown in electron micrographs (Walker et al. 2000) and can be used to determine the actin polarity when bound M5-HMM is stationary. However, the AFM movie could not resolve detailed stepping behaviors because the step completed within a frame.

To slow down the step motion, we additionally attached streptavidin molecules on the substrate surface as moderate obstacles to the movement, which successfully allowed us to directly visualize the detailed molecular processes during a step (Fig. 5.3b). After trailing head detachment from actin, the nearly straight leading head appeared to spontaneously swing from the reverse arrowhead orientation toward the arrowhead orientation, showing, for the first time, the swinging lever-arm motion initially proposed for muscle myosin (Huxley 1969). The detached trailing head rotationally diffused around the advancing neck–neck junction and then bound to a frontward actin site as a new leading head, completing one step. Thus, the detailed molecular behavior during hand-over-hand movement was directly visualized at unprecedented high spatiotemporal resolution. In the AFM movie, we did not observe other behaviors of M5-HMM expected from the "inchworm" (Hua et al. 2002) model proposed for kinesin and the "biased diffusion" (Okada et al. 2007) model proposed for myosin, at least with the temporal resolution used.

5.3.1.3 Tension Generation and Foot Stomp

The seemingly spontaneous rotation of the leading head following trailing head detachment suggests that intramolecular tension for the advance has already existed in the two-headed bound molecule. In this bound state, the trailing head is in the arrowhead orientation. This orientation is a natural one at least for the ADP-bound and nucleotide-free myosin heads, and hence, the intramolecular tension is not produced by the bound trailing head. On the other hand, the leading head is not in the natural orientation (i.e., in the reverse arrowhead orientation) and hence pays energy cost to generate the intramolecular tension. In fact, the bound leading head is slightly curved outwards while the trailing head is straight (Fig. 5.3a). Upon trailing head detachment, the constraint keeping the bound leading head in the unnatural orientation is removed and hence the leading head spontaneously rotates forwards, meaning that the bound leading head is in the strained prestroke state and thus the lever arm swing is not accompanied by chemical transitions. This inferred view was reinforced by the observation that the short coiled-coil of two-headed bound M5-HMM was sometimes unfolded, and then the leading head rotated from the reverse arrowhead orientation to the arrowhead orientation (Fig. 5.4a), very much similar to the swinging lever arm motion observed as powerstroke in the hand-over-hand movement.

In general, during the actomyosin ATPase cycle, the strained prestroke state has been thought to be always formed through Pi release from an ADP-Pi bound head weakly interacting with actin (Goldman 1987; Geeves and Holmes 1999). Thus, it may be concluded that the strained prestroke state, which is formed directly by the binding of ADP-bound leading head to actin, does not participate in the forward step in ATP. However, we observed a molecular behavior in ATP that indicates that it does participate. In two-headed bound M5-HM, both of the motor domains frequently exhibited brief dissociation and reassociation on the same actin filament (or a brief translocation by around ±5 nm along the actin filament), whereas M5-HMM remained at approximately the same position on the filament (Fig. 5.4b). We have termed this behavior "foot stomp." The briefly detached leading head does not carry bound Pi because Pi is already released from an ADP-Pi-bound M5 head immediately after its initial binding to actin (De La Cruz et al. 1999). Nevertheless, the detached leading head with only ADP bound rebinds to actin, still in the reverse arrowhead orientation, and then swings forward following trailing head detachment. The detachment of leading head during foot stomp is not caused by ATP binding to the leading head. As described below, a nucleotide-free leading head is often sharply bent. We never observed a sharply bent leading head before its foot stomp.

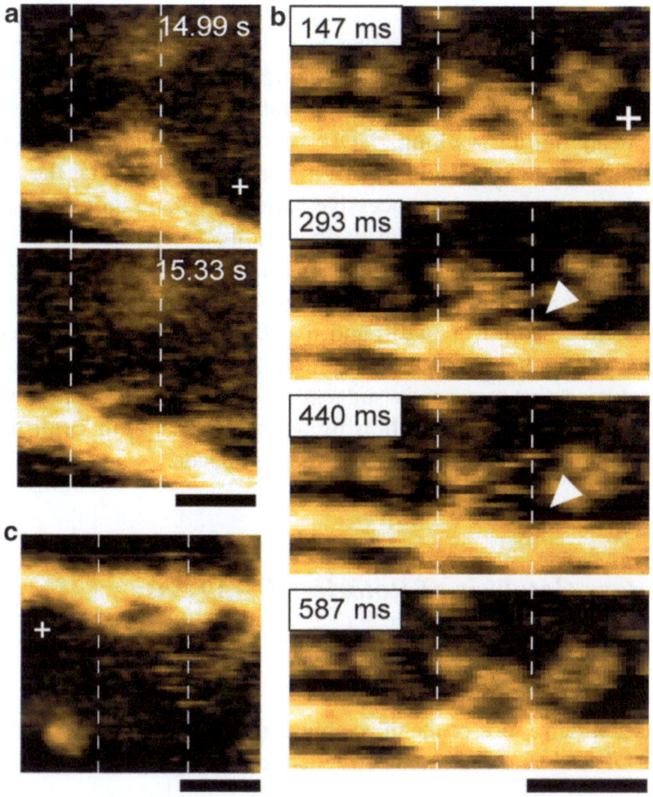

Fig. 5.4 Other dynamic events of M5 HMM observed by HS-AFM. (**a**) Unfolding of coiled-coil tail of two-headed bound M5-HMM in 50 μM ADP. *Top image*, before unfolding; *bottom image*, after unfolding. (**b**) Successive AFM images showing foot stomp at leading head observed in 1 μM ATP. The events of leading head detachment from actin are indicated with the *arrow heads*. Imaging rate, 7 fps. (**c**) AFM image showing sharply bent leading head of two-headed bound M5-HMM in the nucleotide free condition. The marks "*plus*" indicate the plus ends of actin filaments

5.3.1.4 Mechanism for Hand-Over-Hand Movement

The leading head of two-headed bound M5-HMM was always straight (slightly curved outwards) in ADP or ATP, while it was often sharply bent in the nucleotide-free condition (Fig. 5.4c). Therefore, just by looking at the conformation of leading head, we can judge whether or not the leading head contains nucleotides. To estimate the ADP dissociation rate constant at the leading head, M5-HMM bound to actin in various concentrations of ADP was observed, and then, the lifetime of straight leading head was analyzed as a function of ADP concentration. The ADP dissociation rate constant at the leading head was estimated to be 0.1/s. This means that ADP is released from the leading head every 10 s, on average. However, M5-HMM walks many steps for 10 s, meaning that ADP does not dissociate from the leading head. That is, before ADP dissociation, the leading head is switched to the trailing head during walking. ADP dissociation, and the subsequent ATP binding, and the resulting detachment from actin solely occur at the trailing head. This is the basis underlying the processive hand-over-hand movement. This mechanism was inferred previously from various indirect experiments (Rosenfeld and Sweeney 2004; Purcell et al. 2005; Veigel et al. 2005; Forgacs et al. 2008; Oguchi et al. 2008; Sakamoto et al. 2008) but is now clearly observed by our high-speed AFM experiments. Moreover, we found that the foot stomps occurred more frequently at the leading head than at the trailing head. This asymmetry of foot stomp frequency between the two heads suggests that the leading head would be less susceptible to the catalytic activation by actin, which might contribute to some extent to the kinetic asymmetry between the two heads.

5.3.2 Rotary Catalysis of Rotorless F_1-ATPase

5.3.2.1 General Properties and Cooperativity of F_1-ATPase

ATP synthase, which is one of the most abundant proteins in organisms, synthesizes ATP from ADP and Pi using the electrochemical energy of proton gradient across membranes (Boyer 1993, 1997). It consists of two coupled rotary motors, called F_0 and F_1. F_0 is embedded in membranes while water-soluble F_1 is exposed to the solution. They can be separated and studied individually. F_1 is composed of five different subunits with a composition of $\alpha_3\beta_3\gamma\delta\varepsilon$ (Todd et al. 1980; Walker et al. 1985). The three α and β subunits are arranged alternately to form a hexameric ring, and the central cavity of the hexamer is occupied by the γ subunit. The minimum rotary unit of F_1 is the $\alpha_3\beta_3\gamma$ subcomplex, in which the central γ shaft rotates within the $\alpha_3\beta_3$ stator ring when ATP is hydrolyzed by the ring (Matsui and Yoshida 1995). The catalytic site of F_1 is formed at the interface between the α and β subunits. The catalytic residues mainly reside in the β subunit (Abrahams et al. 1994). The rotational motion of γ has been demonstrated by single-molecule optical microscopy in which optical markers are attached to the γ subunit (Noji et al. 1997; Yasuda et al. 1998).

In the earliest obtained crystal structure (Fig. 5.5a), three catalytic sites are in different nucleotide-bound states; one binds to an ATP analog (β_{TP} and α_{TP}), another binds to ADP (β_{DP} and α_{DP}), and the third binds to none (β_E and α_E). Both β_{TP} and β_{DP} swing the C-terminal domain toward γ, assuming an inwardly retracted conformation, whereas β_E swings the domain away from γ, assuming an outwardly extended conformation. Because the two conformational states seem to push γ or be pushed by γ as shown in Fig. 5.5b and c, it was considered that the β–γ

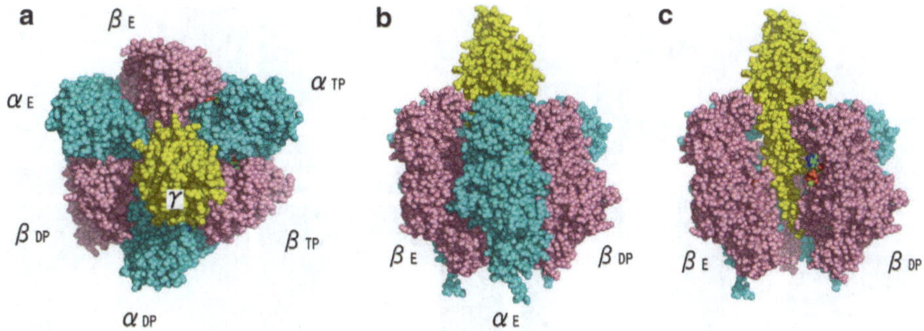

Fig. 5.5 Crystal structure of nucleotides-bound F_1-ATPase (PDB code: 1BMF). (**a**) *Top view*. (**b**) *Side view*. (**c**) *Side view* in which the α subunit on the nearest face is removed

interaction controls the conformational and catalytic states of individual β subunits and thus is essential for sequential torque generation (Wang and Oster 1998). Some biochemical studies are considered to suggest that the $\alpha_3\beta_3$ ring alone does not possess intrinsic cooperativity and γ mediates interplays among β's (Kaibara et al. 1996; Capaldi 1998; Nakamoto et al. 1999). On the other hand, although very few, there are biochemical studies that suggest that the intrinsic cooperativity is engendered by the $\alpha_3\beta_3$ ring alone (Yoshida and Allison 1990; Aloise et al. 1991).

Recently, it was reported that even when the interaction sites between β and γ are abolished, the γ still rotates unidirectionally (Hossain et al. 2006, 2008; Furuike et al. 2008). However, single-molecule optical microscopy does not allow us to directly examine whether the intrinsic cooperativity in the $\alpha_3\beta_3$ ring is responsible for sequential torque generation since attachment of a probe to the rotary shaft is required for visualization. We addressed this issue by directly imaging the ATP-driven conformational transition of β's in the isolated $\alpha_3\beta_3$ ring using high-speed AFM (Uchihashi et al. 2011).

5.3.2.2 AFM Observation of $\alpha_3\beta_3$ Subcomplex Without ATP

The $\alpha_3\beta_3$ subcomplex with Lys_7-tag at the N terminus of β was covalently immobilized on a mica surface treated first with 3-aminopropyl-triethoxysilane and then with glutaraldehyde. Therefore, the C-terminal domain of the $\alpha_3\beta_3$ subcomplex is at the top side. As shown in Fig. 5.6a, a ring structure with six high protrusions was observed under the nucleotide free conditions. Figure 5.6b shows an averaged high-resolution image of the $\alpha_3\beta_3$ subcomplex without nucleotide. This image shows a pseudo-sixfold symmetric ring in which three subunits alternately arranged in the ring are higher than the other three. The simulated AFM image of the C-terminal side of the ring constructed from the crystal structure of nucleotide-free $\alpha_3\beta_3$ subcomplex (Shirakihara et al. 1997) well reproduced the observed image

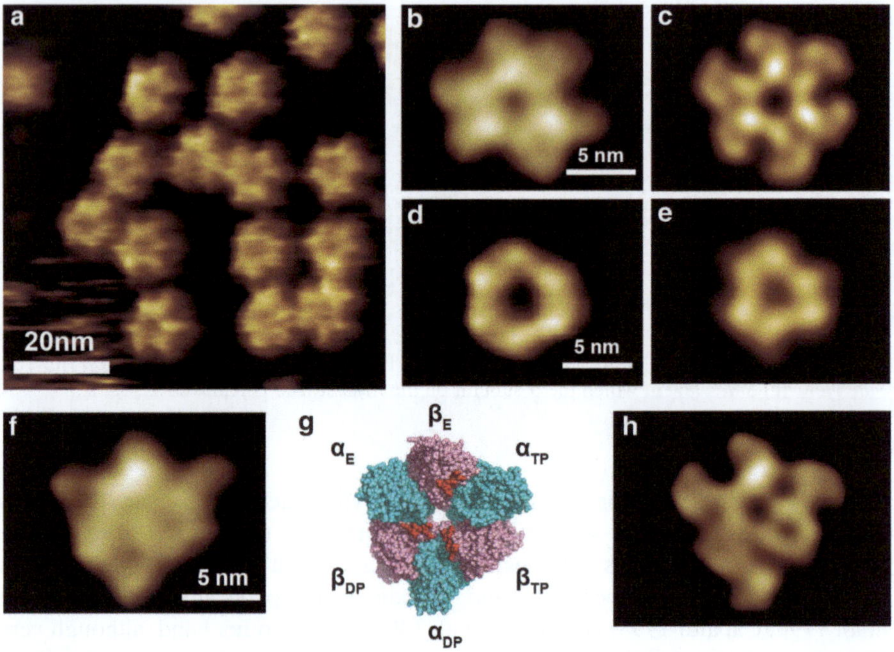

Fig. 5.6 (**a**) Wide-area AFM image of $\alpha_3\beta_3$ subcomplex without nucleotide. (**b**) and (**c**) are averaged experimental and simulated AFM images of C-terminal side of the $\alpha_3\beta_3$ subcomplex without nucleotide, respectively. (**d**) and (**e**) are averaged experimental and simulated AFM images of N-terminal side of the $\alpha_3\beta_3$ subcomplex without nucleotide. (**f**) Averaged AFM image of C-terminal side of the $\alpha_3\beta_3$ subcomplex in 1 mM AMPPNP. (**g**) Atomic structure of $\alpha_3\beta_3$ subcomplex with bound nucleotides. This structure is obtained by removing γ from a crystal structure of F_1. (**h**) Simulated AFM image constructed from the crystal structure shown in (**g**)

(Fig. 5.6c), indicating that the N-terminal side was in fact selectively attached to the mica, although a smaller crown corresponding to the N-terminal side was occasionally observed (Fig. 5.6d). The image was similar to a simulated image (Fig. 5.6e). Three β's, all assuming the outwardly extended conformation in the crystal structure of the $\alpha_3\beta_3$, gave three protrudent peaks in the simulated image as bright spots, and thus the three subunits showing higher protrusions seen in the AFM image were identified as β's.

When AMPPNP, which is a nonhydrolyzable ATP analog, was added, the ring became triangular and the central hole was obscure as shown in Fig. 5.6f. Three α's having the lower protrusions retained the same conformation as those under the nucleotide-free condition and two of the three β's retracted towards the center and at the same time lowered their protrusions. This structure did not change during the observation. A simulated image of $\alpha_3\beta_3$ with bound nucleotides (Fig. 5.6h), which was constructed using the crystal structure (Abrahams et al. 1994) of F_1 without γ (Fig. 5.6g), showed an asymmetric ring very similar to the observed image.

This excellent agreement indicates that only two β's can assume the inwardly retracted conformation even in saturating AMPPNP, which is consistent with the previous observation that three β's do not assume the inwardly retracted conformation simultaneously (Menz et al. 2001).

5.3.2.3 HS-AFM Observation of $\alpha_3\beta_3$ Subcomplex in ATP

In the presence of ATP, the ring shape was very similar to that observed in the presence of AMPPNP but the ring showed distinct conformational dynamics; each β underwent conformational transition between the outwardly extended high state (O-state) and the retracted low state (C-state) (Fig. 5.7a); the outwardly extended and inwardly retracted conformations are well correlated with the high and low protrusion states, respectively (Fig. 5.7b). Prominent features observed are that when the O-to-C transition occurs at one β, the opposite C-to-O transition occurs simultaneously at its counterclockwise neighbor β in most cases as shown in Fig. 5.7a. By this cooperative effect, the CCO conformational state propagates in the counterclockwise direction as seen in the time evolution of the cumulated angle of the open-β position relative to the center of the ring (Fig. 5.7c).

The rate of rotational propagation of the CCO conformational state increased with increasing ATP concentration (2–4 μM), indicating that ATP binding is rate limiting. The rate constant of the counterclockwise O-to-C shift in β at each ATP concentration was comparable with the corresponding initial rate of ATP hydrolysis measured biochemically. Thus, ATP hydrolysis is well coupled with the O-to-C transition of β.

The interplay between β subunits in $\alpha_3\beta_3$ responsible for the rotational propagation of the CCO conformational state was further supported by the following observation. A single subunit was occasionally dissociated from $\alpha_3\beta_3$ during observation (Fig. 5.7d), due presumably to the imperfect covalent binding of the subcomplex to the mica surface and to transient strong tip–sample interaction. Once one β is lost, the rotary propagation completely ceased (Fig. 5.7e), although some cooperative conformational transitions occurred between the remaining β's.

These results provide clear evidence that the structural basis of unidirectional rotary catalysis is intrinsically programmed in the stator ring. Thus, the "γ-dictator" model (Furuike et al. 2008), which proposes that only the interaction with γ determines the conformational and catalytic states of β's (Adachi et al. 2007; Watanabe et al. 2010), does not seem to be valid. On the other hand, the ATP-binding rate and the efficiency of unidirectionality of the CCO state propagation are distinctly lower than those of F_1 (Uchihashi et al. 2011). Thus, the interaction with γ is dispensable but still important for the rapid and precise rotary catalysis. Our findings are not inconsistent with the observations that the rate of the catalytic reaction are apparently under the control of the rotary angle of γ (Itoh et al. 2004; Rondelez et al. 2005; Iko et al. 2009). The intrinsic interplay among β's would reinforce the catalytic control by γ; even if γ tightly interacts with only one β, it still can act on all β's through β–β interplay.

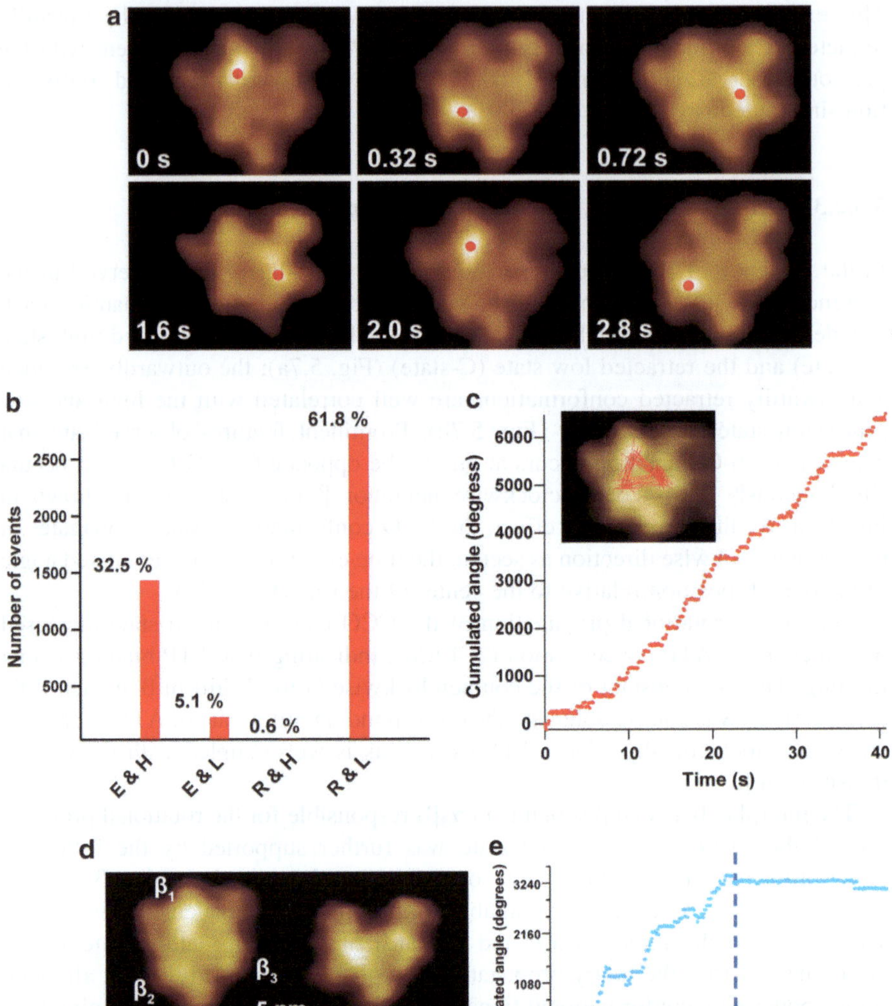

Fig. 5.7 (**a**) Successive AFM images showing conformational change in $\alpha_3\beta_3$ subcomplex in 2 μM ATP. The highest pixel in each image is indicated by the *red circle*. Frame rate, 12.5 fps. (**b**) Tight correlation between two types of conformational changes in β observed in ATP; conformations with different-height protrusions ("H," high; "L," low) and outwardly extended (E) and inwardly retracted (R) conformations of the distal region. The higher protrusion is mostly accompanied by the E-conformation, whereas the lower protrusion is mostly accompanied by the R-conformation. (**c**) Time evolution of cumulated angle of highest pixel. The inset shows a trajectory, superimposed on an AFM image, of the highest pixels corresponding to the high protrusions of β's. The center of rotation is defined by the averaged *x*- and *y*-positions of the highest pixels, and the cumulated angles are calculated relative to the first frame. (**d**) and (**e**) Cessation of rotary propagation of conformational states upon loss of a single subunit. (**d**) AFM images before (*left*) and after (*right*) loss of β_1. (**e**) Cumulated angles of open β measured using the highest pixel position in each frame. ATP concentration, 2 μM; frame rate, 12.5 fps. β_1 is lost at 16.2 s

5.3.3 Processive Movement and Traffic Jams of Cellulase on Cellulose Surface

5.3.3.1 General Issues of Cellulose Hydrolysis by Cellulase

Cellulase, which is produced mainly by cellulolytic fungi and bacteria, is a group of enzymes that degrade the plant polysaccharide called cellulose, a linear polymer connected by β-1,4 linkages (Ryu and Mandels 1980; Hon 1994). Although all types of cellulases can hydrolyze amorphous cellulose, only a limited number of cellulases can hydrolyze crystalline cellulose (Teeri 1997). The enzymes that hydrolyze crystalline cellulose are called cellobiohydrolases (CBHs) because the major product of the reaction is cellobiose, a soluble β-1,4-glucosidic dimer (Teeri et al. 1998). CBHs share a similar two-domain structure consisting of a catalytic domain (CD) and a cellulose-binding domain (CBD) (Abuja et al. 1988a, b, 1989). The two domains are connected by a highly *O*-glycosylated linker region (Rouvinen et al. 1990; Divne et al. 1994, 1998). At the initial step of the reaction, the cellulase enzymes adsorb onto the surface of crystalline cellulose via the CBD, and then glucosidic linkages are hydrolyzed by the CD (Johansson et al. 1989; Ståhlberg et al. 1991). Loss of the CBD causes a significant decrease in crystalline cellulose degradation but has less effect on the hydrolysis of soluble or amorphous cellulose, suggesting that adsorption of the enzymes on the surface via the CBD is essential for the effective hydrolysis of crystalline cellulose (Ståhlberg et al. 1991). The CD has a relatively long tunnel formed by surface loops with catalytic amino acids extending from the central fold of the CD. A cellulose chain threads through the tunnel and is hydrolyzed (Divne et al. 1994, 1998). Therefore, it has been considered that CBHs hydrolyze crystalline cellulose chains in a processive manner, making consecutive cuts without releasing the chain (Davies and Henrissat 1995; Reverbel-Leroy et al. 1997), but this idea has not been experimentally verified until recent HS-AFM imaging studies (Igarashi et al. 2009, 2011). The CBHs are categorized into two main types; CBH I hydrolyzes cellulose from the reducing end of cellulose, while CBH II does from the nonreducing end (Henrissat and Bairoch 1993). The CD of CBH I belongs to the glycoside hydrolase (GH) family 7 and the CD of CBH II belongs to the GH family 6.

Although the kinetics of crystalline cellulose hydrolysis by CBHs has been investigated extensively, the mechanism of crystalline cellulose degradation by CBHs is still elusive. For example, although it is known that the hydrolysis rate decreases rapidly with time (Eriksson et al. 2002), its mechanism is not fully understood. Furthermore, several studies indicate that CBH I and CBH II types display strong synergy for the hydrolysis of cellulose (Henrissat et al. 1985; Boisset et al. 2000), but the reason for the synergy effect is not yet revealed. The main reason for the difficulty in understanding these issues is the lack of analytical methods to monitor the enzymatic reaction at the solid/liquid interface.

5.3.3.2 Processive Run of Cellulase on Cellulose I_α

The dynamic behavior of CBHs on a crystalline cellulose substrate was visualized using HS-AFM (Igarashi et al. 2009, 2011). We first observed a CBH I (*Tr*Cel7A) secreted from the industrially important cellulolytic ascomycete fungus *Trichoderma reesei*. *Tr*Cel7A is one of the best-studied CBHs (Abuja et al. 1989; Henrissat 1991; Henrissat and Bairoch 1993; Divne et al. 1994; Linder 1997; Henrissat et al. 1998; Boraston et al. 2004). We used a crystalline cellulose prepared from the cell wall of green algae, which consists of mainly cellulose I_α (Hon 1994). As a substrate surface, we chose highly ordered pyrolytic graphite (HOPG) because its hydrophobic surface has a high affinity for the hydrophobic surface of the crystalline cellulose. Figure 5.8a shows HS-AFM images of *Tr*Cel7A molecules moving unidirectionally and processively on the cellulose I_α surface. This movement of *Tr*Cel7A was observed only on the top surface of the cellulose crystal where the individual molecules were moving in a line. Interestingly, as shown in Fig. 5.8b, the movement of molecules was often halted due probably to the presence of an obstruction on the crystalline cellulose surface. When the movement of a molecule was halted, the following molecules were blocked and/or slowed down (Igarashi et al. 2011). These phenomena resemble the movement of traffic on a road.

To evaluate the velocity of the movement, we tracked the center of each molecule. Figure 5.8c shows the time courses of displacement for seven molecules. Some molecules move continuously without stopping, whereas the movement of some molecules was occasionally halted or slowed down (the open green and closed red plots in Fig. 5.8c). Figure 5.8d shows a histogram of the measured velocity values. The histogram comprises two Gaussian distributions of the average velocity with mean $\pm$ SD values of -0.32 ± 3.4 and 7.1 ± 3.9 nm/s. This is consistent with the idea that *Tr*Cel7A has two modes of adsorption on a cellulose surface; the productive adsorption mode in which both CD and CBD contribute to the binding and the nonproductive adsorption mode in which the enzyme binds only via the CBD (Ståhlberg et al. 1991). The overall average velocity for many molecules was 5.3 ± 4.9 nm/s. From the velocity, the hydrolysis rate is estimated to be 5/s taking into account that one hydrolysis drives the movement of the molecule by 1 nm, which is the length of cellobiose. Surprisingly, this value is 500 times larger than that (0.01/s) estimated biochemically (Nidetzky et al. 1994). This inconsistency is presumably because in the biochemical estimation of the hydrolysis rate, it is assumed that all molecules productively work on the cellulose surface. However, the number of reduced ends available for *Tr*Cel7A binding is limited and the movement of many molecules is hampered by the traffic jams.

5.3.3.3 Processive Run of Cellulase on Cellulose III_I

The fabric form of cellulose I_α can be converted to cellulose III_I polymorph by supercritical ammonia treatment (Igarashi et al. 2007). Although it is known that this conversion dramatically enhances the cellulose hydrolysis by *Tr*Cel7A

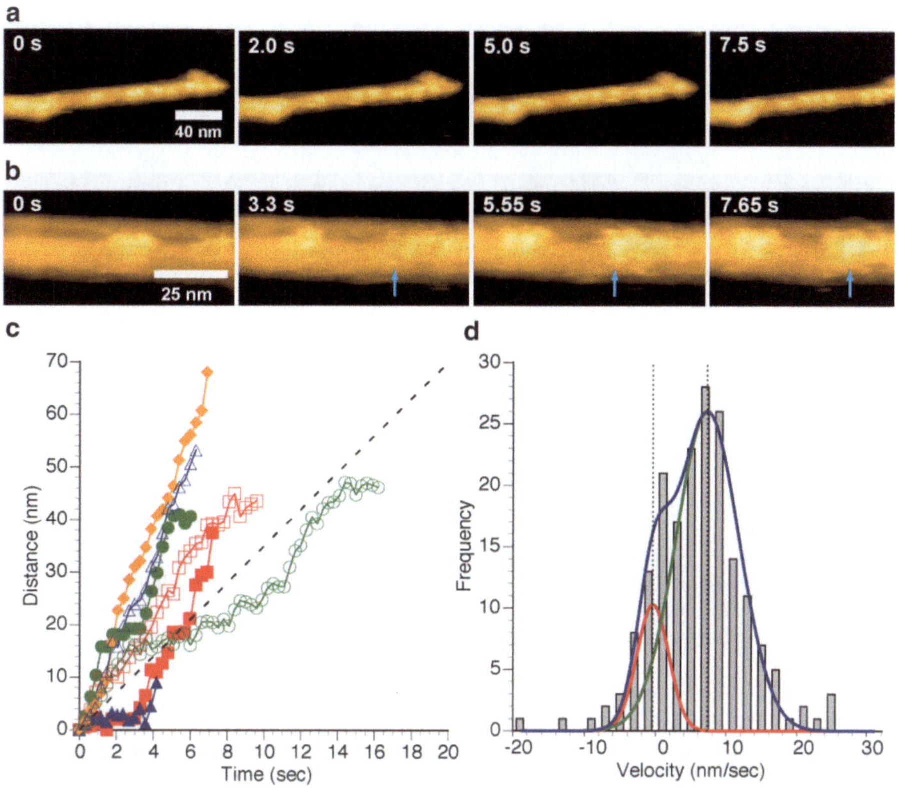

Fig. 5.8 Successive AFM images showing processive and unidirectional movement of *Tr*Cel7A on crystalline cellulose I_α (**a, b**). Frame rate, 1.1 fps. (**a**) The enzyme molecules move from the *right* to *left* on the crystal. (**b**) The movement of one molecule is halted and then following molecules are blocked at a position indicated by *arrows*. (**c**) Time course of displacement for individual *Tr*Cel7A molecules. The *different symbols* indicate different molecules. (**d**) Velocity distribution of linearly moving *Tr*Cel7A molecules on crystalline cellulose I_α. The histogram was fitted by the combination (*blue line*) of two Gaussian distribution curves with mean ± SD values of −0.32 ± 3.4 (*red*) and 7.1 ± 3.9 (*green*)

(Igarashi et al. 2007), its mechanism is unknown. We investigated this issue by visualizing the degradation of cellulose III_I (Igarashi et al. 2011). Figure 5.9a shows successive images of *Tr*Cel7A molecules processively moving on a cellulose III_I surface. The number of molecules on the cellulose III_I surface appears larger than that on the cellulose I_α surface shown in Fig. 5.9a although the velocity of the molecules on both surfaces is similar. In the case of cellulose I_α, the molecules only slide along limited lanes, whereas on cellulose III_I *Tr*Cel7A molecules moved over almost the whole exposed surface. This is due probably to the difference of affinity for *Tr*Cel7A between cellulose I_α and III_I. Comparing the crystalline shapes of cellulose I_α and III_I (Nishiyama et al. 2003; Wada et al. 2004), cellulose I_α has a hydrophobic surface (where the enzyme molecule can bind) at (110) surface (Fig. 5.9c), whereas cellulose III_I has a hydrophobic surface not only at (110)

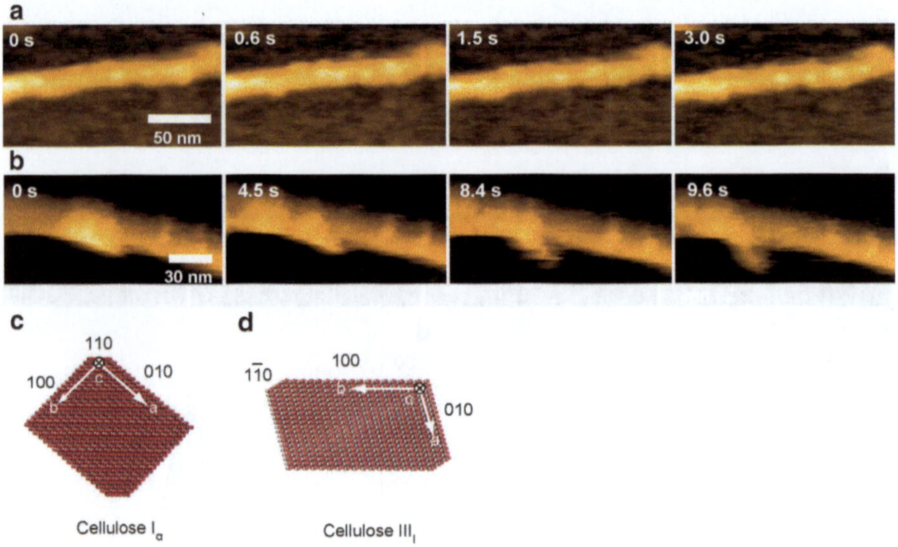

Fig. 5.9 (**a**) Successive AFM images of *Tr*Cel7A molecules on crystalline cellulose III_I. Frame rate, 3.3 fps. (**b**) Fibrillation of crystalline cellulose III_I after congestion of molecules. Frame rate, 3.3 fps. Schematics of sections of crystalline cellulose I_α (**c**) and cellulose III_I (**d**)

surface but also at (100) surface as shown in Fig. 5.9d. Therefore, cellulose III_I has more lanes available for cellulase movement and thus causes less traffic jams. This is the mechanism underlying the enhanced degradation of cellulose III_I.

The traffic jams occurring on cellulose III_I showed a unique feature that did now appear on cellulose I_α. After several additional molecules were blocked, the enzyme molecules started to move again on the surface and at the same time a cellulose bundle was peeled off from the crystalline cellulose as shown in Fig. 5.9b. This phenomenon suggests that one enzyme molecule is unable to overcome an obstacle to the advance but accumulation of molecules behind the blocked molecule overcomes the obstacle. As a result, the blocked molecules restart linear movement from the point where the first molecule had stopped.

5.3.3.4 Synergetic Hydrolysis by *Tr*Cel7A and *Tr*Cel6A

As shown in Fig. 5.10a, two major CBHs, *Tr*Cel7A and *Tr*Cel6A, synergistically increase the efficiency of cellulose III_I hydrolysis. This effect has been called exo-exo synergy (Nidetzky et al. 1994). Previous transmission electron microscopic observations of a crystalline cellulose after partial hydrolysis by both *Tr*Cel6A and *Tr*Cel7A showed that the two enzymes degrade the cellulose from different ends of the crystal; *Tr*Cel7A degrades the crystal from the reducing end, leading to fibrillation, thinning of the crystal, or narrowing of the crystal end, whereas *Tr*Cel6A

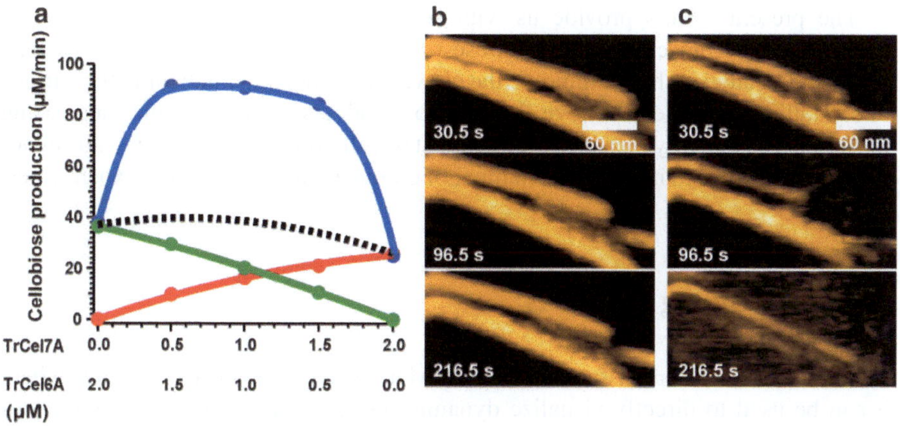

Fig. 5.10 (**a**) Synergy between *Tr*Cel6A and *Tr*Cel7A in cellobiose production from cellulose III_I. The *green* and *red plots* show the rates of cellobiose production by *Tr*Cel6A and *Tr*Cel7A, respectively, and the *blue plot* shows the synergy between the two enzymes. The *dotted line* indicates the simple sum of cellobiose production calculated from the *green* and *red plots*. (**b**) Successive AFM images after injection of *Tr*Cel6A to cellulose III_I at 0 s. The cellulose was not degraded even at 8 min. (**c**) Successive AFM images after *Tr*Cel7A was additionally added at 0 s (at 8.4 min after the initial addition of *Tr*Cel6A). The cellulose was rapidly degraded and almost disappeared at ~4 min. Frame rate, 3.3 fps for (**b**) and (**c**)

hydrolyzes the crystal from the nonreducing end, less processively than *Tr*Cel7A, thereby sharpening the crystal tip (Chanzy and Henrissat 1985; Imai et al. 1998). It is however difficult to explain this type of synergistic action based on their exoenzymatic activity. We imaged the hydrolytic process with crystalline cellulose III_I in the presence of both *Tr*Cel6A and *Tr*Cel7A to understand the mechanism underlying the synergetic degradation (Igarashi et al. 2011).

Cellulose III_I was first imaged in the presence of *Tr*Cel6A alone. The appearance of the cellulose crystals did not significantly change during incubation over ~8 min, even though many enzyme molecules were observed on the cellulose surfaces (Fig. 5.10b). After *Tr*Cel7A was additionally added, enzyme molecules started moving from many points on the cellulose surfaces, and the degradation of crystalline cellulose was dramatically faster than the case with *Tr*Cel7A alone as shown in Fig. 5.10c. Eventually, the cellulose crystals almost disappeared at ~4 min after the addition of *Tr*Cel7A. The observed synergy was independent of the order of addition of *Tr*Cel6A and *Tr*Cel7A. A very similar synergy effect was also observed when both *Tr*Cel6A and *Tr*Cel7A were added simultaneously. These results suggest that *Tr*Cel6A is also an endoenzyme and the synergy observed is therefore endo–exo synergy (Wood and McCrae 1978), although *Tr*Cel6A is generally defined as an exo-glucanase (cellobiohydrolase). If so, *Tr*Cel6A generates nicks in the middle of crystalline cellulose, and these nicks become starting and ending points for *Tr*Cel7A activity.

The present results provide us with some clues to improving the efficiency of conversion from crystalline cellulose to cellobiose by cellulase enzymes. The following strategies should improve the conversion efficiency: flattening the surface of crystalline cellulose to remove obstacles to the advance of the enzymes; increasing the number of lanes by means of pretreatment of cellulose; combining exo- and endo-cellulase enzymes to increase the number of entrances and exits for the exoenzymes.

5.4 Concluding Remarks

In this chapter, we demonstrate that HS-AFM has been developed to the stage where it can be used to directly visualize dynamic processes and structure dynamics of proteins and other biological molecules with unprecedented temporal and spatial resolution. This capability allows us to visually and directly understand how biomolecules operate and function. As demonstrated by our recent results highlighted in this chapter, HS-AFM can provide unique new insights into the functional mechanism of proteins unattainable with previous single-molecule techniques.

Other than the examples describe here, our HS-AFM instrument has already been applied to studying dynamics of various proteins and macromolecular assemblies (Table 5.1); for instances, alternate binding of GroES to the two rings of GroEL (Yamamoto et al. 2009), dynamics of protein 2D crystals (Yamamoto et al. 2008; Yamashita et al. 2009), wiggling motion of intrinsically disordered regions of FACT protein (Miyagi et al. 2008), and photoinduced conformational change of bacteriorhodopsin (Shibata et al. 2010, 2011). Our HS-AFM apparatus is now commercialized by RIBM (Tsukuba, Japan) and has found many other applications (Yokokawa et al. 2006a, b; Crampton et al. 2007; Shinozaki et al. 2009; Casuso et al. 2010; Milhiet et al. 2010; Sugimoto et al. 2010; Wickham et al. 2011). The application of this new type of microscopy is now expanding and should become a tool commonly used for biophysical studies in the near future.

Acknowledgments The authors thank Daisuke Yamamoto, Hayato Yamashita, Mikihiro Shibata, Ryota Iino, Hiroyuki Noji, Kiyohiko Igarashi, Hideki Kandori, and many present and past students in Ando lab for collaborations or contributions to the work described in this chapter. Long-term financial support by NEDO, JST (CREST project), JSPS (Grant-in-Aid for Basic Research (S)), Knowledge Cluster Initiative, and the Mitsubishi Foundation is gratefully acknowledged.

References

Abrahams JP, Leslie AGW, Lutter R, Walker JE (1994) Structure at 2.8 Å resolution of F1-ATPase from bovine heart mitochondri. Nature 370:621–628

Abuja PM, Pilz I, Claeyssens M, Tomme P (1988a) Domain structure of cellobiohydrolase II as studied by small angle X-ray scattering: close resemblance to cellobiohydrolase I. Biochem Biophys Res Commun 156:180–185

Abuja PM, Schmuck M, Pilz I, Tomme P, Claeyssens M, Esterbauer H (1988b) Structural and functional domains of cellobiohydrolase I from trichoderma reesei. Eur Biophys J 15:339–342

Abuja PM, Pilz I, Tomme P, Claeyssens M (1989) Structural changes in cellobiohydrolase I upon binding of a macromolecular ligand as evident by SAXS investigations. Biochem Biophys Res Commun 165:615–623

Adachi K, Oiwa K, Nishizaka T, Furuike S, Noji H, Itoh H, Yoshida M, Kinosita K (2007) Coupling of rotation and catalysis in F_1-ATPase revealed by single-molecule imaging and manipulation. Cell 130:309–321

Aloise P, Kagawa Y, Coleman PS (1991) Comparative Mg^{2+}-dependent sequential covalent binding stoichiometries of 3'-O-(4-benzoyl)benzoyl adenosine 5'-diphosphate of MF_1, TF_1, and the alpha 3 beta 3 core complex of TF_1. The binding change motif is independent of the F1 gamma delta epsilon sub. J Biol Chem 266:10368–10376

Ando T (2012) High-speed atomic force microscopy coming of age. Nanotechnology 23:062001 (27 pp)

Ando T, Kodera N, Maruyama D, Takai E, Saito K, Toda A (2002) A high-speed atomic force microscope for studying biological macromolecules in action. Jpn J Appl Phys 41:4851–4856

Ando T, Kodera N, Takai E, Maruyama D, Saito K, Toda A (2001) A high-speed atomic force microscope for studying biological macromolecules in action. Proc Natl Acad Sci USA 98:12468–12472

Ando T, Kodera N, Uchihashi T, Miyagi A, Nakakita R, Namashita H, Matada K (2005) High-speed atomic force microscopy for capturing dynamic behavior of protein molecules at work. e-J Surf Sci Nanotech 3:384–392

Ando T, Uchihashi T, Fukuma T (2008) High-speed atomic force microscopy for nano-visualization of dynamic biomolecular processes. Prog Surf Sci 83:337–437

Bezanilla M, Drake B, Nudler E, Kashlev M, Hansma PK, Hansma HG (1994) Motion and enzymatic degradation of DNA in the atomic force microscope. Biophys J 67:2454–2459

Binnig G (1992) Force microscopy. Ultramicroscopy 42–44:7–15

Binnig G, Quate C, Gerber C (1986) Phys Rev Lett 56:930–933

Boisset C, Fraschini C, Schulein M, Henrissat B, Chanzy H (2000) Imaging the enzymatic digestion of bacterial cellulose ribbons reveals the endo character of the cellobiohydrolase *Cel*6A from humicola insolens and its mode of synergy with cellobiohydrolase *Cel*7A. Appl Environ Microbiol 66:1444–1452

Boraston AB, Bolam DN, Gilbert HJ, Davies GJ (2004) Carbohydrate-binding modules: fine-tuning polysaccharide recognition. Biochem J 382:769–781

Boyer PD (1993) The binding change mechanism for ATP synthase–some probabilities and possibilities. Biochim Biophys Acta 1140:215–250

Boyer PD (1997) The ATP synthase–a splendid molecular machine. Annu Rev Biochem 66:717–749

Burgess S, Walker M, Wang F, Sellers JR, White HD, Knight PJ, Trinick J (2002) The prepower stroke conformation of myosin V. J Cell Biol 159:983–991

Capaldi RA (1998) Unisite Catalysis without Rotation of the γ - ε Domain in *Escherichia coli* F_1 - ATPase. Biochemistry 273:15940–15945

Casuso I, Kodera N, Le Grimellec C, Ando T, Scheuring S (2009) High-resolution high-speed contact mode atomic force microscopy movies of purple membrane. Biophys J 97:1354–1361

Casuso I, Sens P, Rico F, Scheuring S (2010) Experimental evidence for membrane-mediated protein-protein interaction. Biophys J 99:L47–L49

Chanzy H, Henrissat B (1985) Unidirectional degradation of Valonia cellulose microcrystals subjected to cellulase action. FEBS Lett 184:285–288

Crampton N, Yokokawa M, Dryden DTF, Edwardson JM, Rao DN, Takeyasu K, Yoshimura SH, Henderson RM (2007) Fast-scan atomic force microscopy reveals that the type III restriction enzyme EcoP15I is capable of DNA translocation and looping. Proc Natl Acad Sci USA 104:12755–12760

Davies G, Henrissat B (1995) Structures and mechanisms of glycosyl hydrolases. Structure 3:853–859

Divne C, Stahlberg J, Reinikainen T, Ruohonen L, Pettersson G, Knowles J, Teeri T, Jones T (1994) The three-dimensional crystal structure of the catalytic core of cellobiohydrolase I from Trichoderma reesei. Science 265:524–528

Divne C, Ståhlberg J, Teeri TT, Jones TA (1998) High-resolution crystal structures reveal how a cellulose chain is bound in the 50 A long tunnel of cellobiohydrolase I from Trichoderma reesei. J Mol Biol 275:309–325

Drake B, Prater C, Weisenhorn A, Gould S, Albrecht T, Quate C, Cannell D, Hansma H, Hansma P (1989) Imaging crystals, polymers, and processes in water with the atomic force microscope. Science 243:1586–1589

Endo M, Sugiyama H (2009) Three-dimensional DNA nanostructures constructed by folding of multiple rectangles. Nucleic Acids Symp Ser 53:81–82

Endo M, Hidaka K, Kato T, Namba K, Sugiyama H (2009) DNA prism structures constructed by folding of multiple rectangular arms. J Am Chem Soc 131:15570–15571

Endo M, Hidaka K, Sugiyama H (2011) Direct AFM observation of an opening event of a DNA cuboid constructed via a prism structure. Org Biomol Chem 9:2075–2077

Endo M, Katsuda Y, Hidaka K, Sugiyama H (2010a) Regulation of DNA methylation using different tensions of double strands constructed in a defined DNA nanostructure. J Am Chem Soc 132:1592–1597

Endo M, Sugita T, Katsuda Y, Hidaka K, Sugiyama H (2010b) Programmed-assembly system using DNA jigsaw pieces. Chem Eur J 16:5362–5368

Eriksson T, Karlsson J, Tjerneld F (2002) A model explaining declining rate in hydrolysis of lignocellulose substrates with cellobiohydrolase I (*Cel*7A) and endoglucanase I (*Cel*7B) of Trichoderma reesei. Appl Biochem Biotechnol 101:41–60

Fantner GE, Barbero RJ, Gray DS, Belcher AM (2010) Kinetics of antimicrobial peptide activity measured on individual bacterial cells using high-speed atomic force microscopy. Nat Nanotechnol 5:280–285

Forgacs E, Cartwright S, Sakamoto T, Sellers JR, Corrie JET, Webb MR, White HD (2008) Kinetics of ADP dissociation from the trail and lead heads of actomyosin V following the power stroke. J Biol Chem 283:766–773

Forkey JN, Quinlan ME, Shaw MA, Corrie JET, Goldman YE (2003) Three-dimensional structural dynamics of myosin V by single-molecule fluorescence polarization. Nature 422:399–404

Fukuma T, Okazaki Y, Kodera N, Uchihashi T, Ando T (2008) High resonance frequency force microscope scanner using inertia balance support. Appl Phys Lett 92:243119 (3 pp)

Furuike S, Hossain MD, Maki Y, Adachi K, Suzuki T, Kohori A, Itoh H, Yoshida M, Kinosita K (2008) Axle-less F_1-ATPase rotates in the correct direction. Science 319:955–958

Geeves MA, Holmes KC (1999) Structura mechanism of muscle contraction. Annu Rev Biochem 68:687–728

Gilmore JL, Suzuki Y, Tamulaitis G, Siksnys V, Takeyasu K, Lyubchenko YL (2009) Single-molecule dynamics of the DNA-*EcoR*II protein complexes revealed with high-speed atomic force microscopy. Biochemistry 48:10492–10498

Giocondi M-C, Yamamoto D, Lesniewska E, Milhiet P-E, Ando T, Le Grimellec C (2010) Surface topography of membrane domains. Biochim Biophys Acta—Biomembranes 1798:703–718

Goldman YE (1987) Kinetics of the actomyosin ATPase in muscle fibers. Annu Rev Physiol 49:637–654

Guthold M, Bezanilla M, Erie DA, Jenkins B, Hansma HG, Bustamante C (1994) Following the assembly of RNA polymerase-DNA complexes in aqueous solutions with the scanning force microscope. Proc Natl Acad Sci USA 91:12927–12931

Häberle W, Hörber JKH, Ohnesorge F, Smith DPE, Binnig G (1992) In situ investigations of single living cells infected by viruses. Ultramicroscopy 42–44:1161–1167

Hansma HG, Bezanilla M, Zenhausern F, Adrian M, Sinsheimer RL (1993) Atomic force microscopy of DNA in aqueous solutions. Nucleic Acids Res 21:505–512

Henrissat B (1991) A classification of glycosyl hydrolases based on amino acid sequence similarities. Biochem J 280:309–316

Henrissat B, Bairoch A (1993) New families in the classification of glycosyl hydrolases based on amino acid sequence similarities. Biochem J 280:781–788

Henrissat B, Driguez H, Viet C, Schülein M (1985) Synergism of cellulases from Trichoderma reesei in the degradation of cellulose. Nat Biotechnol 3:722–726

Henrissat B, Teeri TT, Warren RA (1998) A scheme for designating enzymes that hydrolyse the polysaccharides in the cell walls of plants. FEBS Lett 425:352–354

Hon DNS (1994) Cellulose: a random walk along its historical path. Cellulose 1:1–25

Hossain MD, Furuike S, Maki Y, Adachi K, Ali MY, Huq M, Itoh H, Yoshida M, Kinosita K (2006) The rotor tip inside a bearing of a thermophilic F1-ATPase is dispensable for torque generation. Biophys J 90:4195–4203

Hossain MD, Furuike S, Maki Y, Adachi K, Suzuki T, Kohori A, Itoh H, Yoshida M, Kinosita K (2008) Neither helix in the coiled coil region of the axle of F1-ATPase plays a significant role in torque production. Biophys J 95:4837–4844

Hua W, Chung J, Gelles J (2002) Distinguishing inchworm and hand-over-hand processive kinesin movement by neck rotation measurements. Science 295:844–848

Huxley HE (1969) The mechanism of muscular contraction. Science 164:1356–1366

Igarashi K, Koivula A, Wada M, Kimura S, Penttilä M, Samejima M (2009) High speed atomic force microscopy visualizes processive movement of Trichoderma reesei cellobiohydrolase I on crystalline cellulose. J Biol Chem 284:36186–36190

Igarashi K, Uchihashi T, Koivula A, Wada M, Kimura S, Okamoto T, Penttila M, Ando T, Samejima M (2011) Traffic jams reduce hydrolytic efficiency of cellulase on cellulose surface. Science 333:1279–1282

Igarashi K, Wada M, Samejima M (2007) Activation of crystalline cellulose to cellulose III(I) results in efficient hydrolysis by cellobiohydrolase. FEBS J 274:1785–1792

Iko Y, Tabata KV, Sakakihara S, Nakashima T, Noji H (2009) Acceleration of the ATP-binding rate of F1-ATPase by forcible forward rotation. FEBS Lett 583:3187–3191

Imai T, Boisset C, Samejima M, Igarashi K, Sugiyama J (1998) Unidirectional processive action of cellobiohydrolase *Cel*7A on Valonia cellulose microcrystals. FEBS Lett 432:113–116

Inoue S, Uchihashi T, Yamamoto D, Ando T (2011) Direct observation of surfactant aggregate behavior on a mica surface using high-speed atomic force microscopy. Chem Commun 47:4974–4976

Itoh H, Takahashi A, Adachi K, Noji H, Yasuda R, Yoshida M, Kinosita K (2004) Mechanically driven ATP synthesis by F_1-ATPase. J Jpn Biochem Soc 427:42–45

Johansson G, Stahlberg J, Lindeberg G, Engstrom A, Pettersson G (1989) Isolated fungal cellulose terminal domains and a synthetic minimum analogue bind to cellulose. FEBS Lett 243:389–393

Joo C, Balci H, Ishitsuka Y, Buranachai C, Ha T (2008) Advances in single-molecule fluorescence methods for molecular biology. Annu Rev Biochem 77:51–76

Kaibara C, Matsui T, Hisabori T, Yoshida M (1996) Structural asymmetry of F_1-ATPase caused by the gamma subunit generates a high affinity nucleotide binding site. J Biol Chem 271:2433–2438

Kasas S, Thomson NH, Smith BL, Hansma HG, Zhu X, Guthold M, Bustamante C, Kool ET, Kashlev M, Hansma PK (1997) *Escherichia coli* RNA polymerase activity observed using atomic force microscopy. Biochemistry 36:461–468

Kitazawa M, Shiotani K, Toda A (2003) Batch fabrication of sharpened silicon nitride tips. Jpn J Appl Phys 42:4844–4847

Kodera N, Sakashita M, Ando T (2006) Dynamic proportional-integral-differential controller for high-speed atomic force microscopy. Rev Sci Instrum 77:083704 (7 pp)

Kodera N, Yamamoto D, Ishikawa R, Ando T (2010) Video imaging of walking myosin V by high-speed atomic force microscopy. Nature 468:72–76

Kodera N, Yamashita H, Ando T (2005) Active damping of the scanner for high-speed atomic force microscopy. Rev Sci Instrum 76:053708 (5 pp)

Kokavecz J, Tóth Z, Horváth ZL, Heszler P, Mechler Á (2006) Novel amplitude and frequency demodulation algorithm for a virtual dynamic atomic force microscope. Nanotechnology 17: S173–S177

De La Cruz EM, Wells AL, Rosenfeld SS, Ostap EM, Sweeney HL (1999) The kinetic mechanism of myosin V. Proc Natl Acad Sci USA 96:13726–13731

Lin H, Clegg DO, Lal R (1999) Imaging real-time proteolysis of single collagen I molecules with an atomic force microscope. Biochemistry 38:9956–9963

Linder M (1997) The roles and function of cellulose-binding domains. J Biotechnol 57:15–28

Lindsay SM, Nagahara LA, Thundat T, Knipping U, Rill RL, Drake B, Prater CB, Weisenhorn AL, Gould SA, Hansma PK (1989) STM and AFM images of nucleosome DNA under water. J Biomol Struct Dyn 7:279–287

Matsui T, Yoshida M (1995) Expression of the wild-type and the Cys-/Trp-less alpha 3 beta 3 gamma complex of thermophilic F_1-ATPase in *Escherichia coli*. Biochim Biophys 1231:139–146

Mehta AD, Rock RS, Rief M, Spudich JA, Mooseker MS, Cheney RE (1999) Myosin-V is a processive actin-based motor. Nature 400:590–593

Menz RI, Walker JE, Leslie AG (2001) Structure of bovine mitochondrial F(1)-ATPase with nucleotide bound to all three catalytic sites: implications for the mechanism of rotary catalysis. Cell 106:331–341

Milhiet P-E, Yamamoto D, Berthoumieu O, Dosset P, Le Grimellec C, Verdier J-M, Marchal S, Ando T (2010) Deciphering the structure, growth and assembly of amyloid-like fibrils using high-speed atomic force microscopy. PLoS One 5:e13240 (8 pp)

Miyagi A, Tsunaka Y, Uchihashi T, Mayanagi K, Hirose S, Morikawa K, Ando T (2008) Visualization of intrinsically disordered regions of proteins by high-speed atomic force microscopy. Chemphyschem 9:1859–1866

Müller DJ, Büldt G, Engel A (1995) Force-induced conformational change of bacteriorhodopsin. J Mol Biol 249:239–243

Müller DJ, Engel A (1999) Voltage and pH-induced channel closure of porin OmpF visualized by atomic force microscopy. J Mol Biol 285:1347–1351

Nakamoto RK, Ketchum CJ, Al-Shawi MK (1999) Rotational coupling in the F_0F_1 ATP synthase. Annu Rev Biophys Biomol Struct 28:205–234

Nidetzky B, Steiner W, Hayn M, Claeyssens M (1994) Cellulose hydrolysis by the cellulases from Trichoderma reesei: a new model for synergistic interaction. Biochem J 298:705–710

Nishiyama Y, Sugiyama J, Chanzy H, Langan P (2003) Crystal structure and hydrogen bonding system in cellulose I(alpha) from synchrotron X-ray and neutron fiber diffraction. J Am Chem Soc 125:14300–14306

Noji H, Yasuda R, Yoshida M, Kinosita K (1997) Direct observation of the rotation of F1-ATPase. Nature 386:299–302

Oberleithner H, Schillers H, Wilhelmi M, Butzke D, Danker T (2000) Nuclear pores collapse in response to CO2 imaged with atomic force microscopy. Pflugers Arch Eur J Physiol 439:251–255

Oguchi Y, Mikhailenko SV, Ohki T, Olivares AO, De La Cruz EM, Ishiwata S (2008) Load-dependent ADP binding to myosins V and VI: implications for subunit coordination and function. Proc Natl Acad Sci USA 105:7714–7719

Ohnesorge F, Heckl WM, Häberle W, Pum D, Sara M, Schindler H, Schilcher K, Kiener A, Smith DP, Sleytr UB (1992) Scanning force microscopy studies of the S-layers from Bacillus coagulans E38-66, Bacillus sphaericus CCM2177 and of an antibody binding process. Ultramicroscopy 42–44:1236–1242

Okada T, Tanaka H, Iwane AH, Kitamura K, Ikebe M, Yanagida T (2007) The diffusive search mechanism of processive myosin class-V motor involves directional steps along actin subunits. Biochem Biophys Res Commun 354:379–384

Oke OA, Burgess SA, Forgacs E, Knight PJ, Sakamoto T, Sellers JR, White H, Trinick J (2010) Influence of lever structure on myosin 5a walking. Proc Natl Acad Sci USA 107:2509–2514

Peterman EJG, Sosa H, Moerner WE (2004) Single-molecule fluorescence spectroscopy and microscopy of biomolecular motors. Annu Rev Phys Chem 55:79–96

Purcell TJ, Sweeney HL, Spudich JA (2005) A force-dependent state controls the coordination of processive myosin V. Proc Natl Acad Sci USA 102:13873–13878

Putman CAJ, Van Der Werf KO, De Grooth BG, Van Hulst NF, Greve J (1994) Tapping mode atomic force microscopy in liquid. Appl Phys Lett 64:2454–2456

Rajendran A, Endo M, Katsuda Y, Hidaka K, Sugiyama H (2011) Programmed two-dimensional self-assembly of multiple DNA origami jigsaw pieces. ACS Nano 5:665–671

Reverbel-Leroy C, Pages S, Belaich A, Belaich JP, Tardif C (1997) The processive endocellulase CelF, a major component of the Clostridium cellulolyticum cellulosome: purification and characterization of the recombinant form. J Bacteriol 179:46–52

Rondelez Y, Tresset G, Nakashima T, Kato-Yamada Y, Fujita H, Takeuchi S, Noji H (2005) Highly coupled ATP synthesis by F_1-ATPase single molecules. Nature 433:773–777

Rosenfeld SS, Sweeney HL (2004) A model of myosin V processivity. J Biol Chem 279:40100–40111

Rouvinen J, Bergfors T, Teeri T, Knowles JK, Jones TA (1990) Three-dimensional structure of cellobiohydrolase II from *Trichoderma reesei*. Science 249:380–386

Roy R, Hohng S, Ha T (2008) A practical guide to single-molecule FRET. Nat Methods 5:507–516

Ryu DDY, Mandels M (1980) Cellulases: biosynthesis and applications. Enzyme Microb Technol 2:91–102

Sakamoto T, Amitani I, Yokota E, Ando T (2000) Direct observation of processive movement by individual myosin V molecules. Biochem Biophys Res Commun 272:586–590

Sakamoto T, Webb MR, Forgacs E, White HD, Sellers JR (2008) Direct observation of the mechanochemical coupling in myosin Va during processive movement. Nature 455:128–132

Sannohe Y, Endo M, Katsuda Y, Hidaka K, Sugiyama H (2010) Visualization of dynamic conformational switching of the G-quadruplex in a DNA nanostructure. J Am Chem Soc 132:16311–16313

Schabert FA, Engel A (1994) Reproducible acquisition of *Escherichia coli* porin surface topographs by atomic force microscopy. Biophys J 67:2394–2403

Schäffer TE, Cleveland JP, Ohnesorge F, Walters DA, Hansma PK (1996) Studies of vibrating atomic force microscope cantilevers in liquid. J Appl Phys 80:3622–3627

Sellers JR, Weisman LS (2008) Myosins: a superfamily of molecular motors. In: Coluccio LM (ed) Proteins and cell regulation. Springer, Netherland, pp 289–324

Shao Z, Yang J (1995) Progress in high resolution atomic force microscopy in biology. Q Rev Biophys 28:195–251

Shibata M, Uchihashi T, Yamashita H, Kandori H, Ando T (2011) Structural changes in bacteriorhodopsin in response to alternate illumination observed by high-speed atomic force microscopy. Angew Chem Int Ed 50:4410–4413

Shibata M, Yamashita H, Uchihashi T, Kandori H, Ando T (2010) High-speed atomic force microscopy shows dynamic molecular processes in photoactivated bacteriorhodopsin. Nat Nanotechnol 5:208–212

Shinozaki Y, Sumitomo K, Tsuda M, Koizumi S, Inoue K, Torimitsu K (2009) Direct observation of ATP-induced conformational changes in single P2X4 receptors. PLoS Biol 7:e1000103 (12 pp)

Shirakihara Y, Leslie AG, Abrahams JP, Walker JE, Ueda T, Sekimoto Y, Kambara M, Saika K, Kagawa Y, Yoshida M (1997) The crystal structure of the nucleotide-free alpha 3 beta 3 subcomplex of F1-ATPase from the thermophilic Bacillus PS3 is a symmetric trimer. Structure 5:825–836

Ståhlberg J, Johansson G, Pettersson G (1991) A new model for enzymatic hydrolysis of cellulose based on the two-domain structure of cellobiohydrolase I. Bio/Technology 9:286–290

Sugimoto S, Yamanaka K, Nishikori S, Miyagi A, Ando T, Ogura T (2010) AAA+ chaperone ClpX regulates dynamics of prokaryotic cytoskeletal protein FtsZ. J Biol Chem 285:6648–6657

Suzuki Y, Higuchi Y, Hizume K, Yokokawa M, Yoshimura SH, Yoshikawa K, Takeyasu K (2010) Molecular dynamics of DNA and nucleosomes in solution studied by fast-scanning atomic force microscopy. Ultramicroscopy 110:682–688

Teeri TT (1997) Crystalline cellulose degradation: new insight into the function of cellobiohydrolases. Trends Biotechnol 15:160–167

Teeri TT, Koivula A, Linder M, Wohlfahrt G, Divne C, Jones TA (1998) Trichoderma reesei cellobiohydrolases: why so efficient on crystalline cellulose? Biochem Soc Trans 26:173–178

Todd RD, Griesenbeck TA, Douglas MG (1980) The yeast mitochondrial adenosine triphosphatase complex Subunit stoichiometry and physical characterization. J Biol Chem 255:5461–5467

Uchihashi T, Iino R, Ando T, Noji H (2011) High-speed atomic force microscopy reveals rotary catalysis of rotorless F_1-ATPase. Science 333:755–758

Veigel C, Schmitz S, Wang F, Sellers JR (2005) Load-dependent kinetics of myosin-V can explain its high processivity. Nat Cell Biol 7:861–869

Vianni MB, Pietrasanta LI, Thompson JB, Chand A, Gebeshuber IC, Kindt JH, Richter M, Hansma HG, Hansma PK (2000) Probing protein-proein interactions in real time. Nature Struct Biol 7:644–647

Wada M, Chanzy H, Nishiyama Y, Langan P (2004) Cellulose III I crystal structure and hydrogen bonding by synchrotron X-ray and neutron fiber diffraction. Macromolecules 37:8548–8555

Walker JE, Fearnley IM, Gay NJ, Gibson BW, Northrop FD, Powell SJ, Runswick MJ, Saraste M, Tybulewicz VL (1985) Primary structure and subunit stoichiometry of F1-ATPase from bovine mitochondria. J Mol Biol 184:677–701

Walker ML, Burgess SA, Sellers JR, Wang F, Hammer JA, Trinick J, Knight PJ (2000) Two-headed binding of a processive myosin to F-actin. Nature 405:804–807

Walz T, Tittmann P, Fuchs KH, Müller DJ, Smith BL, Agre P, Gross H, Engel A (1996) Surface topographies at subnanometer-resolution reveal asymmetry and sidedness of aquaporin-1. J Mol Biol 264:907–918

Wang H, Oster G (1998) Energy transduction in the F1 motor of ATP synthase. Nature 396:279–282

Watanabe R, Iino R, Noji H (2010) Phosphate release in F1-ATPase catalytic cycle follows ADP release. Nat Chem Biol 6:814–820

Weisenhorn AL, Hansma PK, Albrecht TR, Quate CF (1989) Forces in atomic force microscopy in air and water. Appl Phys Lett 54:2651 (3 pp)

Wendel M, Lorenz H, Kotthaus JP (1995) Sharpened electron beam deposited tips for high resolution atomic force microscope lithography and imaging. Appl Phys Lett 67:3732 (3 pp)

Wickham SFJ, Endo M, Katsuda Y, Hidaka K, Bath J, Sugiyama H, Turberfield AJ (2011) Direct observation of stepwise movement of a synthetic molecular transporter. Nat Nanotechnol 6:166–169

Wood TM, McCrae SI (1978) The cellulase of Trichoderma koningii purification and properties of some endoglucanase components with special reference to their action on cellulose when acting alone and in synergism with the cellobiohydrolase. Biochem J 171:61–72

Yamamoto D, Nagura N, Omote S, Taniguchi M, Ando T (2009) Streptavidin 2D crystal substrates for visualizing biomolecular processes by atomic force microscopy. Biophys J 97:2358–2367

Yamamoto D, Uchihashi T, Kodera N, Ando T (2008) Anisotropic diffusion of point defects in a two-dimensional crystal of streptavidin observed by high-speed atomic force microscopy. Nanotechnology 19:384009 (9 pp)

Yamamoto D, Uchihashi T, Kodera N, Yamashita H, Nishikori S, Ogura T, Shibata M, Ando T (2010) High-speed atomic force microscopy techniques for observing dynamic biomolecular processes. Methods Enzymol 475:541–564

Yamashita H, Voïtchovsky K, Uchihashi T, Contera SA, Ryan JF, Ando T (2009) Dynamics of bacteriorhodopsin 2D crystal observed by high-speed atomic force microscopy. J Struct Biol 167:153–158

Yanagida T, Ishii Y (eds) (2008) Single molecule dynamics in life science. Wiley, Weinheim

Yasuda R, Noji H, Kinosita K, Yoshida M (1998) F_1-ATPase is a highly efficient molecular motor that rotates with discrete 120 degree steps. Cell 93:1117–1124

Yildiz A, Forkey JN, McKinney SA, Ha T, Goldman YE, Selvin PR (2003) Myosin V walks hand-over-hand: single fluorophore imaging with 1.5-nm localization. Science 300:2061–2065

Yokokawa M, Carnally SM, Henderson RM, Takeyasu K, Edwardson JM (2010) Acid-sensing ion channel (ASIC) 1a undergoes a height transition in response to acidification. FEBS Lett 584:3107–3110

Yokokawa M, Wada C, Ando T, Sakai N, Yagi A, Yoshimura SH, Takeyasu K (2006a) Fast-scanning atomic force microscopy reveals the ATP/ADP-dependent conformational changes of GroEL. EMBO J 25:4567–4576

Yokokawa M, Yoshimura SH, Naito Y, Ando T, Yagi A, Sakai N, Takeyasu K (2006b) Fast-scanning atomic force microscopy reveals the molecular mechanism of DNA cleavage by ApaI endonuclease. IEE Proc Nanobiotechnol 153:60–66

Yoshida M, Allison WS (1990) The ATPase activity of the alpha 3 beta 3 complex of the F_1-ATPase of the thermophilic bacterium PS_3 is inactivated on modification of tyrosine 307 in a single beta subunit by 7-chloro-4-nitrobenzofurazan. J Biol Chem 265:2483–2487

Zhong Q, Inniss D, Kjoller K, Elings VB (1993) Fractured polymer/silica fiber surface studied by tapping mode atomic force. Surf Sci Lett 290:L688–L692

Chapter 6
Enzyme Catalysis at the Single-Molecule Level

Raul Perez-Jimenez and Jorge Alegre-Cebollada

6.1 Introduction

Enzyme catalysis is one of the most fundamental processes in living organisms. Since Eduard Buchner's influential work in 1897 on the fermentation of sucrose, the study of enzyme catalysis has been one of the main subjects in Biochemistry.

Enzymes exhibit several features that make them extraordinary molecules. First, enzymes are capable of accelerating chemical reactions up to 10^{23} times and, in many cases, chemical reactions in biological systems are simply not possible without the intervention of enzymes (Kraut et al. 2003). Second, enzymes show high specificity in both substrate specificity and in the type of reaction that they catalyze. Finally, enzymes have developed complex tertiary structures that are mainly responsible for these characteristics. These tertiary structures determine the physical contact between enzyme and substrate and also establish specific interactions between critical residues that generate chemical modifications and conformational fluctuations necessary for catalysis (Eisenmesser et al. 2002, 2005).

Over the course of the twentieth century, the mechanisms of a large number of enzymes were studied using biochemical assays, leading to a comprehensive depiction of the kinetics of enzyme catalysis. Different kinetic models have been described, from the simple steady-state Michaelis–Menten model to more complex models such as those entailing multi-substrate reactions or numerous enzymatic states. All this information has been crucial to understanding the basic chemistry as well as the biological role of enzymes. During the second half of the century, the refinement of molecular biology tools and particularly the development of structural techniques such as X-ray crystallography and nuclear magnetic resonance (NMR) greatly expanded the knowledge of enzyme catalysis (Eisenmesser et al. 2005; Dai et al. 2007; Henzler-Wildman et al. 2007; Faham et al. 2008). Using NMR and X-ray

R. Perez-Jimenez (✉) • J. Alegre-Cebollada
Department of Biological Sciences, Columbia University, New York, NY, USA
e-mail: raulpjc@biology.columbia.edu

A.F. Oberhauser (ed.), *Single-molecule Studies of Proteins*,
Biophysics for the Life Sciences 2, DOI 10.1007/978-1-4614-4921-8_6,

crystallography to solve the structure of enzymes in the free and/or bound state has been of great importance to understanding more deeply the catalytic mechanism of enzymes. These techniques have demonstrated that binding between enzyme and substrate is not as simple as represented by the "lock and key" model, but rather enzyme and substrate undergo structural changes to facilitate mutual interaction and to stabilize the transition state of the reaction (Garcia-Viloca et al. 2004).

While the combination of structural techniques with biochemical assays and theoretical investigations has provided most of the current information about enzymatic catalysis, some aspects of enzyme catalysis remain elusive. For instance, X-ray crystallography does not provide direct information about the dynamics of proteins but rather relies on computational studies (Kohn et al. 2010). Furthermore, X-ray crystallography depends on atomistic models that may introduce uncertainties, which is clearly detrimental when resolving fine details of the enzyme-substrate complex (DePristo et al. 2004). NMR also presents difficulties. First, NMR is limited to relatively small protein systems ($\leq$60,000 Da). Second, although NMR spectroscopy can reach the sub-Ångstrom scale, it does not provide dynamic information at this scale (Hammes-Schiffer and Benkovic 2006). In addition, computational tools are currently used to study enzyme catalysis; however, reaching the timescale in which most enzymatic reactions occur is still a challenge (Klepeis et al. 2009). New experimental methods, as well as new theoretical and computational tools, are necessary to further expand our understanding of enzyme catalysis.

Newly developed single-molecule techniques represent a powerful approach for studying enzymes. These techniques have been successfully applied to the study of biological systems and have provided novel insights. The main advantage of single-molecule techniques is that chemical and biological processes can be observed individually. These methods avoid the uncertainty generated when observing an average of a given process in which different states are possible. The study of enzyme catalysis has been one of the niches in which these single-molecule techniques have succeeded at providing new and unprecedented information. The conversion of a single substrate molecule, the enzymatic rupture of a single covalent bond, and the concerted dynamic during catalysis are just a few examples of processes that can be rigorously examined using single-molecule techniques.

In this chapter, we cover some of the most significant achievements of the single-molecule methodologies in elucidating enzyme catalysis. We pay special attention to AFM techniques, a field in which we ourselves have made significant contributions in recent years.

6.2 Single-Molecule Techniques

There are at least three different single-molecule techniques that have been successfully and extensively implemented to offer insight into enzyme catalysis: optical tweezers (OT), single-molecule fluorescence resonance energy transfer (sm-FRET), and atomic force microscopy (AFM). Each technique has its own strengths and weaknesses and, in general, the results they provide do not overlap with each other, but rather offer insight into different phenomena. OT and AFM rely on the application of mechanical force to probe the system under study, while FRET is based on transference of energy from a donor to an acceptor through dipole–dipole interaction (Cornish and Ha 2007). These techniques provide details about the dynamics of enzyme catalysis reaching the sub-Ångstrom scale and are capable of resolving motions occurring in the sub-second scale (Wiita et al. 2007; Liu et al. 2010).

6.2.1 Single-Molecule Fluorescence Techniques

Single-molecule fluorescence techniques were made possible by the ability to detect the fluorescence of a single fluorophore. Perhaps the most informative modality within this field is sm-FRET. This technique is based on the transference of nonradiative energy between two fluorescent dye molecules, donor and acceptor. The macromolecules of interest, proteins and/or nucleic acids, are labeled with a pair of dyes in specific positions. Upon excitation of the donor at a given wavelength (generally around 500 nm), the light emission associated with the transference of energy from the donor to the acceptor can be monitored in real time (Fig. 6.1a, b). The efficiency of the energy transference can be expressed as $E = (1 + (R/R_0)^6)^{-1}$, where R is the distance between the dyes and R_0 is the distance at which $E = 0.5$, generally known as Förster radius. Therefore, the transmission of energy is very sensitive to the distance between the dyes. This particular fact makes sm-FRET very effective in determining conformational changes and dynamics in proteins and enzymes as well as nucleic acids at the nanometer scale (Roy et al. 2008).

One of the earliest studies reporting the activity of a single enzyme measuring fluorescence emission was on cholesterol oxidase (Lu et al. 1998). In this case, a single fluorescent probe was used. Cholesterol oxidase is a flavin enzyme that catalyzes the oxidation of cholesterol by oxygen using FAD as a cofactor. The reduction of fluorescent FAD generates nonfluorescent $FADH_2$, which is further oxidized by oxygen. Each catalytic cycle produces an on–off fluctuation in the fluorescence signal. Statistical analyses of catalytic turnovers showed that single cholesterol oxidase enzymes exhibit fluctuations in the reaction rate instead of a single, time-independent reaction rate. This phenomenon was attributed to conformational changes in the individual enzyme. These catalytic fluctuations appear to be a general phenomenon (English et al. 2006). These experiments also demonstrated that the Michaelis–Menten model holds at the single-enzyme level (Kou et al. 2005).

In general, sm-FRET is a particularly powerful technique to determine conformational changes; however, due to the distance sensitivity of the energy transference, the dyes must be placed in close proximity, which limits the observed dynamics to relatively small alterations (Roy et al. 2008).

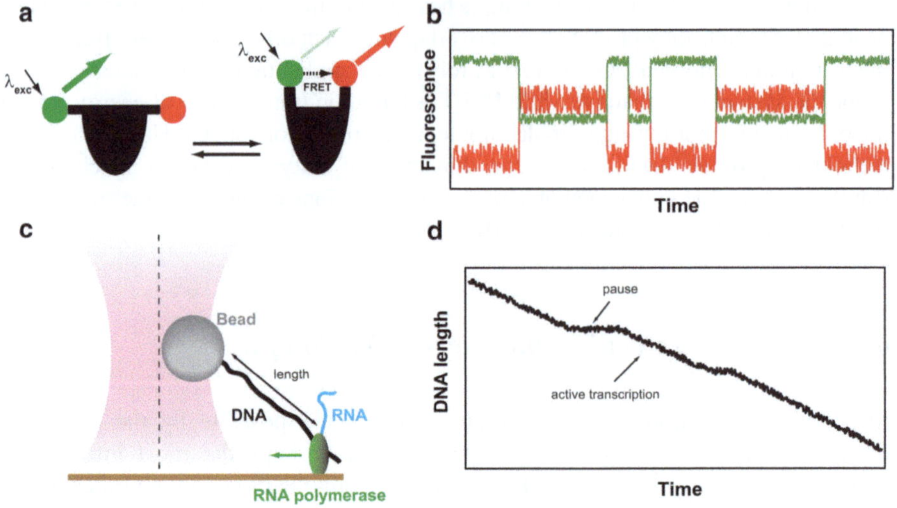

Fig. 6.1 (**a**) Schematic representation of a sm-FRET experiment. The molecule of interest is labeled with a pair of dyes, donor (*green*) and acceptor (*red*). The donor chromophore is excited at λ_{exc} and its emission can be detected (*green*). When conformational changes within the molecule occur and the two dyes are brought in close proximity to one another, the transference of energy from the donor to the acceptor occurs. The emission of the acceptor can be now detected at a different wavelength (*red*). (**b**) A typical two-color sm-FRET experimental trace. The *green trace* corresponds to the donor emission and the *red trace* is the acceptor emission upon FRET. (**c**) Representation of an OT setup to study the activity of RNA polymerase. One end of a DNA molecule is attached to the bead and serves as a template for RNA polymerase, which is specifically attached to the surface. As the enzyme decodes the information producing RNA, the DNA molecule is shortened causing the displacement of the bead. (**d**) Idealized example of an OT trace of RNA polymerase transcription. As DNA is decoded its length becomes shorter (active transcription). Pauses during transcription can be easily detected since no change in length occurs

6.2.2 *Optical Tweezers*

In the mid-1980s, Ashkin and colleagues realized that microscopic dielectric particles can be optically trapped by focusing a laser through a high numerical aperture microscope objective (Ashkin et al. 1986). This observation opened the door to the application of optical trapping techniques to biological systems (Ashkin and Dziedzic 1987; Ashkin et al. 1987). Since then, the term optical tweezers (OT) has become increasingly popular to describe any experiment that exploits optical trapping to study biological processes. To obtain information at the single-molecule level, OT makes use of micrometer-sized beads coated with the molecule of interest. The experimental design is such that one end of the molecule binds to the trapped bead, whereas the other end is attached to a surface or to a second bead. Using this approach, a single-molecule tether can be formed (Fig. 6.1c). When trapped, beads behave as minute springs. Thus, any displacement from their equilibrium position is accompanied by a restoring force acting on the bead.

OT setups are able to measure the position of trapped beads with sub-nanometer resolution, allowing the simultaneous determination of force and displacement (Neuman and Nagy 2008; Moffitt et al. 2008; Fazal and Block 2011).

The inspection of short molecules (e.g., most proteins) by OT is hampered, in general, by nonspecific interactions between the surfaces at both ends of the tether (Cecconi et al. 2008). However, OT are extraordinarily versatile when it comes to the study of micrometer-long molecules, such as nucleic acids or large multidomain proteins. Consequently, OT can examine the activity of enzymes acting on this type of substrates (Fig. 6.1d). The last two decades have seen an explosion in the number of reports using OT to study enzymes acting on DNA or RNA, including polymerases (Yin et al. 1995; Wang et al. 1998; Larson et al. 2011; Wuite et al. 2000), endonucleases (Gemmen et al. 2006), and helicases (Cheng et al. 2011; Dumont et al. 2006). More recently, using protein substrates conjugated to DNA handles, OT has also been shown to capture proteolytic activity (Zhang et al. 2009; Aubin-Tam et al. 2011; Maillard et al. 2011).

The study of enzymes at work using optical trapping has provided new perspectives on their mechanism of action. Like any other single-molecule technique, OT can capture heterogeneity among individual enzyme molecules (Davenport et al. 2000; Adelman et al. 2002). In addition, enzymes working on long DNA or RNA molecules usually move along their substrates in discrete steps, the size of which can be determined using OT (Cheng et al. 2011; Dumont et al. 2006; Abbondanzieri et al. 2005). The majority of these enzymes are true molecular motors that use the energy derived from hydrolysis of metabolites such as ATP to power movement along their substrates.

6.2.3 Single-Molecule Atomic Force Microscopy

We will dedicate the rest of this chapter to discuss the application of AFM techniques to the study of enzyme catalysis. AFM seems to be the latest addition to the toolbox of single-molecule techniques that can examine enzymatic reactions. Similarly to OT, AFM relies on the application of mechanical force to study the mechanism of action of enzymes. In an AFM, a single protein molecule is mechanically stretched by placing it between a sharp cantilever tip and a surface attached to a piezoelectric actuator (Fig. 6.2). The cantilever tip acts as a spring that obeys Hooke's law with a spring constant of 15–40 pN/nm. The piezoelectric actuator can accurately generate displacements in the nanometer scale. Therefore, force and distance at the pN and nm scale, respectively, can be easily resolved in an AFM. One of the main advantages of AFM over OT is that the sharpness of the cantilever tip minimizes surface interactions and, as a consequence, shorter molecules such as protein domains can be studied.

AFM's ability to measure force makes it ideal to study molecular systems that are under mechanical force in their environment. Force is a natural perturbation occurring in biological systems. At the molecular level, mechanical forces not

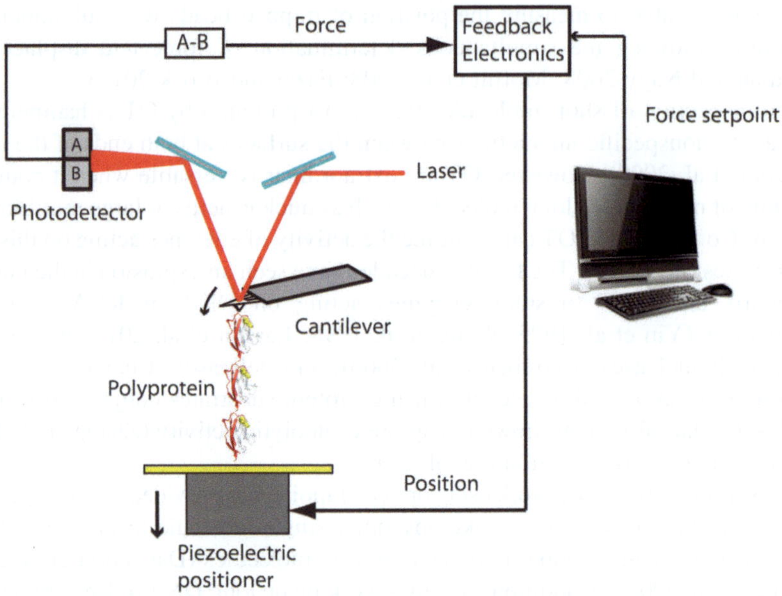

Fig. 6.2 Schematic diagram of an AFM apparatus operating in force-clamp mode. Polyprotein molecules are absorbed on a coverslip placed on top of a piezoelectric actuator. A silicon-nitride cantilever is pressed against the coverslip allowing a polyprotein to attach to it. By retracting the piezoelectric actuator, the polyprotein can be mechanically stretched. When the cantilever deflects, a change in the laser trajectory is detected in the two-part photodetector. This change in the signal, (A–B), is proportional to the change in force. In the force-clamp mode, the measured force is compared to a setpoint force. If the measured force is different than the setpoint force, the feedback system triggers a displacement of the piezoelectric actuator to restore the force

only modulate muscle proteins, they also activate ion channels and mediate cell adhesion, cell signaling, and bacterial division (Zhang et al. 2009; del Rio et al. 2009; Vogel and Sheetz 2006, 2009; Martinac 2004; Bershadsky et al. 2003). In addition, numerous proteins that may be exposed to mechanical tension such as membrane proteins, surface receptors, or cell adhesion molecules (CAM) also undergo enzymatic transformations (Merinen et al. 2005; Salmi and Jalkanen 2005, 2006). While the effect of force on cell adhesion, motility, and signaling has been documented both at the cellular and molecular level, much less effort has been devoted to understanding the effect of mechanical force on enzymatic reactions.

AFM techniques provide an excellent platform to study enzymatic reaction occurring under mechanical loads. Yet, the study of mechanoenzymatic reactions requires the ability to control the force applied to the system. This allows us to have a direct and controlled intervention over the enzymatic reaction, which is possible thanks to the development of *force-clamp* spectroscopy. This technique allows the application of a well-calibrated force to a single protein molecule (Fernandez and Li 2004). The ability to control force makes force-clamp spectroscopy highly suitable to studying processes such as protein folding, as it allows us for the capture of the different phases that a protein undergoes before acquiring its folded state (Garcia-Manyes et al. 2009). Although protein (un)folding was the first application studied by force-clamp spectroscopy, its ability to study enzyme catalysis greatly expanded the capabilities of the technique.

In the following sections, we describe recent developments of a force-clamp single-molecule assays to study the chemistry of thioredoxin enzymes (Trx), a class of oxidoreductases present in all known living organisms that act as electron donors for reducing disulfide bonds (Arner and Holmgren 2000). These experiments demonstrated that force can reveal the dynamics during enzyme catalysis with sub-Ångstrom resolution.

6.3 Basic Concepts of Force-Clamp Spectroscopy

Force-clamp spectroscopy takes fundamental concepts from the voltage-clamp technique used in electrophysiology, in which the current passing through ion channels can be controlled (Sakmann and Neher 1984). In the case of force-clamp, mechanical force can be programmed to be held constant over time by adjusting the displacement of the piezoelectric actuator. The heart of the force-clamp technique is an electronic feedback that corrects any alteration produced in the programmed force (Oberhauser et al. 2001). These alterations can be produced, for instance, when a protein maintained under a constant stretching force unfolds or when a disulfide bond, also under force, is cleaved by a reducing agent. In both cases, the force will drop upon unfolding or reduction as a consequence of the increase in the contour length of the polymer. The electronic feedback will rapidly trigger a displacement in the piezoelectric positioner that will restore the programmed force. The time response of the piezoelectric actuator is critical. Current state-of-the-art piezos can respond well below 1 ms, which is more than adequate to study the (un)folding of a multitude of proteins as well as bond ruptures. Therefore, in force-clamp, the unfolding of a protein or the reduction of disulfide bond is detected as steps of increment in length as a function of time (Fig. 6.3a, b).

A fundamental aspect when working in force-clamp, and force spectroscopy in general, is to create signatures or fingerprints that unambiguously confirm that only the protein of interest is being examined. This is important since spurious interactions between the cantilever tip and the surface often occur. One approach to generate molecular fingerprint is the construction of polyproteins made of equal repeats of the protein of interest. Generally, polyproteins are made of four to ten repeats and current molecular biology techniques make the construction easy and efficient (Carrion-Vazquez et al. 1999, 2000; Perez-Jimenez et al. 2006). When a polyprotein is stretched and unfolded, a number of steps of exact equal length corresponding to the length of each domain are detected. The probability that unspecific interactions produce such a highly repetitive pattern is virtually zero.

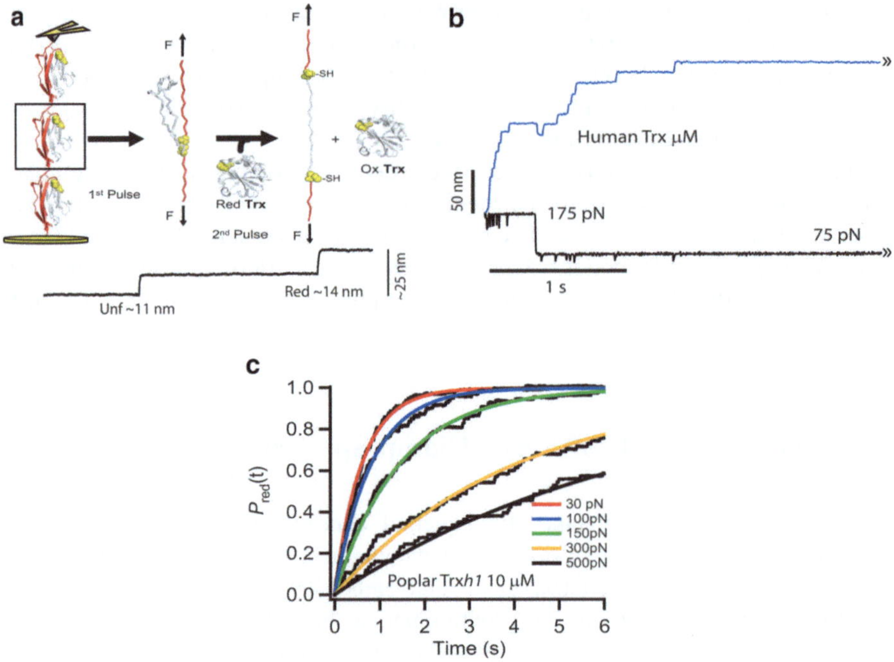

Fig. 6.3 (**a**) Schematic representation of a single-molecule force-clamp experiment for disulfide bond reduction. The I27 substrate, containing a disulfide bond in positions 32 and 75 ($I27_{SS}$), is first stretched to trigger unfolding of the titin domains. Upon unfolding, the disulfide bond is exposed to the solution and can be reduced by a single Trx enzyme. The unfolding is registered as an 11 nm increment in length and the reduction of the disulfide bond as a 14 nm step. (**b**) Experimental force-clamp trace of $(I27_{SS})_8$ reduction by human Trx. A two-force pulse protocol is used to separate unfolding events (175 pN) from reductions of disulfide bonds (75 pN). (**c**) Summed and averaged traces of disulfide bond reduction events at different forces. An exponential fit is used to estimate the reduction rate at each force ($r = 1/\tau$). Modified from Perez-Jimenez et al. Nat Struct Mol Biol. (2009)

The first application of the force-clamp technique was the study of protein unfolding and folding. In a pioneering work by Fernandez and Li (2004), polyubiquitin was first mechanically unfolded and subsequently refolded by quenching the force. These experiments allowed the identification of the different stages that a protein undergoes during its folding trajectory. Importantly, the same stages were observed when using a single ubiquitin monomer instead of a polyprotein (Garcia-Manyes et al. 2007), ruling out the possibility that the use of polyproteins alters the behavior of individual protein domains. However, working with monomers is complicated by a lesser frequency of molecule pick-ups, which makes this option much more time consuming. For these reasons, the use of polyproteins is preferred.

Shortly after the first application of force-clamp, the technique was used to study the kinetics of disulfide bond reduction by dithiothreitol (DTT) under the effect of a stretching force. In this case, a polyprotein with engineered disulfide bonds was used, and it was demonstrated that the application of mechanical force to the disulfide bond accelerates the reduction by DTT (Wiita et al. 2006). These experiments were the prelude for the study of Trx catalysis by force-clamp spectroscopy.

6.4 Force-Clamp Assay for Studying Trx Catalysis

Thioredoxin enzymes are present in virtually all known organisms. Their main function is to donate electrons in reactions that generally involve the reduction of disulfide bonds (Arner and Holmgren 2000). Trx enzymes contain a highly conserved pair of cysteines that are generally located at positions 32 and 35. Both cysteines have anomalous pK_a values of ~7 and ~9.5, respectively (Dyson et al. 1997), far from the normal pK_a for a standard cysteine residue (8.5). These pK_a values are, in part, responsible for the unique characteristics of Trx enzymes.

The basic mechanism of Trx for the reduction of disulfide bonds has been known for years. The cysteine in position 32 attacks the substrate disulfide forming a mix disulfide. This intermediate disulfide is resolved by the attacking of Cys35 with the interplay of other residues such as Asp26 (Chivers and Raines 1997), releasing the reduced substrate and the oxidized enzyme. The enzyme is reactivated to the reduced form by NADPH and a second enzyme, Trx reductase. All together these components form part of the so-called thioredoxin system (Holmgren 1985).

The experimental assay utilized to determine the chemistry of Trx enzymes in bulk relies on a coupled reaction using insulin or DTNB as substrates (Holmgren 1979; Perez-Jimenez et al. 2008). In the insulin assay, Trx molecules reduce insulin, which aggregates and precipitates upon reduction. The turbidity generated is measured spectrophotometrically at 600 nm. The disulfide reduction rate is estimated from the rate of insulin precipitation. In the case of the DTNB assay, Trx reduces this compound to TNB which absorbs at 412 nm. The reduction rate is estimated by determining the change in absorbance at this wavelength. Although these assays have provided a good understanding of Trx catalysis, they are far from providing detailed information on the dynamics occurring during catalysis at the Ångstrom level.

To study the chemistry of Trx enzymes using force-clamp spectroscopy, the first step is to engineer a substrate that provides a clear fingerprint. For that, a polyprotein made of eight repeats of the 27th module of human cardiac titin (I27) was created, and a disulfide bond was introduced in each module by mutating residues in positions 32 and 75 to Cys $(I27_{SS})_8$. The disulfide bond between these two residues remains buried within the core of the protein. The advantage of using I27 is that it is a well-known protein that has been extensively studied by force spectroscopy (Oberhauser et al. 2001; Li et al. 2000). The disulfide bond will remain intact in the force regime applied in a typical force-clamp experiment, generally bellow 1 nN. Upon mechanical unfolding, the disulfide bonded substrate will extend only ~11 nm per module leaving the rest of the residues sequestered behind the disulfide bond. Upon reduction of the disulfide bond by Trx enzymes, the release of the trapped residues produces and extra elongation of ~14 nm giving rise to the total extension of 25 nm characteristic of wild-type I27 (Fig. 6.3a, b).

In a typical force-clamp experiment, the substrate is deposited on top of a gold-coated cover slip, and a drop containing the Trx system, i.e., Trx enzymes, Trx reductase, and NADPH. The final concentration of Trx is typically between 5 and 10 μM. When the polyprotein substrate is stretched, the domains unfold and the disulfide bonds are exposed to the solution where they can be reduced by Trx enzymes. Typically, a two-pulse force protocol is used to separate unfolding from reduction events. The unfolding events of 11 nm are collected during a short pulse of ~180 pN and 0.3 s. After this pulse the force is switched to a different value ranging from 25 to 600 pN. In this second pulse, the only traces selected are those showing the same number of reduction events (14 nm) as unfolding events (11 nm) during the first pulse. A typical trace of unfolding and disulfide bond reduction by Trx enzymes is shown in Fig. 6.3b, where the events are clearly differentiated. Numerous traces of the second pulse containing only reduction events can be collected at a given force. These traces are summed, averaged, and fitted to a single exponential to obtain the reduction rate at a particular force (Fig. 6.3c). This allows us obtaining the force dependency of disulfide bond reduction (Fig. 6.4a, b).

6.5 Understanding the Chemistry of Thioredoxin

In the case of DTT and other simple reducing agents, the force-dependency for disulfide bond reduction follows a single exponential, indicating that the reaction is exponentially accelerated as the force increases (Wiita et al. 2006; Ainavarapu et al. 2008). However, when using Trx enzymes from *Escherichia coli* and human, the force dependency is clearly different from that of DTT and a biphasic behavior is observed. When the substrate is pulled at forces between 25 and 200 pN, both Trx enzymes show a negative force dependency of the disulfide bond reduction rate. Interestingly, when the substrate is pulled at forces over 200 pN, *E. coli* Trx shows an exponential increase in the reduction rate, whereas human Trx remains essentially force independent (Fig. 6.5a). Surprisingly, this marked difference between human and *E. coli* Trx can be generalized to bacterial and eukaryotic Trx enzymes. By testing a number of Trx enzymes from the domains Bacteria and Eukaryota, the exponential increase of the reduction rate at forces over 200 pN is observed in all bacterial Trx enzymes and the force independent behavior is common to all eukaryotic Trxs (Fig. 6.4a, b) (Perez-Jimenez et al. 2009).

To understand this complex chemistry for disulfide bond reduction, we developed a kinetic model that considers different chemical mechanisms. Common to all Trx enzymes is the negative force dependency for disulfide reduction observed at low forces. This mechanism is unique to Trx enzymes and is not observed when using simple reducing agents such us DTT, glutathione, or L-cys (Perez-Jimenez et al. 2009). This behavior can be explained with a Michaelis–Menten mechanism in

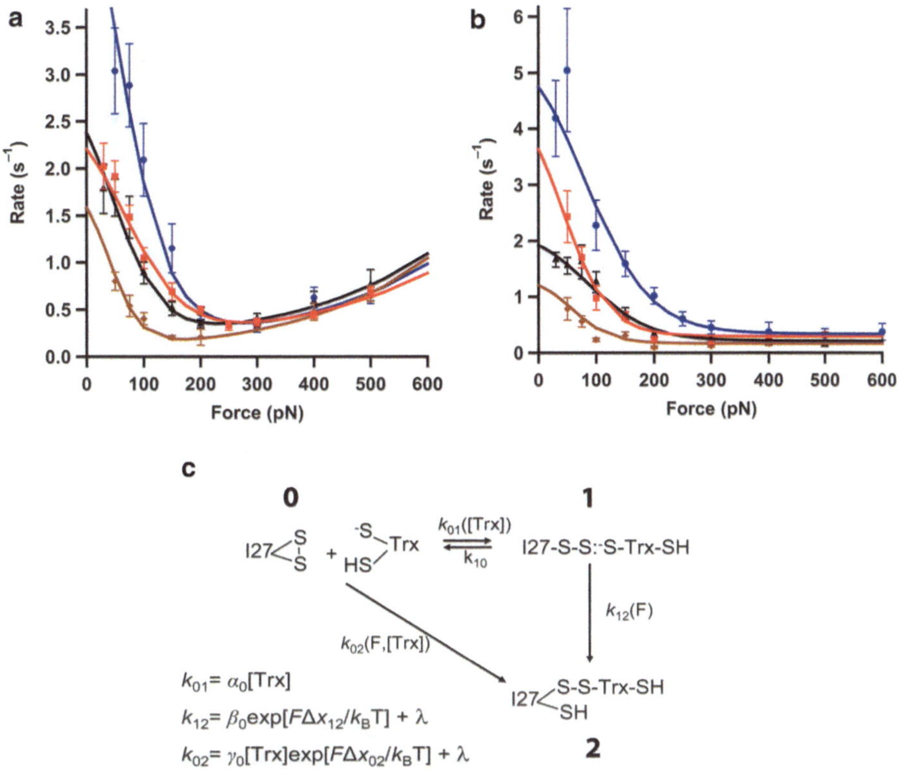

Fig. 6.4 Force-dependence of disulfide bond reduction for bacterial-origin Trxs (**a**) and eukaryotic origin Trx enzymes (**b**). The bacterial Trxs are *E. coli* Trx1 (*red*), *E. coli* Trx2 (*brown*), human mitochondrial Trx2 (*blue*), and pea chloroplastic Trx*m* (*black*). The eukaryotic Trxs are human Trx1 (*blue*), poplar Trx*h*1 (*black*), *Plasmodium falciparum* Trx (*red*), and poplar Trx*h*3 (*brown*). All Trx enzymes show a negative force dependency at forces below 200 pN. This pathway is attributed to a Michaelis–Menten mechanism. At forces above 200 pN, bacterial Trxs show a S_N2 mechanism, whereas eukaryotic Trxs show a force-independent single-electron transfer mechanism. (**c**) Kinetic model of force dependent reduction by Trx enzymes. Three possible states are considered in this model, Michaelis–Menten S_N2 (pathway 0–1–2). In this pathway, the formation of the enzyme-substrate complex is determinant. The second pathway does not require formation of the complex (pathway 0–2). In the case of bacterial Trx, this pathway is exponentially accelerated consistent with an S_N2 reaction mechanism. In eukaryotic Trx, the reduction occurs following a force-independent single-electron transfer mechanism considered by the parameter λ. We speculate that this contribution is ubiquitous to all Trx enzymes. For eukaryotic Trx, only the force-independent contribution is observed and, therefore, the exponential contribution in k_{02} may be eliminated, leaving $k_{02} = \lambda$. From Perez-Jimenez et al. Nat Struct Mol Biol. (2009)

which the formation of the enzyme-substrate complex is essential. Within this complex, the disulfide bond is reoriented in the binding pocket of Trx in order to achieve the correct orientation for reduction, that is, an angle of 180° between the three sulfur atoms participating in the S_N2 reaction (Wiita et al. 2007; Perez-Jimenez et al. 2009). This reorientation requires the rotation of the disulfide bond, which is

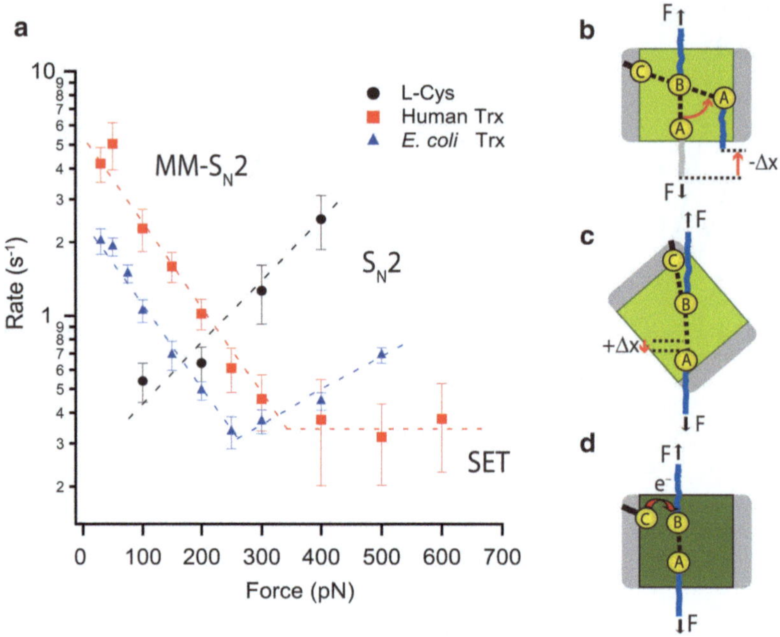

Fig. 6.5 Chemical mechanisms for disulfide bond reduction by Trx enzymes (human and *E. coli*) and simple reducing agent such as L-Cys detected by force-clamp spectroscopy. (**a**) Force-dependence for disulfide bond reduction for several reducing agents (human Trx, *E. coli* Trx, and L-Cys), showing the different behaviors observed. (**b**) Michaelis–Menten chemical mechanisms (MM-SN2) detected in all Trx enzymes. In this mechanism, the active site of the enzyme regulates the position of the three participating sulfur atoms. A rotation of the substrate disulfide bond is required to achieve the 180° required for an SN2 reaction. (**c**) Representation of the SN2 reaction mechanisms resulting by alignment of the three sulfur atoms with no active site regulation. This configuration is achieved by simple collisions in the correct orientation. This mechanism is favored in Trx enzymes with a shallow grove such as those of bacterial origin. (**d**) Single-electron transfer mechanism occurring in all enzymes but highly visible in eukaryotic Trxs. For this mechanism, no specific orientation of the sulfur atoms is required. From Perez-Jimenez et al. Nat Struct Mol Biol. (2009)

accompanied by a shortening of the polypeptide chain (Fig. 6.5b). Both the rotation of the angle and the shortening of the polypeptide can be determined from the kinetic model. This shortening is generally below 1 Å (Δx_{12} in the kinetic diagram in Fig. 6.4c) and represents a direct measurement of the dynamics occurring during catalysis. Intuitively, the applied force is against the rotation of the disulfide bond and therefore it will negatively impact the reduction rate, explaining thus the negative force dependency observed at low forces.

When the substrate is pulled at forces above 200 pN, two distinct behaviors are observed. The explanation for these two mechanisms can be found by analyzing the structure of the binding grove of Trx enzymes. A detailed inspection of pdb files of different Trx enzymes demonstrates that Trx enzymes from bacterial origin have a shallow grove, whereas enzymes from eukaryotic organisms have a deep grove.

Computational calculations using the pdb files show that the difference in depth between bacterial and eukaryotic origin Trx enzymes is about two times (Perez-Jimenez et al. 2009). A shallow grove will allow a bacterial Trx enzyme to search the space for the appropriate conformation for an S_N2 reaction similar to that observed with simple compounds (Fig. 6.5c). A deep grove will freeze the substrate in a specific conformation avoiding the possibility of acquiring the orientation necessary for an S_N2 reaction. In fact, molecular dynamic simulations of the mobility of substrate trapped in the binding grove of several bacterial and eukaryotic Trx enzymes demonstrate that in the case of bacterial Trx the substrate has a high mobility, whereas in the case of eukaryotic Trx the substrate is rather static. Therefore, in eukaryotic Trxs, a nucleophilic attack is prevented and the reduction of disulfide bonds occurs throughout a single-electron transfer reaction (SET) that seems to be force independent (Fig. 6.5a, d). This is further demonstrated by the fact the metallic zinc that reduces disulfide bond by transferring electrons also displays a force-independent reaction for disulfide bond reduction in the force-clamp assay (Perez-Jimenez et al. 2009). The SET mechanism is incorporated in the kinetic model as a constant contribution λ, i.e., force independent (Fig. 6.4c). Although it is more noticeable in eukaryotic Trx, this mechanism is probably present in all Trx enzymes.

The reason why Trx enzymes have developed these different chemical mechanisms is not obvious but an interesting possibility could be related to the different functions, environments, and conditions that Trx enzymes support. For instance, the S_N2 reaction utilized by bacterial Trx when pulling at elevated forces could be related to the ability of bacteria to live under extreme situations in which elevated forces can be found such as high pressures or extreme salinity conditions which may generate drastic changes in intracellular osmotic pressures causing cells to swell or shrink (Sharma et al. 2002; La Duc et al. 2007; Bartlett 1992; Koch 1984). Strikingly, Trx has been shown to promote high-pressure resistance in *E. coli* (Malone et al. 2006). Nevertheless, it is important to note that force is used to restrict the orientation and conformation of the substrate disulfide bond. It is reasonable to speculate that similar situations might happen in vivo without the need of reaching elevated forces. For instance, this might be the case of DsbD, a member of the Trx superfamily present in *E. coli* that interacts with cytoplasmic Trx (Katzen and Beckwith 2000). DsbD is associated to the membrane and that might restrict its mobility originating a situation in which the adaptability of *E. coli* Trx might be beneficial for the transmembrane electron flow.

6.6 Exploring the Origin of Trx Catalysis: Resurrecting the Past

It is obvious that Trx enzymes have developed a much more complex chemistry than simple reducing agents such as DTT, L-Cys, or glutathione. First, Trx enzymes are more efficient than the mentioned compounds requiring concentrations in the order of micromolar, whereas simple reducing agents require concentration in the

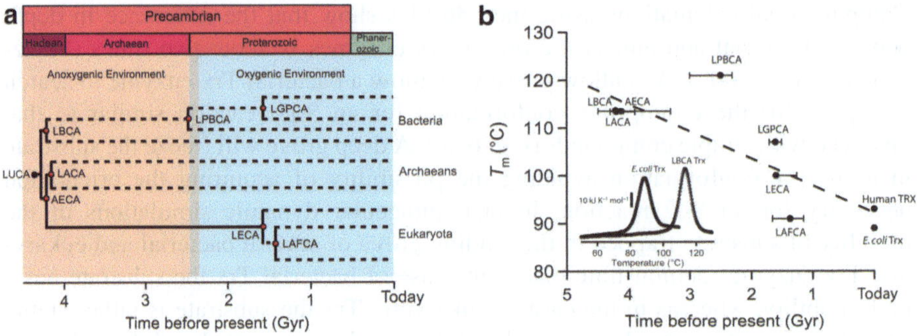

Fig. 6.6 (**a**) Schematic representation of thioredoxin phylogeny showing different extinct organisms (see Sect. 6.6 for details). Divergence times are also indicated. (**b**) Denaturation temperatures (T_m) of extinct and extant Trx enzymes versus geological time. The *dashed line* represents a linear fit with a slope of about −6°C/Myr. The inset represents experimental DSC thermograms for *E. coli* Trx and LBCA Trx. From Perez-Jimenez et al. Nat Struct Mol Biol (2011)

millimolar scale. Second, the Michaelis–Menten mechanism entails an orchestrated reorganization of the cysteine substrates conducted by the active site of the enzyme. In addition, the marked difference between eukaryotic and bacterial Trx enzymes suggests that evolution has operated in the development of this complex and diverse chemistry. The main question now is to understand the origin of the complex chemistry of Trx.

One might be tempted to speculate that primordial Trx enzymes were similar in their chemistry to simple compounds. As life evolved ancient Trx enzymes developed the binding grove that originated their complex chemistry. But is it possible to investigate the chemistry of primitive Trx enzymes? The answer is yes, but not by examining fossil remains. The oldest fossil records that contain organic material are from animals and do not go very far back in time, ~300 million years. However, using computational procedures, it is possible to infer the sequence of ancient proteins if we know the sequence of their descendants (Liberles 2007). This is a technique called ancestral sequence reconstruction (ASR), which is made possible thanks to the extensive number of sequences deposited in databases such as Uniprot (http://www.uniprot.org/) or PFAM (http://pfam.sanger.ac.uk/). ASR relies on statistical methodologies based on maximum parsimony, maximum likelihood, and Bayesian analysis. The reconstruction of ancient proteins and genes requires a number of steps equally important for error-free and reliable inference.

First, it is necessary to perform a smart selection of extant sequences that are diverse and contrasted. Once the sequences have been selected, they must be aligned. A multiple sequence alignment (MAS) is an arrangement of the sequences site by site, to correlate regions that are identical or similar. The MAS represents the basis for the construction of the phylogeny that is, perhaps, the most crucial step for the reconstruction of ancestors.

How can we estimate the sequences of the genes or proteins represented by the nodes in the phylogeny represented in Fig. 6.6a. For a particular site in the protein sequence, the ancestral state that diverged on two modern residues is estimated by calculating the posterior probability for the twenty natural amino acids, given that we know the residue that occupies that position in the descendants. The process is applied to every site for every node in the tree. Those residues with the highest probability are assigned to each site. Therefore, one can estimate the most probabilistic ancestral sequence for each node (Perez-Jimenez et al. 2011; Gaucher et al. 2008). Nowadays a number of programs are available that make the process relatively easy (Liberles 2007).

This methodology was applied to resurrect Trx enzymes from extinct organisms in order to study their properties as well as their chemistry. Trx sequences are abundant in databases since Trx is a universal enzyme present in all living organisms. Therefore, it is possible to retrieve numerous sequences to build a tree with a generous representation of every domain of life. After building the tree, seven internal nodes were selected for sequence reconstruction (Fig. 6.6a). These nodes correspond to the last common ancestor of bacteria (LCBA), last common ancestor of archaea (LACA), archaea/eukaryotes common ancestor (AECA), last common ancestor of photosynthetic bacteria (LPBCA), last common ancestor of gammaproteobacteria (LGPCA), last common ancestor of eukaryotes (LECA), and finally the last common ancestor of animal and fungi (LAFCA). Their age ranges from ~4.2 to ~1.6 billion years old (Fig. 6.6a).

After the computational reconstruction has been performed, the resulting ancestral genes can be synthesized and cloned onto a vector for expression in *E. coli* cells. The purified ancestral Trx enzymes were tested using different approaches. Firstly, a measurement of the denaturation temperature of the ancestral Trx enzymes (T_m) was carried out using differential scanning calorimetry (DSC). This measurement demonstrated that the T_m followed a paleotemperature trend similar to that previously observed with other ancestral proteins and chemical analysis of isotopes (Gaucher et al. 2008). This trend demonstrated the progressive cooling of the planet at a rate of ~6 °C/Myr (Fig. 6.6b). The oldest enzymes, LBCA, LACA, and AECA, demonstrated a T_m of ~112 °C, which is about ~25 °C higher than the T_m of representative Trx enzymes such as human or *E. coli* Trx. This high value suggests that these organisms were likely thermophiles living in hot environments with temperatures over 60°C. As mentioned before, the capability of force-clamp techniques to discern complex details of the chemistry of Trx enzymes makes them perfect to investigate early Trx enzymes as well as their chemical evolution.

Surprisingly, the seven ancestral enzymes showed very similar chemical mechanisms than those of modern ones (see Fig. 6.7a, b for examples). All of them showed the typical negative force-dependent Michaelis–Menten pathway present in modern Trx enzymes, thus refuting the hypothesis that primitive Trx enzymes would be similar to simple reducing compounds. Interestingly, all the ancestral bacterial Trx (LBCA, LPBCA, and LGPCA) also show the force-dependent nucleophilic mechanism (S_N2 reaction) at elevated forces, whereas the eukaryotic and archaeal origin Trxs (AECA, LACA, LECA, and LAFCA) operate with a force-independent mechanism at forces above 200 pN (Perez-Jimenez et al. 2011).

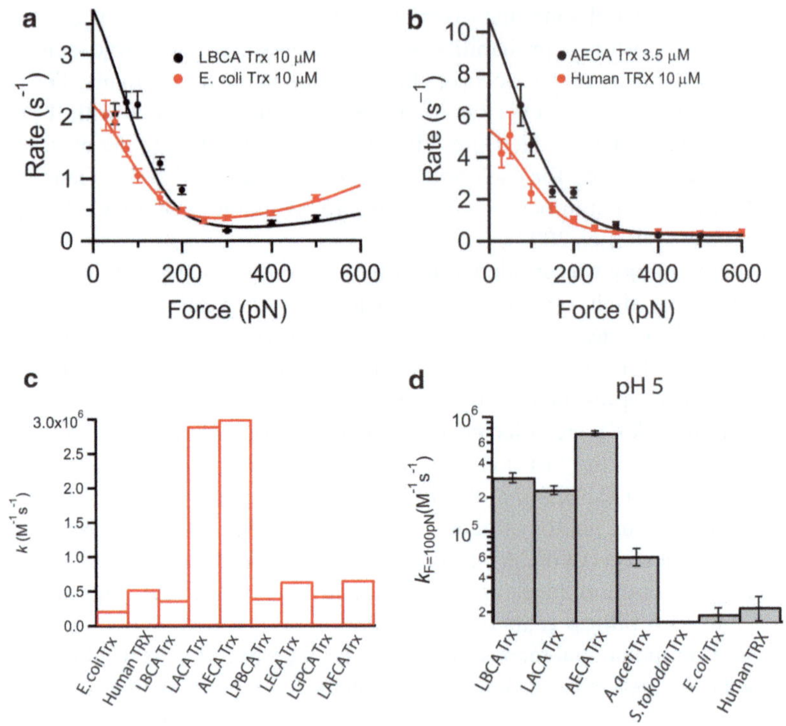

Fig. 6.7 Force-dependence for disulfide bond reduction by ancestral Trx enzymes LBCA Trx (**a**) and AECA Trx (**b**). Modern counterpart, *E. coli* and human Trxs are also shown for comparative purposes. The same chemical mechanisms that were observed in modern Trx seem to be present in ancestral enzymes, suggesting that the chemistry of Trx catalysis is highly conserved. Nevertheless, some differences were observed such as the high rate constant obtained for LACA and AECA ancestral Trxs (**c**). This rate constant is estimated by extrapolating to zero force the fitting to the kinetic model in Fig. 6.3c. (**d**) Rate constant at $F = 100$ pN obtained at pH 5 for ancestral Trx enzymes belonging to LBCA, AECA, and LACA and some modern organisms. Ancestral Trx 4 billion years old show considerable activity when compared with modern organisms. Only Trx from *A. aceti*, an organism that lives under acidic conditions and does not regulate internal pH, shows also activity at pH 5. The ability of ancestral Trx enzymes to work under acidic conditions might be correlated with the high content on CO_2 in the Hadean Earth (4–4.5 billion years ago). From Perez-Jimenez et al. Nat Struct Mol Biol (2011)

This striking conservation of the enzymatic mechanisms might be highly significant in terms of the origin of life. The step from simple compounds to well-structured and functional enzymes occurred very early in evolution and very rapidly. Perhaps, this was a condition necessary for life rather than evolving from poorly efficient enzymes. Another intriguing hypothesis might be that these enzymes, and maybe others, were not created in our planet but came in meteorites or asteroids. This hypothesis is the so-called Panspermia, and was first proposed by notable scientists such as Kelvin, Arrhenius, Berzelius, or Helmholtz. This hypothesis is still under investigation (Wickramasinghe 2010).

There is still an interesting observation about the activity of these ancestral Trx enzymes. Two of the oldest ones, AECA and LACA Trx enzymes, show an activity much higher than the other ancestral Trx but also much higher than modern Trx (up to ten times) (Fig. 6.7c). The activity of Trx enzymes is largely determined by the pK_a of the reactive cysteine. This cysteine needs to be deprotonated to be active. Thanks to the low pK_a of this cysteine this is possible even at physiological pH. However, the activity is very sensitive to pH and drops abruptly at low pH. Surprisingly, LBCA, AECA, and LACA Trx demonstrate a considerable activity at acidic pH (Fig. 6.7 d). This activity under acidic conditions might be explained by the fact that in the Hadean Earth (4.5–4 billion years ago) the atmospheric content of O_2 was pretty low as compared with the content of CO_2 (Walker 1983; Russell and Hall 1997). This CO_2 content made oceans to be acid (pH ~ 5.5) as well as hot. Therefore, organisms living under these conditions were acidophilic and thermophilic. In addition, the pH tolerance of these enzymes might also reflect that primitive organisms did not regulate internal pH.

6.7 Conclusions

In this chapter, we have presented a general description of the current single-molecule techniques used to study enzymes catalysis with special emphasis on AFM methods. It is safe to say that traditional biochemical tools are reaching their limits, and therefore it seems necessary and desirable to expand our knowledge about enzymes throughout the use of novel single-molecule techniques.

The direct observation of the dynamics occurring during the enzymatic process and the discovery of new chemical mechanisms utilized by enzymes or tracking the reaction carried out by single enzymes are some of the phenomena that are now possible to access. Perhaps, the future directions on enzyme catalysis have their guidelines in the single-molecule techniques, but there are still a number of obstacles need to be overcome. For instance, time resolution continues to be a limitation as well as distance resolution. Nevertheless, it is reasonable to envision a future in which the reaction and dynamics of enzymes could be resolved with atomic resolution. This will only be possible throughout the use of single-molecule techniques.

References

Abbondanzieri EA et al (2005) Direct observation of base-pair stepping by RNA polymerase. Nature 438(7067):460–465

Adelman K et al (2002) Single molecule analysis of RNA polymerase elongation reveals uniform kinetic behavior. Proc Natl Acad Sci USA 99(21):13538–13543

Ainavarapu SRK et al (2008) Single-molecule force spectroscopy measurements of bond elongation during a bimolecular reaction. J Am Chem Soc 130(20):6479–6487

Arner ES, Holmgren A (2000) Physiological functions of thioredoxin and thioredoxin reductase. Eur J Biochem 267(20):6102–6109
Ashkin A, Dziedzic JM (1987) Optical trapping and manipulation of viruses and bacteria. Science 235(4795):1517–1520
Ashkin A et al (1986) Observation of a single-beam gradient force optical trap for dielectric particles. Opt Lett 11(5):288
Ashkin A, Dziedzic JM, Yamane T (1987) Optical trapping and manipulation of single cells using infrared laser beams. Nature 330(6150):769–771
Aubin-Tam ME et al (2011) Single-molecule protein unfolding and translocation by an ATP-fueled proteolytic machine. Cell 145(2):257–267
Bartlett DH (1992) Microbial life at high pressures. Sci Prog 76(301–302 Pt 3–4):479–496
Bershadsky AD, Balaban NQ, Geiger B (2003) Adhesion-dependent cell mechanosensitivity. Annu Rev Cell Dev Biol 19:677–695
Carrion-Vazquez M et al (1999) Mechanical and chemical unfolding of a single protein: a comparison. Proc Natl Acad Sci USA 96(7):3694–3699
Carrion-Vazquez M et al (2000) Mechanical design of proteins studied by single-molecule force spectroscopy and protein engineering. Prog Biophys Mol Biol 74(1–2):63–91
Cecconi C et al (2008) Protein-DNA chimeras for single molecule mechanical folding studies with the optical tweezers. Eur Biophys J 37(6):729–738
Cheng W et al (2011) Single-base pair unwinding and asynchronous RNA release by the hepatitis C virus NS3 helicase. Science 333(6050):1746–1749
Chivers PT, Raines RT (1997) General acid/base catalysis in the active site of *Escherichia coli* thioredoxin. Biochemistry 36(50):15810–15816
Cornish PV, Ha T (2007) A survey of single-molecule techniques in chemical biology. ACS Chem Biol 2(1):53–61
Dai S et al (2007) Structural snapshots along the reaction pathway of ferredoxin-thioredoxin reductase. Nature 448(7149):92–96
Davenport RJ et al (2000) Single-molecule study of transcriptional pausing and arrest by *E. coli* RNA polymerase. Science 287(5462):2497–2500
del Rio A et al (2009) Stretching single talin rod molecules activates vinculin binding. Science 323 (5914):638–641
DePristo MA, de Bakker PI, Blundell TL (2004) Heterogeneity and inaccuracy in protein structures solved by X-ray crystallography. Structure 12(5):831–838
Dumont S et al (2006) RNA translocation and unwinding mechanism of HCV NS3 helicase and its coordination by ATP. Nature 439(7072):105–108
Dyson HJ et al (1997) Effects of buried charged groups on cysteine thiol ionization and reactivity in *Escherichia coli* thioredoxin: structural and functional characterization of mutants of Asp 26 and Lys 57. Biochemistry 36(9):2622–2636
Eisenmesser EZ et al (2002) Enzyme dynamics during catalysis. Science 295(5559):1520–1523
Eisenmesser EZ et al (2005) Intrinsic dynamics of an enzyme underlies catalysis. Nature 438 (7064):117–121
English BP et al (2006) Ever-fluctuating single enzyme molecules: Michaelis-Menten equation revisited. Nat Chem Biol 2(2):87–94
Faham S et al (2008) The crystal structure of a sodium galactose transporter reveals mechanistic insights into Na+/sugar symport. Science 321(5890):810–814
Fazal FM, Block SM (2011) Optical tweezers study life under tension. Nat photonics 5:318–321
Fernandez JM, Li H (2004) Force-clamp spectroscopy monitors the folding trajectory of a single protein. Science 303(5664):1674–1678
Garcia-Manyes S et al (2007) Force-clamp spectroscopy of single-protein monomers reveals the individual unfolding and folding pathways of I27 and ubiquitin. Biophys J 93(7):2436–2446
Garcia-Manyes S et al (2009) Direct observation of an ensemble of stable collapsed states in the mechanical folding of ubiquitin. Proc Natl Acad Sci USA 106(26):10534–10539

Garcia-Viloca M et al (2004) How enzymes work: analysis by modern rate theory and computer simulations. Science 303(5655):186–195
Gaucher EA, Govindarajan S, Ganesh OK (2008) Palaeotemperature trend for Precambrian life inferred from resurrected proteins. Nature 451(7179):704–707
Gemmen GJ, Millin R, Smith DE (2006) Tension-dependent DNA cleavage by restriction endonucleases: two-site enzymes are "switched off" at low force. Proc Natl Acad Sci USA 103(31):11555–11560
Hammes-Schiffer S, Benkovic SJ (2006) Relating protein motion to catalysis. Annu Rev Biochem 75:519–541
Henzler-Wildman KA et al (2007) Intrinsic motions along an enzymatic reaction trajectory. Nature 450(7171):838–844
Holmgren A (1979) Reduction of disulfides by thioredoxin. Exceptional reactivity of insulin and suggested functions of thioredoxin in mechanism of hormone action. J Biol Chem 254 (18):9113–9119
Holmgren A (1985) Thioredoxin. Annu Rev Biochem 54:237–271
Katzen F, Beckwith J (2000) Transmembrane electron transfer by the membrane protein DsbD occurs via a disulfide bond cascade. Cell 103(5):769–779
Klepeis JL et al (2009) Long-timescale molecular dynamics simulations of protein structure and function. Curr Opin Struct Biol 19(2):120–127
Koch AL (1984) Shrinkage of growing *Escherichia-Coli*-cells by osmotic challenge. J Bacteriol 159(3):919–924
Kohn JE et al (2010) Evidence of functional protein dynamics from X-ray crystallographic ensembles. PLoS Comput Biol 6(8):pii: e1000911
Kou SC et al (2005) Single-molecule Michaelis-Menten equations. J Phys Chem B 109 (41):19068–19081
Kraut DA, Carroll KS, Herschlag D (2003) Challenges in enzyme mechanism and energetics. Annu Rev Biochem 72:517–571
La Duc MT et al (2007) Isolation and characterization of bacteria capable of tolerating the extreme conditions of clean room environments. Appl Environ Microbiol 73(8):2600–2611
Larson MH, Landick R, Block SM (2011) Single-molecule studies of RNA polymerase: one singular sensation, every little step it takes. Mol Cell 41(3):249–262
Li H et al (2000) Point mutations alter the mechanical stability of immunoglobulin modules. Nat Struct Biol 7(12):1117–1120
Liberles DA (2007) Ancestral sequence reconstruction, vol xiii. Oxford University Press, Oxford, p 252
Liu S et al (2010) Initiation complex dynamics direct the transitions between distinct phases of early HIV reverse transcription. Nat Struct Mol Biol 17(12):1453–1460
Lu HP, Xun L, Xie XS (1998) Single-molecule enzymatic dynamics. Science 282 (5395):1877–1882
Maillard RA et al (2011) ClpX(P) generates mechanical force to unfold and translocate its protein substrates. Cell 145(3):459–469
Malone AS, Chung YK, Yousef AE (2006) Genes of *Escherichia coli* O157:H7 that are involved in high-pressure resistance. Appl Environ Microbiol 72(4):2661–2671
Martinac B (2004) Mechanosensitive ion channels: molecules of mechanotransduction. J Cell Sci 117(Pt 12):2449–2460
Merinen M et al (2005) Vascular adhesion protein-1 is involved in both acute and chronic inflammation in the mouse. Am J Pathol 166(3):793–800
Moffitt JR et al (2008) Recent advances in optical tweezers. Annu Rev Biochem 77:205–228
Neuman KC, Nagy A (2008) Single-molecule force spectroscopy: optical tweezers, magnetic tweezers and atomic force microscopy. Nat Methods 5(6):491–505
Oberhauser AF et al (2001) Stepwise unfolding of titin under force-clamp atomic force microscopy. Proc Natl Acad Sci USA 98(2):468–472

Perez-Jimenez R et al (2006) Mechanical unfolding pathways of the enhanced yellow fluorescent protein revealed by single molecule force spectroscopy. J Biol Chem 281(52):40010–40014
Perez-Jimenez R et al (2008) Force-clamp spectroscopy detects residue co-evolution in enzyme catalysis. J Biol Chem 283(40):27121–27129
Perez-Jimenez R et al (2009) Diversity of chemical mechanisms in thioredoxin catalysis revealed by single-molecule force spectroscopy. Nat Struct Mol Biol 16(8):890–896
Perez-Jimenez R et al (2011) Single-molecule paleoenzymology probes the chemistry of resurrected enzymes. Nat Struct Mol Biol 18(5):592–596
Roy R, Hohng S, Ha T (2008) A practical guide to single-molecule FRET. Nat Methods 5 (6):507–516
Russell MJ, Hall AJ (1997) The emergence of life from iron monosulphide bubbles at a submarine hydrothermal redox and pH front. J Geol Soc Lond 154:377–402
Sakmann B, Neher E (1984) Patch clamp techniques for studying ionic channels in excitable membranes. Annu Rev Physiol 46:455–472
Salmi M, Jalkanen S (2005) Cell-surface enzymes in control of leukocyte trafficking. Nat Rev Immunol 5(10):760–771
Salmi M, Jalkanen S (2006) Developmental regulation of the adhesive and enzymatic activity of vascular adhesion protein-1 (VAP-1) in humans. Blood 108(5):1555–1561
Sharma A et al (2002) Microbial activity at gigapascal pressures. Science 295(5559):1514–1516
Vogel V, Sheetz M (2006) Local force and geometry sensing regulate cell functions. Nat Rev Mol Cell Biol 7(4):265–275
Vogel V, Sheetz MP (2009) Cell fate regulation by coupling mechanical cycles to biochemical signaling pathways. Curr Opin Cell Biol 21(1):38–46
Walker JCG (1983) Possible limits on the composition of the Archean Ocean. Nature 302 (5908):518–520
Wang MD et al (1998) Force and velocity measured for single molecules of RNA polymerase. Science 282(5390):902–907
Wickramasinghe C (2010) The astrobiological case for our cosmic ancestry. Int J Astrobiol 9(2):119–129
Wiita AP et al (2006) Force-dependent chemical kinetics of disulfide bond reduction observed with single-molecule techniques. Proc Natl Acad Sci USA 103(19):7222–7227
Wiita AP et al (2007) Probing the chemistry of thioredoxin catalysis with force. Nature 450 (7166):124–127
Wuite GJ et al (2000) Single-molecule studies of the effect of template tension on T7 DNA polymerase activity. Nature 404(6773):103–106
Yin H et al (1995) Transcription against an applied force. Science 270(5242):1653–1657
Zhang X et al (2009) Mechanoenzymatic cleavage of the ultralarge vascular protein von Willebrand factor. Science 324(5932):1330–1334

Chapter 7
Single-Molecule Studies of Amyloidogenic Proteins

Miklós S.Z. Kellermayer, Árpád Karsai, Ünige Murvai, Szilvia Erdélyi-Bótor, József Kardos, and Ricardo H. Pires

7.1 Introduction

Amyloidosis is a severe, usually fatal clinical disorder in which insoluble filamentous aggregates called amyloid fibrils are deposited in the extracellular space of various tissues (Selkoe 1997, 2003; Mattson 2004; Pepys 2006). Amyloid fibrils are self-associating filamentous structures formed from various misfolded peptides and proteins (Greenwald and Riek 2010; Friedman 2011; Eisenberg and Jucker 2012). Because amyloid is thought to form as a consequence of improper folding of the particular protein, amyloidosis and its related pathological conditions are also called misfolding diseases (Dobson 1999; Selkoe 2003). Several dozens of unrelated peptides have been reported to generate amyloid fibrils in vivo (Westermark et al. 2005; Greenwald and Riek 2010; Friedman 2011; Eisenberg and Jucker 2012). Although oligomeric intermediates of amyloidogenic protein species are now considered to be mostly responsible for the cytotoxic behavior (Fandrich 2012), the accumulation of amyloid fibrils is still considered to be an important factor in the pathogenesis of the disease. The exact structure, dynamics, and the mechanisms of the amyloid fibrils' formation and cellular effects are still not clearly understood.

M.S.Z. Kellermayer (✉) • Ü. Murvai • R.H. Pires
Department of Biophysics and Radiation Biology, Semmelweis University, Budapest, Hungary
e-mail: kellermayer.miklos@med.semmelweis-univ.hu

Á. Karsai
Department of Biophysics and Radiation Biology, Semmelweis University, Budapest, Hungary

Department of Neurobiology, Physiology, and Behavior, University of California-Davis, Davis, CA, USA

S. Erdélyi-Bótor
Department of Biophysics and Radiation Biology, Semmelweis University, Budapest, Hungary

University of Pécs, Pécs, Hungary

J. Kardos
Department of Biochemistry, Eötvös Loránd University, Budapest, Hungary

A.F. Oberhauser (ed.), *Single-molecule Studies of Proteins*,
Biophysics for the Life Sciences 2, DOI 10.1007/978-1-4614-4921-8_7,

The structure of amyloid fibrils has for long been difficult to explore because of insoluble aggregate formation that precludes the use of standard structural methods. Furthermore, the details in higher-order structure and the structural dynamics of amyloid fibrils remain to be resolved.

The first protein to be extracted from amyloid deposits was β-protein, the main component of Alzheimer's amyloid β-fibrils (Aβ-fibrils) (Pepys 2006). Amyloidogenic proteins are considered to be the etiological factors in other notable degenerative diseases as well: Parkinson's disease, type 2 diabetes mellitus, Huntington's disease, transmissible spongiform encephalopathies, amyotrophic lateral sclerosis (ALS), and systemic amyloidoses (Greenwald and Riek 2010; Friedman 2011; Eisenberg and Jucker 2012). In recent years several, sometimes unexpected, proteins have also been shown to form amyloid fibrils (Fandrich et al. 2001). Furthermore, it is increasingly recognized that amyloids may be utilized in diverse biological functions ranging from surface adhesion to biofilm formation (Maury 2009; Mostaert et al. 2009; Otzen 2010; Blanco et al. 2012). Although Alzheimer's disease (Alzheimer 1907; Castellani et al. 2010; Ballard et al. 2011) is not considered to be classical amyloidosis per se (Pepys 2006), Aβ-fibrils represent probably the most frequent and popular model system for exploring the mechanisms of amyloid fibrillogenesis. Systemic amyloidoses are caused by circulating proteins or peptides that are deposited in viscera, blood vessel walls, and connective tissues (Falk et al. 1997; Buxbaum 2003; Pepys 2006). These include conditions caused by plasma cell disorders, familial amyloidosis, chronic inflammation, and hemodialysis. The most notable amyloidogenic proteins of systemic amyloidoses are β_2-microglobulin (Kad et al. 2001, 2003; Kardos et al. 2004, 2005; Myers et al. 2006; Pepys 2006; Eichner and Radford 2011), serum amyloid A (Falk et al. 1997; Buxbaum 2003), transthyretin, a transport protein synthesized by the liver and the choroid plexus (Tan and Pepys 1994; Saraiva 1995, 2002; Damas and Saraiva 2000; Buxbaum 2003; Damas and Saraiva 2005), and the immunoglobulin light chain (Tan and Pepys 1994; Bellotti et al. 2000; Buxbaum 2003).

Although the number of proteins that can form amyloid in vivo is large, it is increasingly realized that essentially every protein may, under certain circumstances, be converted to amlyoid fibrils (Dobson 2003; Greenwald and Riek 2010; Friedman 2011; Eisenberg and Jucker 2012). There are a number of common structural and functional features of amyloid fibrils regardless of the primary sequence of the constituent peptide subunit, and the amyloid fibrillogenetic pathway may be generalized (Fig. 7.1). One of the common architectural features is cross-β structure, which is characterised by the arrangement of the peptide unit β-strands perpendicularly to the main axis of the fibril (Sunde et al. 1997; Eisenberg and Jucker 2012). Another main common structural feature is protofilamentous architecture. Protofilaments are composed of a varying number of β-sheets, and several protofilaments wind together helically to form the mature fibril (Serpell et al. 2000; Petkova et al. 2006). These common structural features strongly suggest recurring motifs in the mechanisms of amyloid assembly. One mechanistic motif is the presence of partially unfolded intermediates and oligomeric and protofibrillar species along the path from the native protein to the amyloidogenic form. Such intermediates are being identified in an increasing number of proteins (Eakin et al. 2006; Fandrich 2012).

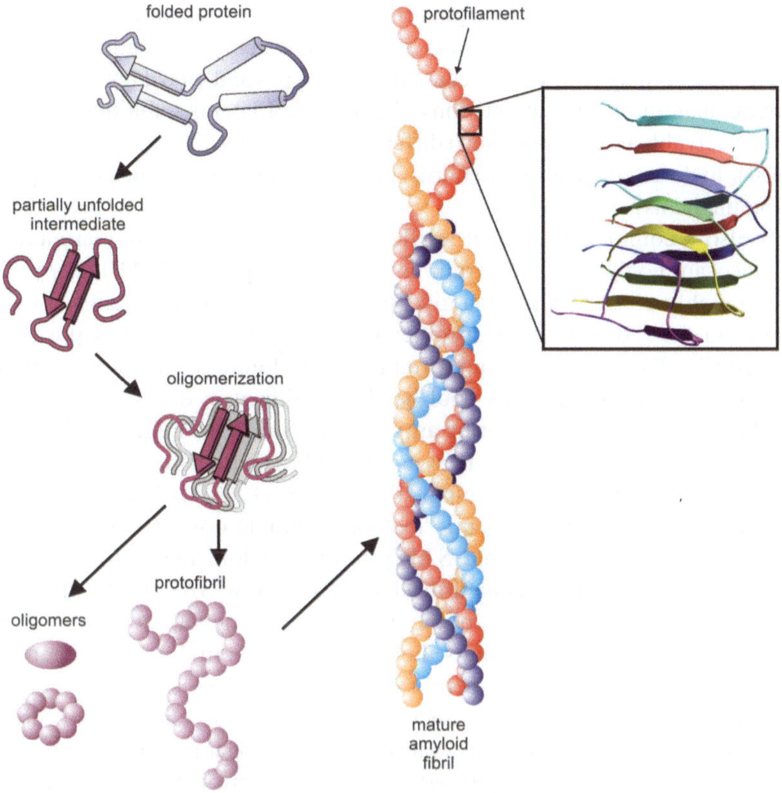

Fig. 7.1 Generic model of amyloid fibrillogenesis. The β-sheet structure of the mature amyloid fibril is based on adaptation from Petkova et al. (2006)

During amyloidogenesis the formation of fibrils is preceded by the appearance of globular aggregates, which are thought to fuse, by not fully understood mechanisms, into fibrillar structures (Rochet and Lansbury 2000). Curved, beaded, ~200-nm-long, and ~6–8-nm-wide fibrillar precursors were described to appear in the amyloidogenetic pathway, which are called protofibrils (Harper et al. 1997; Walsh et al. 1999; Nichols et al. 2002). The protofibrils are thought to go through a structural transition on their way to form amyloid fibril. Another mechanistic motif is the assembly of amyloid fibrils, which is thought to follow nucleation-dependent polymerization mechanisms (Lomakin et al. 1997). In support of this notion, the cooperative, polymerization-like growth of individual Aß fibrils was observed directly with total internal reflection fluorescence microscopy (TIRFM) (Ban et al. 2003, 2004). However, the details of the assembly and propagation dynamics of amyloid fibrils are far from being understood. Because of the structural hierarchy in amyloid fibrils, it is tempting to envision a temporally hierarchical assembly process that entails longitudinal growth and the evolution of complex lateral associations between β-sheets, protofilaments, and eventually fibrils. However, it is unclear whether and how one process (e.g., longitudinal growth) influences the other processes (e.g., lateral associations) and vice versa, just as it is unknown how many intermediates there might be along the assembly pathway. Because different types and sets of bonds are involved in longitudinal and lateral interactions, the axial and lateral stabilities of the fibrils might conceivably be different. Such differences may determine the method by which it is easier to disassemble amyloid fibrils: unzip the lateral associations or depolymerize protofilaments?

It is also a common feature of amyloid fibrils that in vivo they are associated with glucosaminoglycans and serum amyloid P component (SAP) (Pepys 2006). SAP is a highly conserved plasma glycoprotein of the pentraxin family that binds, in a calcium-dependent manner, specifically to all types of amyloid deposits (Pepys et al. 2002). The dynamic equilibrium of amyloid-associated SAP with the circulating plasma SAP plasma pool has been utilized in developing an isotope-based diagnostic tool for monitoring amyloid deposits in the human body (Hawkins et al. 1988). SAP apparently stabilizes amyloid fibrils and thereby it might facilitate amyloid fibrillogenesis. Pharmacological depletion of SAP is being seriously considered as a prospective therapeutic method in amyloidosis (Pepys et al. 2002). Although it has been shown that SAP protects amyloid fibrils from proteolysis (Tennent et al. 1995), the detailed mechanisms of how SAP might influence the stability and assembly dynamics of the fibrils are not known.

The numerous pressing questions about the details of amyloid fibrillogenesis call for experimental strategies that permit high-resolution assessment of fibril structure, dynamics, mechanics, and fluctuations. The most common experimental methodologies utilized in the investigation of amyloidogenic proteins are structural techniques: electron microscopy (EM), scanning transmission electron microscopy (STEM), atomic force microscopy (AFM), small-angle X-ray scattering (SAXS), X-ray crystallography, and NMR (Friedman 2011). Because amyloidogenic proteins are typically difficult to crystallize and amyloid fibrils cannot be investigated with solution NMR, high-resolution structure, and particularly intrafibrillar structural dynamics of amyloid assemblies, pose special problems for exploration. In recent years, single-molecule methodologies have been adapted in the investigation of amyloid fibrillar systems (Kellermayer et al. 2005; Karsai et al. 2005, 2006) and it has been suggested that they may provide unique insights into the structural dynamics and mechanics of amyloid fibrils.

In the past decade, partly due to instrumental development, emerging interesting problems particularly in biology, and ingenious experimental arrangements and technical solutions, the progress in the single-molecule field has been enormous (Mehta et al. 1997, 1998, 1999a, b; Nie and Zare 1997; Gimzewski and Joachim 1999; Moerner and Orrit 1999; Wang 1999; Weiss 1999; Bustamante et al. 2000a, b; Ishii and Yanagida 2000; Zlatanova et al. 2000, Noy, 2011 #3270; Ishii et al. 2001; Michalet and Weiss 2002; Muller et al. 2002; Best et al. 2003; Peterman et al. 2004; Kellermayer 2005; Garcia-Saez and Schwille 2007; Yanagida et al. 2008; Harriman and Leake 2011). The methods employed in the exploration of biomolecular systems have classically been standard biochemical and structural techniques, which examine a bulk sample that contains a large number of molecules.

These ensemble methods provide information about the *average* properties of the protein studied. Single-molecule methods, by contrast, provide an insight about the *distribution* of the properties across many individual molecules. Single-molecule methods have yielded unprecedented insights into the various aspects of protein behavior and function, and certain features, for example mechanical properties, can be explored only by these approaches. In single-molecule techniques, the measurements are carried out in such a way that the molecules are in an aqueous environment that more or less mimics the native, in situ, and ideally the physiological, in vivo conditions. Highly specialized and sophisticated methods are required to observe, grab, and handle individual protein molecules or their ligands. Proteins and ligands also come in many flavors and sizes. For example, an ATP molecule, the substrate of motor proteins and a wide range of enzymes, is a mere 500 Da, whereas the giant titin molecule is up to 3.7 million Daltons. Thus, different methods are required to either visualize or manipulate individual molecules and to follow their interactions with relevant ligands, and different types of information (structure, dynamics, mechanics) are obtained about the molecules in these investigations.

The knowledge obtained through single-molecule experiments has been fundamental, novel and far-reaching. Often the observations provided essential proofs of existing principles. In other instances, however, unprecedented insights were obtained. These include observations that would otherwise not be obtained in ensembles, such as stochastic fluctuations, individual trajectories of biological processes such as protein folding, elasticity of biomolecules, the mechanical work generated by motor proteins, and the identification of individuals in molecular crowds. Accordingly, single-molecule methods are particularly important in areas where temporal and spatial averaging is to be avoided, mechanical forces need to be measured, and individuals are to be tracked.

7.2 Single-Molecule Methods

Most of our knowledge in biology, chemistry, and physics is derived from ensembles of molecules with the inherent assumption that under identical conditions molecules of identical composition behave more or less the same way. Recent findings of single-molecule experiments suggest, however, that molecules of the same species may follow quite different paths during transition from one state to another in a given process. Stochastic events may also occur, which remain hidden in an ensemble. In complex interaction pathways, such as signaling, which are highly relevant in biology, there may be very little spatial and temporal synchrony, obscuring the details of the process. Thus, although distinctly different molecular states may be present, the observer recognizes only an ensemble average. Single-molecule methods, by contrast, aid in discovering, identifying, and describing temporally and spatially distinct states of a molecular species or of a complex process. These states may be related to structure (e.g., conformation, configuration), electronic states (e.g., ground, excited), and interactions (e.g., binding, dissociation). Furthermore, there are situations where single-molecule methods have no serious alternatives: the mechanics of a molecule (elasticity, force-generating transitions) can be inferred only from single-molecule experiments. One may thus identify four different areas in which single-molecule methods provide unique information not obtainable with other techniques (1) identification of temporally distinct states that may occur during stochastic processes [e.g., fluorophore blinking (Dickson et al. 1997)]; (2) identification of spatially distinct states that occur during processes that follow different pathways [e.g., protein folding (Zhuang and Rief 2003)]; (3) mechanics of molecules [e.g., elasticity (Bustamante et al. 2000a, b; Kellermayer and Grama 2002), motor proteins (Seidel and Dekker 2007; Yanagida et al. 2008)], and (4) tracking individuals in molecular crowds [e.g., travel of virus particles through the cytoplasm (Seisenberger et al. 2001)]. Single-molecule visualization and manipulation methods often require sophisticated technologies, frequent realignments, painstaking data acquisition, and analysis. However, due to the valuable insights obtained by these methods, they are becoming more widely available.

7.2.1 Atomic Force Microscopy Based Methods

The atomic force microscope (AFM) is a high-resolution scanning-probe device in which a sharp tip at the end of a flexible cantilever is scanned across a surface (Binnig et al. 1986). The atoms at the tip surface and sample surface interact in an attractive or repulsive manner, which deflects the cantilever. The miniscule motions of the cantilever are detected by directing a laser beam on the cantilever, which is reflected and projected onto a position-sensing photodiode (Fig. 7.2a). By monitoring the position of the laser beam, sub-angstrom cantilever deflections can be detected, and a surface topographical image is reconstructed. In the past decade, the AFM has been increasingly used in biological applications owing to improvements in sample preparation, development of different data acquisition modes (tapping or oscillation mode) and experimental geometries (e.g., force spectroscopy, see below), and the ability of functionalizing the AFM tip that interacts with the sample.

The AFM may be used not only for imaging topographical structure but also for manipulating biomolecules and cells. The mechanical manipulation of individual molecules adds new dimensions on the information obtained through visualizing them. By measuring forces, protein unfolding pathways, elastic properties, and intra- and intermolecular interactions can be characterized. The AFM belongs to the mechanical transducer type of molecular manipulators (Bustamante et al. 2000a, b). When mechanically manipulating with the AFM, the cantilever is typically moved vertically, away from the surface (Fig. 7.2b). If a molecule is captured by the cantilever tip, then the cantilever movement results in the generation of force within the molecule or within the biomolecular system it is coupled to. In the recent decade, an entire experimental methodology, called single-molecule

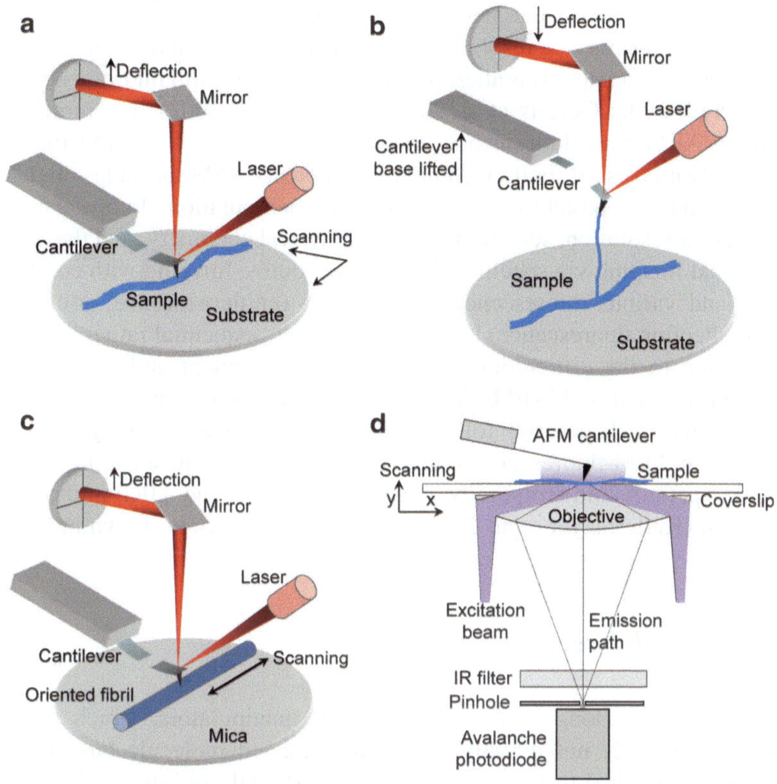

Fig. 7.2 Single-molecule imaging and manipulation methods based on scanning force microscopy. (**a**) Instrumental arrangement of the atomic force microscope (AFM). Scanning direction is indicated with *arrows*. Either the AFM tip or the sample is scanned relative to each other. (**b**) Single-molecule manipulation using AFM. In this force-mode AFM, the cantilever (or, equivalently, the sample) is moved in the vertical direction. Force generated in the manipulated molecular strand is calculated from cantilever deflection multiplied with cantilever stiffness. The end-to-end distance of the strand is obtained from the cantilever base displacement corrected with cantilever bending. (**c**) Scanning force kymography. Scanning occurs along one axis only, while a time-dependent kymogram image is obtained during repetitive scanning along the axis of the investigated fibril. (**d**) Combined AFM and total internal reflection fluorescence (TIRF) microscopy. The sample is simultaneously investigated with AFM and an objective-based TIRF microscope

force spectroscopy, has evolved based on this method (Rief et al. 1997a, b; Rief and Grubmuller 2002). Force spectroscopy requires proper calibration of cantilever stiffness and precise measurements of both the cantilever displacement and deflection. Recently, force-clamp methods have evolved that enable to keep the biomolecular system under constant mechanical load (Fernandez and Li 2004).

A special adaptation of AFM is scanning force kymography (Fig. 7.2c) (Kellermayer et al. 2008). In scanning force kymography, the topographical scanning occurs in one dimension only. Subsequent scans are assembled into a kymogram, which contains temporal information along its vertical *y*-axis. Spatial detail is resolved along the horizontal x-axis. In scanning force kymography, positional changes may be resolved with resolution unaffected by tip convolution.

There have been numerous attempts to combine AFM with fluorescence imaging (Mathur et al. 2000; Gorelik et al. 2002; Nishida et al. 2002). The motivation for combining AFM with fluorescence arises from the desire to exploit the complementary advantages of both techniques. Whereas AFM provides high spatial resolution and the possibility of mechanical manipulation, fluorescence offers high temporal resolution, sensitivity to local physical chemistry, and the possibility of functional imaging via specific labeling procedures. Imaging with combinations of AFM and various fluorescence microscopies (epifluorescence, confocal, total internal reflection fluorescence) has most often been sequential rather than simultaneous, owing to the difficulties in geometric arrangement and grossly different acquisition timescales. TIRFM (Axelrod 2003), due to the narrow excitation field and its ability to detect and follow individual fluorophores, may offer special advantages in AFM-fluorescence combinations. By utilizing scanning-stage AFM geometry, spatially and temporally synchronized AFM and TIRF microscopy can be achieved (Kellermayer et al. 2006), a diagram of which is shown in Fig. 7.2d.

7.2.2 *Optical Tweezers*

Optical tweezers belong to the external field manipulators, which act from a distance and usually not directly on the investigated molecule but on a handle (e.g., a bead) attached to the molecule's end. External field manipulators include optical tweezers, flow fields, and magnetic tweezers (Bustamante et al. 2000a, b). The basis of optical tweezers (synonyms: laser tweezers, optical trap, laser trap) is momentum exchange between photons in a focused beam and a refractile microscopic bead (Ashkin 1980; Ashkin et al. 1987, 1990; Svoboda and Block 1994). In a balance of the various optical forces, most notably the scattering and gradient forces, the bead is captured in an equilibrium position. Any displacement from this position, by external forces, for example, results in a force imbalance that tends to reestablish equilibrium. As a result, the optical trap acts as a virtual spring and can be used as a force transducer in the piconewton force range. The bead captured by the optical trap functions as a handle to which molecules can be attached with different conjugation chemistries. Different optical tweezer geometries have been designed according to the features of the molecule and the type of the experiment. An example for stretching filamentous biomolecules with optical tweezers is shown in Fig. 7.3. The position of the optical trap can be modulated with high speed using galvano mirrors and acousto-optic modulators, with the help of which single-molecule force feedback systems have been implemented (Svoboda and Block 1994). Although optical tweezers have become an important tool in the investigation of various filamentous biomolecular systems and molecular motors (Nishizaka et al. 1995; Mehta et al. 1998; Bustamante et al. 2000a, b; Kellermayer et al. 2001; Pope et al. 2002; Kimura and Bianco 2006; Bianco et al. 2007; Neuman and Nagy

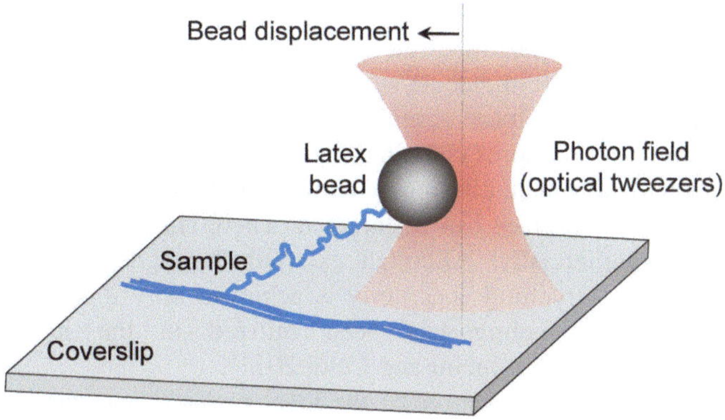

Fig. 7.3 Mechanical manipulation with optical tweezers. One end of the fibrillar sample is attached to a bead, while the other one is attached to the substrate surface. Force is calculated typically from the displacement of the captured bead from the trap center, multiplied by the virtual stiffness of the optical trap

2008; Neuman 2010), their use in the exploration of amyloidogenic protein behavior has been limited, and reports on their applications have been made only recently (Dong et al. 2010; Castro et al. 2011).

7.2.3 *Fluorescence Methods*

The basis of fluorescence is the relaxation of a suitable molecular system from an excited singlet electronic state to the ground state accompanied by radiation (Lakowicz 2006). Individual molecules are excited by the use of different microscope illumination arrangements. The visualization of individual fluorophores became possible due to the development of highly sensitive detectors. Initially, fluorescence spectroscopy of single molecules was carried out in the condensed state (Orrit and Bernard 1990; Ambrose and Moerner 1991). Subsequently, individual immobilized fluorophores were visualized at room temperature (Betzig and Chichester 1993). Currently, there is a multitude of various single-molecule fluorescence techniques available, which permit the exploration of protein structure and dynamics with an ever-increasing detail. Often, the driving force behind novel technologies is the desire to fill gaps in the spatial and temporal resolution scales. In recent years, fluorescence microscopic methods have reached levels of sophistication in the form of superresolution methods in which nanometer resolution has been possible within complex biomolecular and cellular environments (Huang et al. 2009).

The fluorescence of proteins may originate from intrinsic fluorophores (such as tryptophan), but most commonly external fluorophores attached via chemical or genetic means to specific sites are used in single-molecule fluorescence. Intrinsic tryptophan fluorescence has been excessively exploited in the exploration of the structure and dynamics of proteins in ensemble spectroscopic methods (Lakowicz 2006). However, owing to its low extinction coefficient and quantum efficiency it is not ideally suited for single-molecule studies (Kapanidis and Weiss 2002). Extrinsic fluorescent labels are a more common choice in single-molecule fluorescence methods. Extrinsic fluorophores contain a fluorescent dye conjugated with a chemical linker group via which different chemical groups (e.g., sulfhydryl, amino groups) on the protein surface can be targeted. There is a wide range of different dyes available commercially. Although external fluorophores provide relative chemical specificity, structural specificity is achieved only if the binding site is unique. Site-specific labeling has been utilized in the investigation of amyloidogenic proteins (Bertoncini and Celej 2011).

Because fluorescence is sensitive to the local physical and chemical environment, and therefore structural and dynamical informational may be obtained about the investigated biomolecular system, there has been a great effort to search for suitable fluorophores for use in the exploration of amyloidogenic proteins (Bertoncini and Celej 2011). Thioflavin T (ThT) has for long been recognized as a fluorophore specific for amyloid structure (LeVine 1993; Biancalana et al. 2009; Biancalana and Koide 2010). ThT labeling has been used for following the polymerization of individual β_2-microglobulin and amyloid β fibrils (Ban et al. 2003, 2004). More recently, Alexa Fluor 488 and 647 were used to label an N-terminal cystein mutant of SH3 domain from bovine phosphatidylinositol-3-kinase (PI3–SH3), which has been shown to form amyloidogenic oligomers at low pH (Orte et al. 2008). Oligomerization was followed in a novel approach called two-color coincidence detection (TCCD) (Orte et al. 2008, 2010, 2011). The sample is illuminated by two laser beams of the respective excitation wavelengths brought to a common focal point. In a mixture of PI3-SH3 labeled with either of the fluorophores, the coincidence of high-intensity bursts indicated the appearance of amyloidogenic oligomers (Orte et al. 2008).

Commonly used fluorescence microscopic excitation geometries include confocal (Nie et al. 1994) and TIRFM (Axelrod et al. 1984; Axelrod 1989, 2003; Paige et al. 2001). In the confocal microscope a laser beam, with preselected spectral properties, is focused by an objective lens on the sample (Webb 1999). The diffraction-limited focal spot defines the excitation volume, which is on the order of femtoliters (10^{-15} L). The emitted fluorescence is collected by the objective and projected, via a pinhole positioned in the conjugate focal plane, onto the detector. The pinhole serves to avoid the collection of photons generated at locations different from the focal plane. By following the fluctuation of fluorescence intensity due to the diffusion in and out of the focal volume, the diffusive properties, hence the size and associations of molecules, may be investigated in a method known as fluorescence correlation spectroscopy (FCS) (Schwille et al. 1999, 2000; Van Craenenbroeck and Engelborghs 2000). FCS has been used for following the aggregation of amyloidogenic proteins (Tjernberg et al. 1999). The two-color coincidence detection (TCCD) technique (Orte et al. 2008) essentially uses the FCS geometry and data acquisition method. TIRFM has become a widely used method in recent years (see Fig. 7.2d) (Axelrod et al. 1984; Axelrod 1989, 2003;

Paige et al. 2001). The basis of TIRFM is the formation of an excitation field, called evanescent field. An evanescent field is formed in the lower-index-of-refraction phase at a boundary between two phases of different refraction indices, provided that the light beam arrives at this boundary at an angle greater than the critical angle. The intensity of the evanescent field decays exponentially with the distance from the boundary according to

$$I_z = I_0 e^{-z/d}, \tag{7.1}$$

where I_0 is the intensity at the boundary, I_z is the intensity measured at a distance z from the boundary, and d is the decay length that depends on the relative refractive indices (n_1, n_2) and the angle of incidence (θ) and scales with the wavelength (λ) of the light used as

$$d = \frac{\lambda}{4\pi}\left[n_1^2 \sin^2\theta - n_2^2\right]^{-\frac{1}{2}}. \tag{7.2}$$

Recently, TIRFM has been used to follow the assembly of individual amyloid fibrils in real time by using thioflavin-T fluorescence (Ban et al. 2003, 2004).

7.3 Measured Variables

The data obtained in single-molecule manipulation experiments are force and length, whereas in fluorescence methods they are related to either the direct or derived parameters of the emitted photons. Force is generated in the molecule upon mechanical perturbation, and length is the distance between the points of molecular attachment along which the force acts. In the usual experimental arrangements, neither force nor length are obtained directly. Force is obtained by the force probe (cantilever or trapped bead) located at one end of the molecule after proper calibration procedures. Length is obtained by measuring with high precision the displacement of the end of the molecule followed by proper corrections according to the behavior (deflection depending on probe stiffness and applied force) of the force probe.

7.3.1 *Stiffness Calibration*

Single-molecule manipulation devices function as tensiometers that measure tiny forces in the femto- to nanonewton range. In order to obtain the forces, either the stiffness of the probe must be known or a direct force measurement must be carried out. Stiffness calibration procedures rely on either imposing known forces on the probe or measuring its thermal fluctuations. Direct force measurement, developed recently for measuring forces with optical tweezers, utilizes first principles (principle of conservation of momentum).

Calibration with added weights. Known forces may be imposed on the probe in several ways: preadjusted hydrodynamic drag forces, known weights, and cross-calibration against a probe with known stiffness. Preadjusted drag forces are generated according to Stokes' law and are often used to calibrate optical and magnetic tweezers and to impose known forces in flow-field manipulators (Bustamante et al. 2000a, b). In this calibration scheme, a fluid flow of preadjusted rate (v) is generated, and the corresponding bead displacement is measured.

Thermal calibration. The thermal calibration method has become the most widely used calibration scheme in AFM and single-molecule force spectroscopy. The method employs the forces that are constantly applied to the probe (either the cantilever or a trapped bead) in the form of thermal agitation. The basis of the thermal calibration method (Hutter and Bechhoefer 1993) is the equipartition theorem according to which the average potential energy of the probe is equivalent to the imparted thermal energy:

$$\frac{1}{2}\kappa\langle x^2\rangle = \frac{1}{2}k_B T, \tag{7.3}$$

where κ is stiffness, x is the amplitude of displacement, k_B is Boltzmann's constant, and T is the absolute temperature. Fluctuations are dissipated unevenly across frequencies, particularly in the case of resonating objects such as cantilevers. In the case of cantilevers, stiffness can be obtained by fitting a simple harmonic oscillation equation on the fundamental vibrational mode in the power spectral density function of the thermally excited probe (Wang and Uhlenbeck 1945). In case of overdamped systems, such as a microscopic bead in optical tweezers, stiffness can be obtained as

$$\kappa = \omega_c \gamma, \tag{7.4}$$

where ω_c is the corner frequency above which the system does not respond to external stimuli, and γ is the drag coefficient. In either case, the probe displacement needs to be sampled with high bandwidth.

Photon-flux force sensing. The force acting on a bead trapped in the optical tweezers may be directly obtained based on first principles (Smith et al. 1996, 2003). For a beam of light of power W in a medium of refractive index n, the momentum flux (dP/dt) carried by the beam is

$$\frac{dP}{dt} = \frac{nW}{c}, \tag{7.5}$$

where c is the speed of light. If a particle, such as a bead trapped in the optical tweezers, scatters that light in a new direction, then by conservation of momentum the reaction force on the particle is given by

$$F = \frac{\mathrm{d}(P_{\mathrm{in}} - P_{\mathrm{out}})}{\mathrm{d}t}. \tag{7.6}$$

The force exerted on the particle may be obtained as the integral of radiant intensity across all directions. Thus, if all the photons exiting the trap are collected, then the radiant force exerted on the particle may be directly calculated. The photons exiting the trap are collected with a microscope objective and are projected onto a position-sensing device (dual-axis lateral-effect position detector) in such a way that the detector is conjugate with the objective back aperture. In order to collect photons in a relatively wide range of scattering angles, a low incident angle (hence low numerical aperture) optical configuration is used in a dual-beam, counter-propagating arrangement. Force can then be directly calculated based on the intensity-weighted deflection signal (Σ), the detector's responsivity (Ψ), and the ratio of the effective detector (R_{D}) and objective-lens (R_{L}) radii (Smith et al. 2003):

$$F = \frac{\Sigma}{c\Psi} \frac{R_{\mathrm{D}}}{R_{\mathrm{L}}}. \tag{7.7}$$

7.3.2 *Force Measurement Modalities*

Once the molecule under investigation is mounted, it can be mechanically perturbed either by the action of the investigator or by another molecule (e.g., motor protein). The force data, in the case of chain molecules, are typically compared with the predictions of elasticity models. Deviations from the model usually reflect structural transitions that are subject to further analysis and modeling.

Elasticity measurements. To characterize the elasticity of a protein chain, it is necessary to grab each of its ends. The chain is extended by moving one end away from the other, and the force, measured either at the same or opposite end of the chain, is used to characterize the mechanical behavior. In the case of long protein chains, the force versus extension (end-to-end distance) relationship most often displays a nonlinear curve, indicating that the elasticity mechanisms are primarily entropic. An entropic polymer shortens because of thermally driven bending movements that increase the chain's conformational entropy. Conformational entropy can be reduced by stretching the chain, i.e., separating its ends, which requires external force. As the end-to-end distance approaches the chain's contour length, force asymptotically rises. Two models are commonly used to describe the entropic elasticity of chain molecules, the wormlike chain [WLC (Bustamante et al. 1994)] and freely jointed chain [FJC (Smith et al. 1992)] models. The WLC model treats the polymer as a bendable or deformable rod, the bending rigidity of which is described in terms of persistence length, which is a distance along which the thermally driven bending motions (chain orientations) are correlated. The longer the persistence length, the greater the bending rigidity, but smaller is the force required to stretch the chain to a given fractional extension. The WLC equation describes the relationship between stretch force (F), persistence length (A), and fractional extension (z/L, where z is the end-to-end distance and L is the contour length) (Bustamante et al. 1994; Marko and Siggia 1995):

$$\frac{FA}{k_B T} = \frac{z}{L} + \frac{1}{4(1 - z/L)^2} - \frac{1}{4}. \tag{7.8}$$

The FJC model describes the polymer chain as infinitely stiff segments interconnected with frictionless joints or hinges. The average length of the segments is the Kuhn segment length, which, similarly to the persistence length, is related to the bending rigidity of the polymer chain. The FJC equation is the following (Smith et al. 1992):

$$\langle x \rangle = L\left(\coth\frac{Fb}{k_B T} - \frac{k_B T}{Fb}\right), \tag{7.9}$$

where $<x>$ is the average end-to-end distance, and b is the Kuhn-segment length. Based on theoretical considerations, the two models are interrelated and the Kuhn segment length is equal to two times the persistence length (Doi and Edwards 1986).

In the force curve of mechanically manipulated proteins, deviations from elastic behavior can often be observed. The deviations from predicted elasticity are caused by intramolecular structural arrangements, such as unfolding. In multidomain proteins [e.g., titin (Kellermayer et al. 1997; Rief et al. 1997a, b; Tskhovrebova et al. 1997)], individual domain unfolding events may be resolved depending on the method used. Domain unfolding events are manifested in the appearance of a force sawtooth pattern. Whether an individual force sawtooth can be resolved depends on the stiffness of the probe (the greater the stiffness, the greater the resolving power) and the time resolution of the method. In between the transitions, the molecule behaves elastically. This transient elastic behavior can be identified by fitting the WLC equation to local, nonlinear regions of the force data.

Force spectroscopy. Force spectroscopy entails the analysis of transition forces that correspond to mechanically induced rupture of intra- or intermolecular interactions (Rief and Grubmuller 2002). The interactions may be, for example, bonds that hold the native structure of a protein molecule together (Rief et al. 1997a, b), antigen/antibody bonds (Stout 2001), and ligand/binding site interactions [e.g., biotin/streptravidin (Florin et al. 1994; Bianco et al. 2007)]. AFM is commonly used to obtain a high-resolution force response of the protein system. In a long, multidomain protein molecule such as titin, the repetitive domain unfolding events result in a complex function called *force spectrum*. When the interaction between or within molecules is studied, an external load (force) is placed on the bond(s). In a theoretical work, Bell has shown that mechanical force placed on adhesion bonds influences the bond lifetime (Bell 1978). In a single-molecule experiment, the load is increased as a function of time (t) by pulling the molecules apart, which eventually results in bond rupture. The loading rate (r) can be expressed as

$$r = \frac{F}{t} = \kappa v_s, \tag{7.10}$$

where k is the stiffness (combined stiffness of the probe and the molecular system, Nm^{-1}), and v_s is the stretch rate (ms^{-1}). The variables experimentally measured are the rupture (or unbinding) force [force at which unbinding occurs when the bond is loaded with a force ramp (Evans and Ritchie 1997; Strunz et al. 2000)] and the loading rate (obtained from the time-dependent force trace just prior to the unbinding event, considering that the force trace might not be linear throughout). The unbinding force is directly related to the unbinding or dissociation kinetics of the molecular complex under load, and therefore depends on the loading rate (Evans and Ritchie 1997). In theory, it is possible to determine the thermal off-rate (hence lifetime) of a bond from the relationship of unbinding force and loading rate. The method has been called dynamic force spectroscopy (Evans and Ritchie 1997). The rate of bond dissociation, driven by thermal activation (n_0), can be expressed as

$$v_0 = \omega \mathrm{e}^{-\Delta E/k_\mathrm{B}T}, \tag{7.11}$$

where ΔE is the activation energy barrier, $k_\mathrm{B}T$ is the thermal energy, and ω is an attempt frequency set by Brownian dynamics (bond vibration frequency). Conceptually, external load tilts the potential energy landscape (Evans 2001), lowers the energy barrier, and facilitates bond rupture. The rate of bond rupture under external force (n_F) can be expressed as

$$v_\mathrm{F} = \omega \mathrm{e}^{-(\Delta E - F\Delta x)/k_\mathrm{B}T} = v_0 \mathrm{e}^{F\Delta x/k_\mathrm{B}T}, \tag{7.12}$$

where F is the force acting along the reaction coordinate, and Δx is a characteristic distance (along the reaction coordinate) associated with the transition state for bond breaking (width of the energy barrier). Thus, the activation energy barrier is discounted by the mechanical potential ($F\Delta x$) and leads to an exponential amplification of the dissociation kinetics. Bond rupture is a stochastic process, whereby, for a given loading rate (r), a distribution of unbinding forces is observed around a peak unbinding force (F^*). For a bond with a single activation energy barrier, F^* varies linearly with the natural logarithm of the loading rate, with a slope of $k_\mathrm{B}T/\Delta x$, which is expressed as

$$F^* = \frac{k_\mathrm{B}T}{\Delta x} \ln \frac{r}{v_0(k_\mathrm{B}T/\Delta x)}, \tag{7.13}$$

where n_0 is the thermally activated unbinding rate (Strunz et al. 2000). Thus, by measuring the slope in the peak unbinding force versus the loading rate diagram, one may establish the width of the unbinding potential. In the case of multiple linear regimes (Merkel et al. 1999), it has been argued that intermediates of the dissociation may be identified, and the potential energy landscape of the complex (e.g., receptor-ligand) may be constructed.

Dynamic force spectroscopy has been very successfully applied to studying the forced unfolding and mechanical stability of proteins (Rief et al. 1997a, b, 1998a, b; Fisher et al. 1999; Carrion-Vazquez et al. 2000). The notion that single-molecule mechanics can be used in a unique way to understand protein folding has been a driving force in the development and standardization of dynamic force spectroscopy. Criteria for proper experimentation and data analysis are emerging (Best et al. 2003). Such criteria include the following (1) Experiments should ideally be carried out on engineered polyproteins that contain a tandem array of identical domains. (2) The force trace should contain at least three clearly resolved peaks with an even distribution of peak separation, where the peak separation distance can be correlated with the molecular dimensions of the unfolded domain (contour length). (3) The interpeak force should not drop to zero, because unless the chain breaks, there is a residual entropic force present. (4) The force trace should correspond to the mechanical behavior of a single protein chain (see below). Force spectra containing multiple unevenly spaced force peaks should not be considered. Although in most of the force spectroscopy experiments on amyloidogenic protein systems fibrils were directly manipulated, in some cases engineered polyprotein systems were investigated (Sandal et al. 2008; Sorce et al. 2009).

7.3.3 Kymography

In single amyloid fibril kymography, the detected dataset is the kymogram (Kellermayer et al. 2008). In scanning force kymography, the cantilever tip is scanned repetitively along the axis of an amyloid fibril that grows in a direction maintained for a long distance (Fig. 7.2c). The fibril-end position is detected as a sudden change in topographical height. Precision is limited by pixel resolution (typically around 1 nm/pixel) and is unaffected by tip convolution. Consecutive linescans, separated by a period dictated by the scanning frequency (typically a few Hz in commercial AFM instruments), are assembled into a scanning force kymogram (see Fig. 7.7). Kymograms may be converted to datapoints using custom-written algorithms and further analyzed with commercial software (Kellermayer et al. 2008).

7.3.4 Fluorescence

The signal in single-molecule fluorescence microscopy arises from the photons emitted by the fluorophore, which are then segregated according to various parameters such as intensity, wavelength, polarization, and arrival time. Depending on the type of detector used, high spatial (2D detectors: back-thinned CCD cameras, microchannel-plate intensified CCD cameras) or temporal (point detectors: avalanche photodiode, photomultiplier tubes) resolution is obtained.

Determinants of signal quality. Various fluorophore- and instrument-dependent factors influence the quality of the detected signal. Fluorophore-dependent parameters include brightness, fluorescence lifetime, blinking, and photochemical resistance (Kapanidis and Weiss 2002). Instrumentation-dependent parameters are manifested in the signal-to-background (SBR) and signal-to-noise (SNR) ratios (Michalet and Weiss 2002). The brightness of the fluorophore is determined by its extinction coefficient at the wavelength of excitation (e_{exc}) and the quantum yield at the emission wavelength (Q_{F}), which ideally should exceed 20,000 $\mathrm{cm}^{-1}\,\mathrm{M}^{-1}$ and 0.1, respectively (Kapanidis and Weiss 2002). Considering that the number of photons by the fluorophore is determined by the rate of fluorescence, to maximize the emitted photon number fluorophores with relatively short lifetimes (0.5–5 ns) are preferred. Fluorophore blinking refers to the intermittency in the emission intensity, which is due to the adoption of long-lived dark states (Dickson et al. 1997). Blinking may interfere with the observation of relevant biomolecular phenomena. Photochemical destruction or photobleaching is an irreversible chemical process that results in the loss of fluorescence by the fluorophore. Fluorophores used in single-molecule fluorescence measurement typically emit up to 10^6 photons on average prior to photobleaching (Michalet and Weiss 2002). Photobleaching is a well-defined one-step loss in emission intensity, and therefore it is often used as a criterion of singleness.

Measurement of position and distance. Being a diffraction-limited technique, the spatial resolution of light microscopy is fundamentally limited by the Rayleigh criterion (Slayter and Slayter 1992). Accordingly, spatial resolution is fundamentally determined by wavelength, which sets the minimal resolved distance on the order of a couple of hundred nanometers for visible light. A significant driving force behind the development of special imaging techniques has been the long-standing desire to somehow get around the limit set by the Rayleigh criterion. Several methods have recently been developed that provide spatial information at the small scale. There are methods that obtain high positioning accuracy by fitting a Gaussian distribution function on the intensity profile of the fluorophore. A precision of 1.5-nm has been achieved by this method (Yildiz et al. 2003; Snyder et al. 2004). The method is appropriately named "fluorescence imaging with one nanometer accuracy" (FIONA). Although the method provides high positioning accuracy, resolution is not increased, because closely spaced point objects cannot be distinguished. In order to increase resolution, either the point-spread function (PSF) needs to be reduced (Klar et al. 2000), or the overlapping images of the point emitters need to be separated based on the spectral or dynamic properties of the fluorophores. Accordingly, resolution can be increased if the two image points are separated based on their color (~10 nm resolution) (Lacoste et al. 2000). In addition, by taking advantage of the quantal nature of the photobleaching process and carrying out the PSF fitting prior to and following the photobleaching step, resolution may be increased further (to ~5 nm) (Gordon et al. 2004; Qu et al. 2004).

Certain spectroscopic methods provide very high spatial resolution. In single-pair fluorescence resonance energy transfer (spFRET) (Ha et al. 1996, 1999; Deniz et al. 1999), the excited state of a single fluorophore (donor) relaxes via dipole–dipole interaction with another (acceptor) (Lakowicz 2006). The efficiency of the transfer is a very sensitive function of the distance between the two fluorophores in the 2–10 nm range. The FRET efficiency (*E*) is obtained as

$$E = \frac{1}{1 + (R/R_0)^6}, \tag{7.14}$$

where R is the distance between the donor and acceptor fluorophores and R_0 is the so-called Förster radius, which is the distance at which $E = 0.5$. In experimental practice, E can be obtained as

$$E = [1 + \gamma I_d/I_a]^{-1}, \tag{7.15}$$

where I_d and I_a are the donor and acceptor intensities during donor excitation, respectively, and γ is a correction factor that depends on the donor and acceptor quantum yields and the detector efficiencies (Ha et al. 1996). FRET methods have been utilized for following the aggregation of amyloidogenic proteins and their interaction with various ligands (Bertoncini and Celej 2011).

7.4 Examples of Amyloidogenic Protein Systems

Several amyloidogenic protein systems have been investigated by using single-molecule biophysical tools. In the following, a number of these are introduced, focusing primarily on the nanomechanical and force-spectroscopy investigations.

7.4.1 Amyloid β1–42

Alzheimer's amyloid beta fibrils (Aβ-fibrils) are formed from amyloid beta peptides (Aβ-peptides), which are 39- to 43-residue-long enzymatic cleavage products of the amyloid precursor protein (APP) (Serpell 2000). The Amyloid β1–42 peptide (Aβ1–42) is considered to be essentially the full-length Aβ, which is the most neurotoxic type of Aβ (Ling et al. 2003). Although several short fragments of the Aβ peptide have been shown to form filamentous aggregates (Serpell 2000), Aβ1–42 is thought to be primarily responsible for the clinical forms of Alzheimer's disease (Selkoe 2001; Hardy and Selkoe 2002; Adlard and Cummings 2004; Cummings 2004).

Aβ1–42 fibrils appear as gently curved, 4.8 nm wide filamentous structures, when adsorbed to the mica surface (Fig. 7.4a) (Karsai et al. 2006). Helical structure with periodicity ranging between 25 and 110 nm can often be discerned. Force versus extension curves obtained in single-molecule manipulation experiments using AFM are dominated by force plateaus (Fig. 7.4b). A force plateau is characterized by constant force as a function of extension, followed by a force step. Most frequently, several force steps can be observed in the recorded force spectra. Sometimes nonlinear force versus extension curves are observed, onto which transitions with sawtooth shape are superimposed (Fig. 7.4c). The distance between consecutive force peaks typically exceeds the unfolded contour length of the unfolded Aβ1–42 peptide (~16 nm). The plateau force step histogram (Fig. 7.4d) shows multimodal distribution with maxima at ~23, ~45, ~70, ~90, ~110, and ~130 pN. The minimal force is apparently a unit force necessary to evoke the intrafibrillar structural transitions that are manifested in the force plateau. The most plausible explanation for the dominant force plateaus is that filamentous entities most likely corresponding to subunit sheets can be unzipped from the fibril with constant forces in an equilibrium process, indicating that during mechanical relaxation the subunit sheets rapidly rebind to the fibril surface.

7.4.2 Amyloid β1–40

The amyloid β1–40 (Aβ1–40) peptide is also a frequent component of amyloid plaques, and it has become one of the most studied model peptides in amyloid investigation. Site-directed spin labeling (Török et al. 2002) and particularly NMR experiments (Petkova et al. 2002; Tycko 2004; Lührs et al. 2005) have formed the basis of a high-resolution model of the Aβ1–40 fibril: β-hairpins lying perpendicularly to the fibril axis are associated into β-sheets that line up to form protofilaments, which are then assembled in parallel into fibrils. Protofilaments are thus thought to represent a ~2–3-nm-diameter structural unit within the amyloid fibril (Serpell et al. 2000).

Mature Aβ1–40 fibrils appear on AFM images as gently curved filaments often displaying a beaded topographical structure (Fig. 7.5a). In the force spectra of mechanically manipulated Aβ1–40 fibrils, two fundamental types of responses can be observed (Kellermayer et al. 2005): force plateau characterized with a constant force level during stretch and reversible, nonlinear elasticity. Most frequently, the two fundamental processes are combined into complex, hierarchical mechanical responses: force plateaus superimposed onto one another (Fig. 7.5b) or onto a nonlinear force curve (Fig. 7.5c). Superimposed force plateaus usually end with a series of decreasing force steps resulting in a descending staircase pattern, similar to the behavior of Aβ1–42 fibrils.

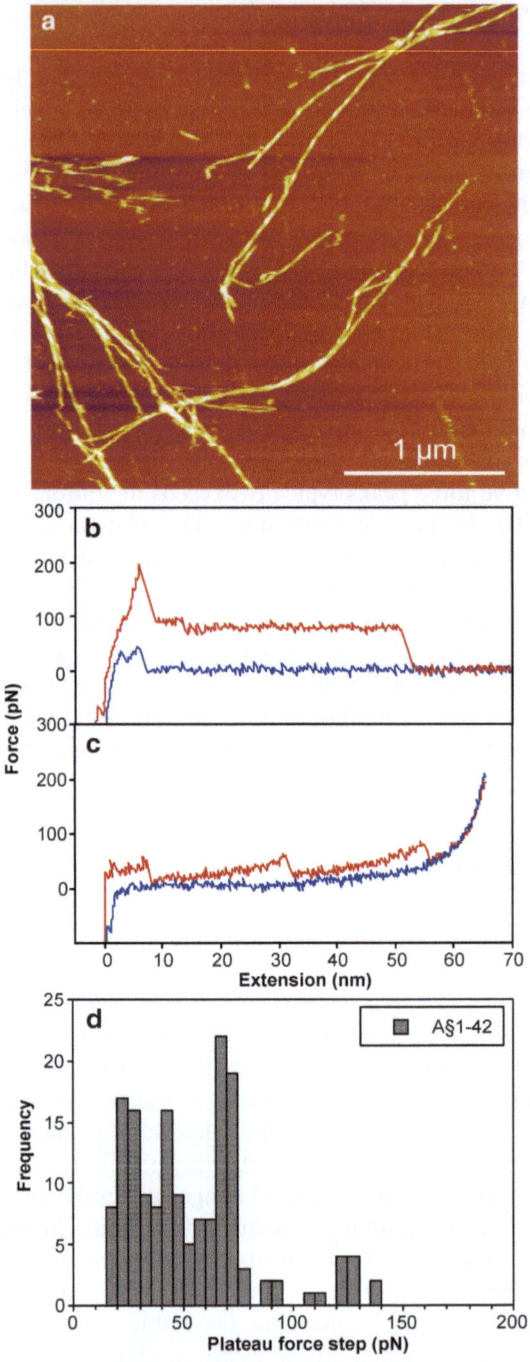

Fig. 7.4 Imaging and manipulation of Aβ1–42 fibrils. (**a**) AFM image of Aβ1–42 fibrils attached to mica surface. Fibrils with helical structure are observed. (**b**) Example of a force versus extension *curve* measured on an Aβ1–42 fibril displaying a force plateau. *Red* and *blue* datapoints correspond to the departure and retraction of the AFM cantilever from and to the sample, respectively. (**c**) Example of a force versus extension *curve* showing nonlinear elasticity on which sawtooth-shaped transitions are superimposed. (**d**) Distribution of plateau force steps. Multimodal histogram is observed with peaks at approximate positions of ~23, ~45, ~70, ~90, ~110, and ~130 pN

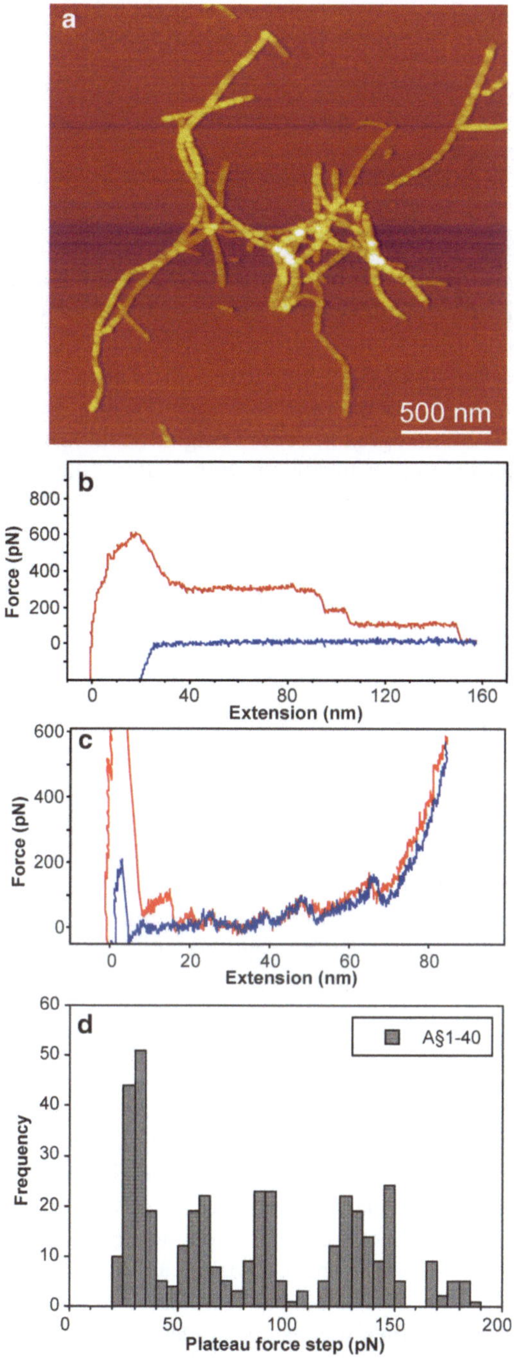

Fig. 7.5 Imaging and manipulation of Aβ1–40 fibrils. (**a**) AFM image of Aβ1–40 fibrils attached to mica surface. Gently curved fibrils with beaded structure are observed. (**b**) Example of a force versus extension *curve* measured on an Aβ1–40 fibril displaying force plateaus interrupted with steps. *Red* and *blue* datapoints correspond to the departure and retraction of the AFM cantilever from and to the sample, respectively. (**c**) Example of a force versus extension *curve* showing nonlinear elasticity on which sawtooth-shaped transitions are superimposed. Notably, the force transitions are reversible indicated by the superposition of the stretch and relaxation *data curves*. (**d**) Distribution of plateau force steps. Multimodal histogram is observed with peaks at approximate positions of ~30, ~60,~90, ~120, ~150, and ~180 pN

The plateau force step histograms display multimodal distribution (Fig. 7.5d). The peaks correspond to integer multiples of the smallest plateau force, 30 pN. The force plateaus correspond, just as in the case of Aβ1–42 fibrils, to the unzipping of peptide subunit sheets from the underlying fibril surface. The plateau transition is often reversible and the force staircases are repeatable. Often the complex pattern of combined staircases and nonlinear force curve is entirely reversible (Fig. 7.5c), persisting through successive stretch and relaxation cycles. The reversible rebinding of subunit-sheets at high loads and loading rates to the underlying fibril surface indicates that the associated state is strongly favored, and a mechanically perturbed amyloid fibril is rapidly recovered by zipping together these sheets. This or a similar process may be important in the structural rearrangements that are thought to occur during the final formation of the amyloid fibril. Notably, oligomeric, prefibrillar intermediates, found in fresh samples of the Aβ1–40 peptide, do not display force plateaus in their force spectra. Rather, repetitive force sawtooth transitions can often be observed (see Fig. 7.8b inset). Conceivably, the observed mechanical behavior describes most likely the capture, unfolding, and release of oligomers of the Aβ1–40 peptide.

7.4.3 Amyloid β25–35

Aβ25–35 is a fragment of the full-length Aβ-peptide comprising a stretch of residues from position 25 to 35. Aβ25–35 has been thought to be the biologically active fragment of Aβ, considering that it forms fibrils and retains the toxicity of the full-length peptide (Forloni et al. 1993; Pike et al. 1995; Yang et al. 1999). It is thought that the basic features of the Aβ25–35 fibril are similar to those formed from other peptides. Thus, adjacent peptides are connected to each other via hydrogen bonds along the axis of the protofilament so as to form a β-sheet, and amino acids side chains lie perpendicular to the axis.

In AFM images, Aβ25–35 fibrils grown in solution for a few days appear as straight, slightly curved filaments with smooth surface (Fig. 7.6a). The fibrils often form bundles. In force spectroscopy experiments, the appearance of force plateaus is particularly pronounced. Single force plateaus (Fig. 7.6b), force steps (Fig. 7.6c), and complex force staircases (Fig. 7.6d) can be observed (Kellermayer et al. 2005). The plateau force step appears multimodal, just as in the case of Aβ1–42 and Aβ1–40 fibrils. The fundamental unitary force, however, is greater in the case of Aβ25–35 fibrils. The histogram peaks appear at integer multiples of ~45 pN. Considering that the role of mechanical force is to rupture the interactions that hold the subunit fibrils together, the results indicate that intrafibrillar interactions in Aβ25–35 fibrils are stronger than in the case of either Aβ1–42 or Aβ1–40 fibrils.

Single-molecule approaches allow the investigation of the interaction of individual fibrils with substrate surfaces as well. It has been shown that charged surfaces interact with Aβ25–35 (Terzi et al. 1994; del Mar Martinez-Senac et al. 1999), suggesting that its binding and, possibly, orientation may be controlled by electrostatic mechanisms.

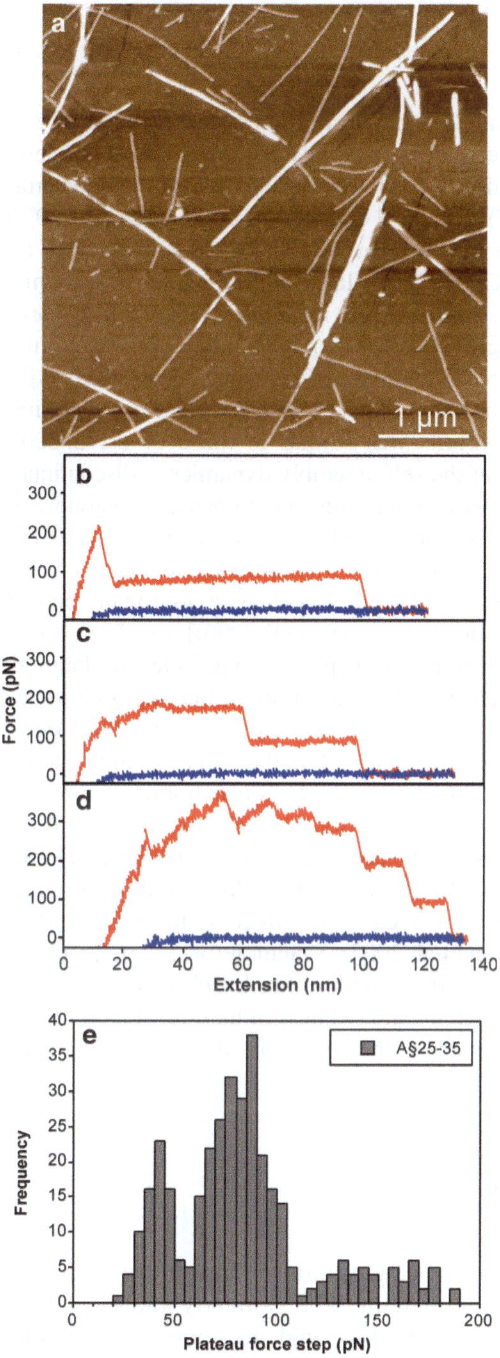

Fig. 7.6 Imaging and manipulation of Aβ25–35 fibrils. (**a**) AFM image of Aβ25–35 fibrils grown in solution for 4 days attached to silanized glass surface. Straight fibrils and bundles are observed with mostly smooth surface. (**b**) Example of a force versus extension *curve* measured on an Aβ25–35 fibril displaying a force plateau. *Red* and *blue* datapoints correspond to the departure and retraction of the AFM cantilever from and to the sample, respectively. (**c**) Example of a force versus extension *curve* measured on an Aβ25–35 fibril displaying two force plateaus interrupted with a step. (**d**) Example of a force versus extension *curve* measured on an Aβ25–35 fibril displaying repetitive force plateaus interrupted with multiple steps. (**e**) Distribution of plateau force steps. Multimodal histogram is observed with peaks at approximate positions of ~45, ~90, and ~135, and ~180 pN

On the surface of mica, a phyllosilicate crystal in which Si, Al, P, and O atoms are organized into layers held together by K^+ (Franzini 1969; Giese 1979; Brigatti et al. 2003), Aβ25–35 fibrils display a highly ordered trigonal arrangement (Fig. 7.7a) (Karsai et al. 2007). The fibril orientations are separated by 60°, reflecting the crystal lattice structure of the mica surface. The oriented binding of the Aβ25–35 fibrils is very sensitive to K^+ ions. Time-lapse AFM imaging and scanning force kymography experiments (Figs. 7.2c and 7.7a) reveal that oriented fibril formation is the result of epitaxial filament growth rather than the binding of fibrils from solution. As revealed by the scanning force kymograms (Kellermayer et al. 2008), Aβ25–35 fibrils grow on both ends, but the net rate is greater on one of them (fast-growing end). The assembly dynamics of such fibrils was similar to those of the underlying one. The most prominent feature of the self-assembly dynamics is discontinuous growth. Bursts of rapid, step-like growth are interrupted with pauses of variable time period. Net fibril growth is achieved via an assembly of consecutive steps and pauses. The step-size histogram (Fig. 7.7b) displays a multimodal distribution with peaks at integer multiples of ~7 nm for both the fast- and slow-growing ends. Considering β-sheet spacing within the fibril [4.7 Å (Serpell 2000)], the 7-nm distance corresponds to a stretch of fibril containing ~15 peptides along its length. Pause time and step size are independent of total peptide concentration, but step rate increases with increasing concentration for both the fast- and slow-growing fibril ends. The concentration dependence of step rate indicates that the rapid assembly step follows second-order kinetics. A critical Aβ25–35 peptide concentration (below which fibril growth is halted) of ~10 μM is obtained for the rapid growth step at both ends. Previously, critical concentrations of 50 μM (Tjernberg et al. 1999) and 100 μM (Lomakin et al. 1996) were measured for the assembly of Aβ1–40 fibrils in solution. The low critical concentration reflects the catalytic effect of the mica surface (Kowalewski and Holtzman 1999; Zhu et al. 2002). Scanning force kymography provides a glimpse at the dynamics of individual amyloid fibril self-assembly with unprecedented detail. Considering its simplicity and richness of high-resolution information, the technique may be used to analyze the assembly dynamics of a wide variety of linear biopolymers.

7.4.4 *Transthyretin*

The precursor protein of various forms of transthyretin amyloidoses is a normal- or mutant-sequence transthyretin (TTR, formerly known as prealbumin), a transport protein synthesized by the liver and the choroid plexus (Tan and Pepys 1994; Saraiva 2002; Buxbaum 2003). TTR is a tetramer of four identical subunits, each 127 amino acid residues long (Damas and Saraiva 2000). Transthyretin is associated with two clinical forms of amyloidosis, senile systemic amyloidosis and hereditary systemic amyloidosis. Senile systemic amyloidosis (SSA) involves the native protein and leads to amyloid deposits in the ventricular myocardium of elderly people (Westermark et al. 1990). The prevalence of TTR amyloidosis

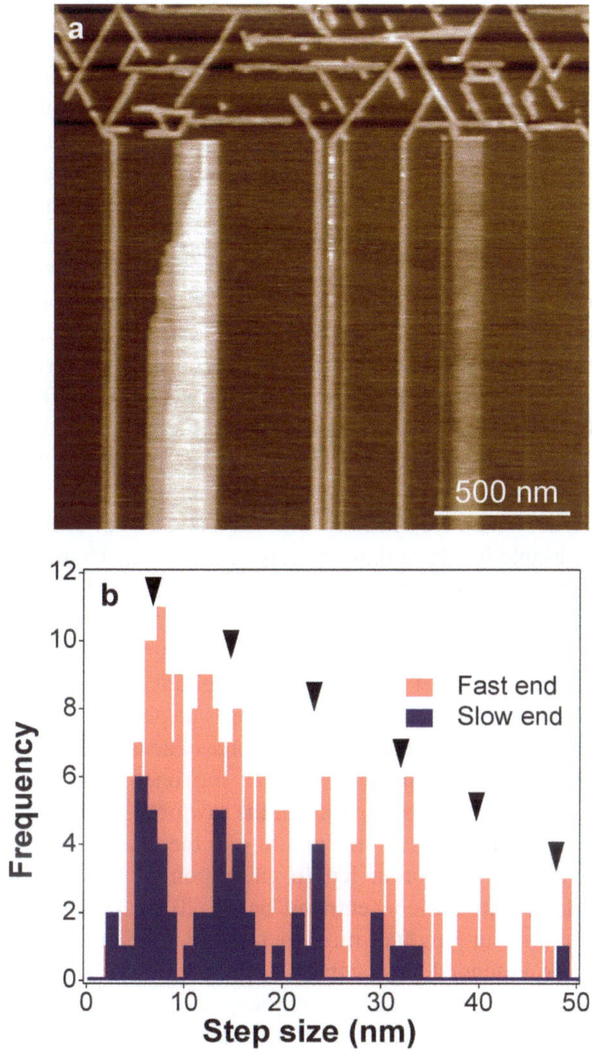

Fig. 7.7 Scanning force kymography. (**a**) Example of a kymogram obtained on an oriented network of Aβ25–35 fibrils evolving on mica. The *upper portion of the image* shows the two-dimensional AFM micrograph, which turns into the kymogram upon stopping the slow scan along the vertical *y*-axis. The time elapsed between consecutive scan lines is 300 ms. The polarized, stepwise growth of a fibril can be clearly seen. (**b**) Distribution of growth step size of epitaxially growing Aβ25–35 fibrils. Fibril growth occurs in discrete steps of ~7 nm or its integer multiples

increases with age, and more than 25% of the population is affected above an age of 80 years (Westermark 1990). Hereditary systemic amyloidosis involves mutant forms of TTR and manifests in familial amyloidotic polyneuropathy. Familial amyloidotic polyneuropathy (FAP) is characterized by amyloid deposits primarily in peripheral nerves, but also in the kidney, spleen, heart, and eye (Saraiva 1995). At least 80 mutations of TTR are known (Saraiva 2001). The most prevalent TTR mutations are Val30Met and Val122Ile (Tan and Pepys 1994). The most aggressive form of FAP is caused by the mutation Leu55Pro (Jacobson et al. 1992). Thermodynamic and kinetic analyses suggested a correlation between the stability of TTR fibrils and the clinical course of the disease (Buxbaum 2003). The exact structure of TTR fibrils is not precisely known. Early electron microscopy analysis suggested a fourfold cross-sectional symmetry in which four protofilaments wind around a hollow core (Serpell et al. 1995). More recent structural investigations reveal a polymorphism in which two, three, four, or even five protofilaments wind together in a more compact arrangement (Cardoso et al. 2002). Each protofilament represents a stack of TTR subunits. There has been much effort to find potential therapeutic agents that might intervene with the formation of TTR amyloid fibrils (Almeida et al. 2004; Morais-de-Sa et al. 2004). Recently, it has been shown that the antibiotic doxycycline successfully disrupts TTR amyloid (Cardoso and Saraiva 2005), paving the way towards possible therapeutic handling of the disease.

TTR protofibrils obtained by mild acidification appear in AFM images as curved filaments with variable length and an axial periodicity of ~15 nm (Fig. 7.8a) (Pires et al. 2011). Notably, when exposed to physiological-pH phosphate-buffered saline solution, the preformed TTR protofibril bundles dissociate into component protofibrils with an axially compacted and radially dilated structure. In single-molecule force spectroscopy experiments on TTR protofibrils, repetitive force sawtooth transitions can be observed (Fig. 7.8b). Thus, TTR protofibrils display a behavior different from mature Aβ fibrils, in the force spectra of which force plateaus dominate. The sawtooth-shaped force transitions are rather similar to the ones observed in the case of prefibrillar Aβ1–40 (Fig. 7.8b inset). Each of the sawtooth-shaped transitions extends the captured chain by a distance corresponding to the spacing between consecutive peaks. The distribution of contour length gain upon such transitions is shown in Fig. 7.8c. The datapoints are concentrated below 20 nm, a distance much shorter than the unfolded contour length of monomeric TTR (~48 nm), but the the captured TTR chain can be extended to distances far exceeding this contour length. The contour-length gain thus corresponds to intramonomeric unfolding transitions evoked by mechanical force. However, the monomers are linked together within the protofibril with sufficiently strong interactions so as to maintain structural continuity in spite of the unfolding transitions. The similarity of the nanomechanical behavior of TTR protofibrils to that of prefibrillar Aβ1–40 points at common structural features in prefibrillar aggregates of amyloidogenic proteins.

7.4.5 β_2-Microglobulin

β_2-Microglobulin amyloidosis is an iatrogenic disease that occurs during long-term hemodialysis. Therefore, the disease is often called "dialysis-related amyloidosis" (DRA) (Tan and Pepys 1994; Falk et al. 1997; Buxbaum 2003). The precursor

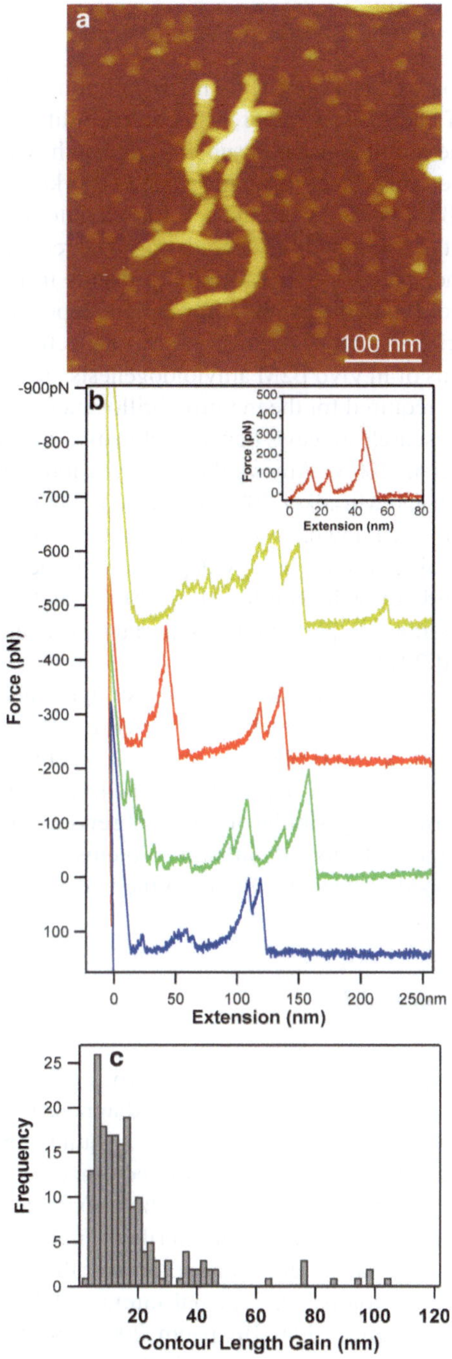

Fig. 7.8 Imaging and manipulation of wild-type transthyretin (TTR) protofibrils. (**a**) AFM image of TTR protofibrils attached to mica surface. *Curved fibrils* with beaded structure are observed. (**b**) Example of force versus extension curves measured on TTR protofibrils. Only datapoints collected during the departure of the AFM cantilever from the sample are shown. Sawtooth-shaped transitions dominate the force spectra. *Inset*, Example, for comparison, of a force *curve* measured on a 1-day-old Aβ1–40 sample in which only prefibrillar aggregates were present. (**c**) Distribution of contour-length gain upon consecutive transitions. Most of the data fall below 20 nm, far below the total unfolded contour length of the TTR monomer

protein of this amyloidosis is a normal β_2-microglobulin (β2M), which is the light chain component of the class I major histocompatibility complex (MHC). The protein belongs to the immunoglobulin superfamily and has a β-sandwich structure stabilized by a single cross-sheet disulphide bond (Björkman et al. 1987). During normal catabolism, β2M is shed from the cell surface into the serum and carried to the kidney, which is the sole site of its degradation. The reduction or abolishment of β2M plasma clearance, caused by renal failure, results in increased plasma β2M levels (by up to 60-fold) and the eventual formation, deposition, and accumulation of β2M amyloid fibrils in various organs (joints, connective tissue) (Pepys 2006). The exact mechanisms of in vivo β2M amyloidogenesis are not clearly understood, as highly acidic pH is required for the in vitro fibrillogenesis of wild type β2M (Kad et al. 2001). Recent research suggests that a combination of synergistic factors may be involved (Myers et al. 2006). Among these, copper ions have been suggested to play a key role (Eakin and Miranker 2005; Eakin et al. 2006). The precise structure of β2M fibrils is not yet known. Systematic structural studies revealed that the fibrils may display a diverse morphology depending on the physical and chemical conditions during fibril growth (Kardos et al. 2005), and the role of non-native folding intermediates may be particularly important in its amyloid fibrillogenesis (Eichner and Radford 2011).

In AFM images of mica-adsorbed β2M fibrils, straight bundles are observed, but in some cases the helical structure can be resolved (Fig. 7.9a). In the force spectra of β2M fibrils (Fig. 7.9b), a force plateau may be seen. In some cases, repetitive sawtooth peaks are also present. The presence of the force plateau reflects structural transitions that are principally similar to those observed for the various types of Aβ-fibrils. Accordingly, sheets of subunit proteins held together by strong interactions are unzipped or peeled away from the underlying fibril surface.

7.4.6 Ex Vivo Amyloid

The term ex vivo amyloid refers to the broad types of amyloids extracted from biological tissue in clinically or postmortem diagnosed cases of amyloidosis (Tennent 1999). Investigation of such samples is quite important because they reflect the properties of amyloid fibrils formed under in vivo conditions. In Fig. 7.10, data obtained on ex vivo amyloid fibrils extracted from human spleen are shown. The protein component of the fibrils is immunoglobulin light chain, which is known to cause a systemic amyloidosis called AL amyloidosis. The precursor protein AL amyloidosis is a clonal immunoglobulin (Ig) light chain or its fragment (Tan and Pepys 1994; Buxbaum 2003). Systemic AL amyloidosis (previously known as primary amyloidosis) is the most common form of clinical amyloidosis in developed countries (Pepys 2006). The disease is a monoclonal plasma cell disorder closely related to myeloma multiplex. A reduced folding stability of the protein appears to be responsible for the amyloidogenic property of Ig light chains (Bellotti et al. 2000). Partially folded Ig light chain intermediates

Fig. 7.9 Visualization and mechanical manipulation of β_2-microglobulin (β2M) fibrils. (**a**) Noncontact mode scanning AFM image (amplitude contrast) of β2M fibrils attached to mica surface. (**b**) Force spectra of β2M fibrils obtained by targeting individual fibrils at specific surface locations. *Red* and *blue* datapoints correspond to the departure and retraction of the AFM cantilever from and to the sample, respectively. The force versus extension data display a force plateau

are critical precursors along the pathway of amyloid oligomer and fibril formation (Bellotti et al.2000; Khurana et al. 2003). Experimental investigation of light-chain amyloid fibrillogenesis has for long been relatively slow because of the lack of recombinant proteins or fragments that permit in vitro studies. Recently, several recombinant immunoglobulin variable domain fragments were generated that opened the way towards detailed biophysical and biochemical analyses (Stevens et al. 1995). Morphological studies on such recombinant fragments suggest that Ig light chain amyloid fibrils acquire a highly intertwined, hierarchical structure (Ionescu-Zanetti et al. 1999). Although the fragments readily form fibrils at acidic pH, the presence of a denaturant is necessary to facilitate fibrillogenesis at

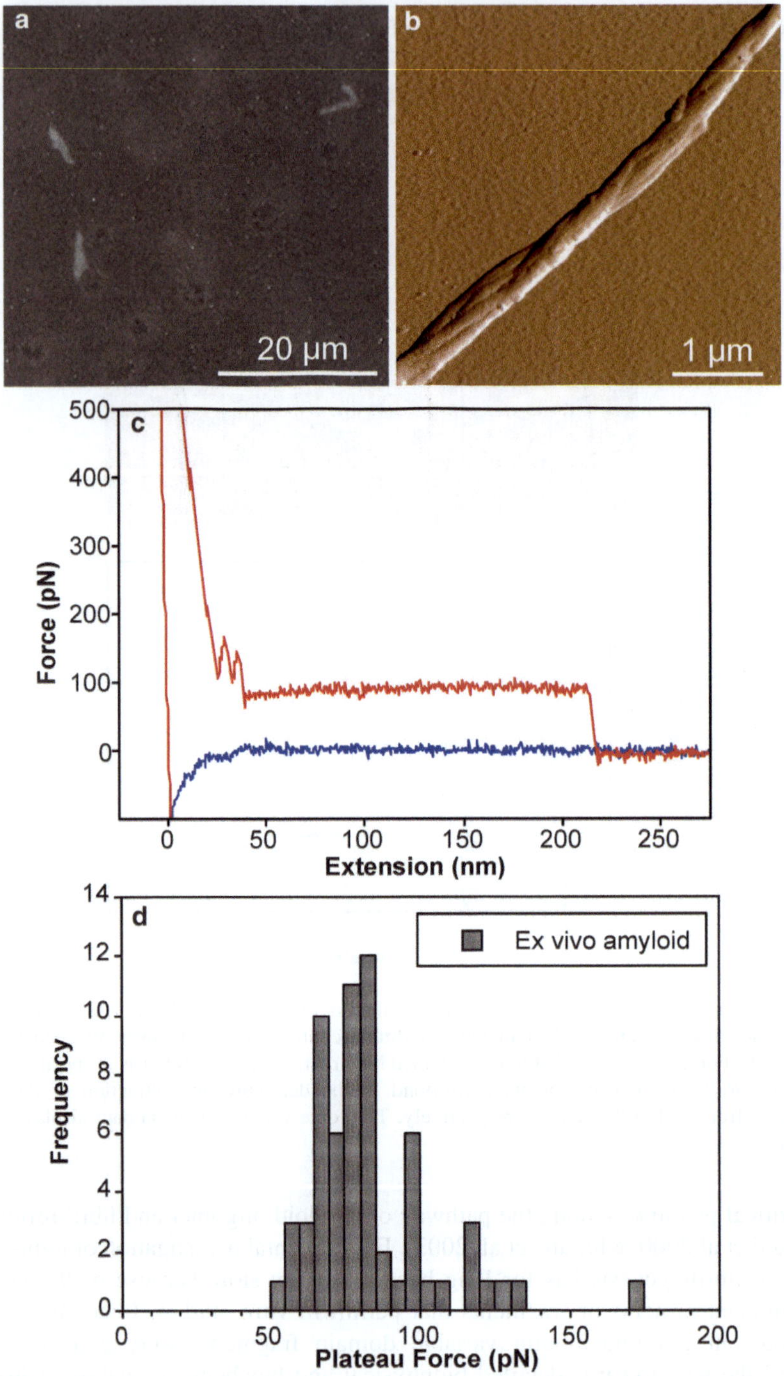

Fig. 7.10 Imaging and manipulation of ex vivo amyloid fibrils extracted from human spleen. The fibrils are composed of immunoglobulin (Ig) light chain. (**a**) Fluorescence microscopic image of ex vivo fibrils labeled with thioflavin-T (ThT). (**b**) AFM image of an ex vivo amyloid fibril bundle attached to mica surface. A thick, several hundred nanometer wide fibril with intertwined structure is observed. (**b**) Example of a force versus extension *curve* measured on the ex vivo amyloid fibril bundle. *Red* and *blue* datapoints correspond to the departure and retraction of the AFM cantilever from and to the sample, respectively. The force versus extension data display a force plateau. (**d**) Distribution of plateau forces. Datapoints fall above 50 pN, with a peak at approximately 80 pN

physiological pH, supporting the notion that amyloid fibril formation proceeds via partially unfolded intermediate states (Souillac et al. 2002a, b).

Ig light-chain ex vivo fibrils appear as thioflavin-T (ThT) positive filamentous structures (Fig. 7.10a) that display an intertwined structure according to the AFM topography (Fig. 7.10b). In situ force spectroscopy experiments, in which the fibril bundle is manipulated at predetermined positions based on the AFM image, reveal distinct force plateaus with plateau lengths exceeding 200 nm (Fig. 7.10c). The presence of the force plateau indicates that subunit sheets can be unzipped from the surface of the fibril bundle. The plateau forces typically exceed 50 pN (Fig. 7.10d), suggesting that the interactions holding together the protofilaments within the fibril are stronger than in the case of Aβ fibrils.

7.4.7 Prion Fibrils

Prions are infectious proteinaceous agents in which the component proteins are in the misfolded or amyloidogenic state (Eisenberg and Jucker 2012). By using the force-spectroscopy approach, the nanomechanical behavior of a recombinant peptide, PrP90–231, of the human prion protein was characterized (Ganchev et al. 2008). The PrP fibrils could be described as wormlike chains, and the forces necessary to extract individual proteins from the fibril backbone were measured. Recently, the nanomechanical properties of the yeast prion Sup35 amyloid fibrils were investigated in elegant experiments by using optical trapping methods (Fig. 7.3) (Dong et al. 2010; Castro et al. 2011). By utilizing the fact that the N-terminal and middle regions of the monomeric Sup35 protein binds with high affinity to fibril ends, Sup35 prion fibrils were captured by the ends and stretched axially. Single Sup35 amyloid fibrils withstood remarkable levels of forces and displayed overstretch behavior only at forces exceeding 200 pN. The overstretch transition is apparently caused by unfolding events within the monomers composing the fibril. Strikingly, in spite of monomer unfolding, the structural continuity of the fibrils is maintained just as in the case of nanomanipulated TTR protofibrils (Fig. 7.8).

7.4.8 Natural Adhesives

It is becoming increasingly recognized that certain amyloids are not pathological agents but play functional roles (Mostaert et al. 2006, 2009; Maury 2009; Otzen 2010; Blanco et al. 2012). Such functional amyloids include natural adhesives found in algae (Mostaert et al. 2006), flatworms (Mostaert et al. 2009), or the barnacle (Sullan et al. 2009). A common feature in the force spectra of these functional amyloid fibrils is the appearance of highly regular, repetitive, and multiple sawtooth-shaped force transitions often superimposed on a force plateau.

The appearance of the sawtooth-shaped transitions is commonly interpreted as unfolding events that are mechanically driven and take place within the monomer component of the amyloid fibril. In the case of covalent polymers, such repetitive sawtooth-shaped transitions are commonly observed (Rief et al. 1997a, b; Kellermayer et al. 2003). Notably, however, in amyloid fibrils the component monomers are linked via secondary chemical bonds. The force spectroscopy data suggest that these interactions are nevertheless strong enough to withstand high external forces. At the same time, via mechanical yield caused by unfolding, the complete rupture of the chain is prevented and an efficient adhesive behavior is established.

7.5 Summary and Perspectives

There are numerous other amyloidogenic protein systems that have been investigated with single-molecule manipulation methods (Smith et al. 2006; Mesquida et al. 2007; del Mercato et al. 2008; Dong et al. 2008; Sweers et al. 2011, 2012), and the list is continuously expanding. Clearly, single-molecule methodologies have yielded unique insights into amyloid behavior. Although the nanomanipulation geometries differ, for the geometries discussed above a common explanation can be drawn for the most prevalent force response: plateau (Fig. 7.11a). During mechanical manipulation of surface-adsorbed fibrils, subunit-sheets are lifted off the fibril surface. As the pulling progresses, the subunit sheets become unzipped, via the sequential rupture of side-chain interactions, which convert the subunit-sheets from an associated to a dissociated state. Because each rupture event increases the length of the unzipped sheet with a uniform and very small distance, the unzipping process proceeds at an apparent constant force that results in a force plateau. Since the sheets within a bundle are coupled in parallel, the plateau force scales with the effective number of subunit-sheets. If a sheet becomes unzipped entirely, the force drops in a unit step. Numerous sequential steps result in a descending force staircase. Most importantly, if the sheet is allowed to retract (prior to becoming completely unzipped), it reassociates to the underlying fibril in a rapid process. The driving force of the rapid rebinding is the very high local concentration of the conjugate binding sites, which increases with each rebinding event. The rapid rebinding is probably further facilitated by the correct orientation of the conjugate binding sites defined by the structural order of the subunit-sheet. As a result, the subunit-sheets are rapidly zippered together. A simulation based on a simple elastically coupled two-state model (Fig. 7.11b) (Rief et al. 1998a, b; Kellermayer et al. 2001) recovers the essential features of the nanomechanical behavior including the ascending and descending force staircases and reversibility (Fig. 7.11c, d). Thus, a reversible zipper behavior is an inherent property of many mature amyloid fibrillar systems. A prerequisite for a mechanical zipper to work is that the interactions holding together the subunit sheet are stronger than the ones binding it sideways to the underlying fibril surface. Notably, the interactions

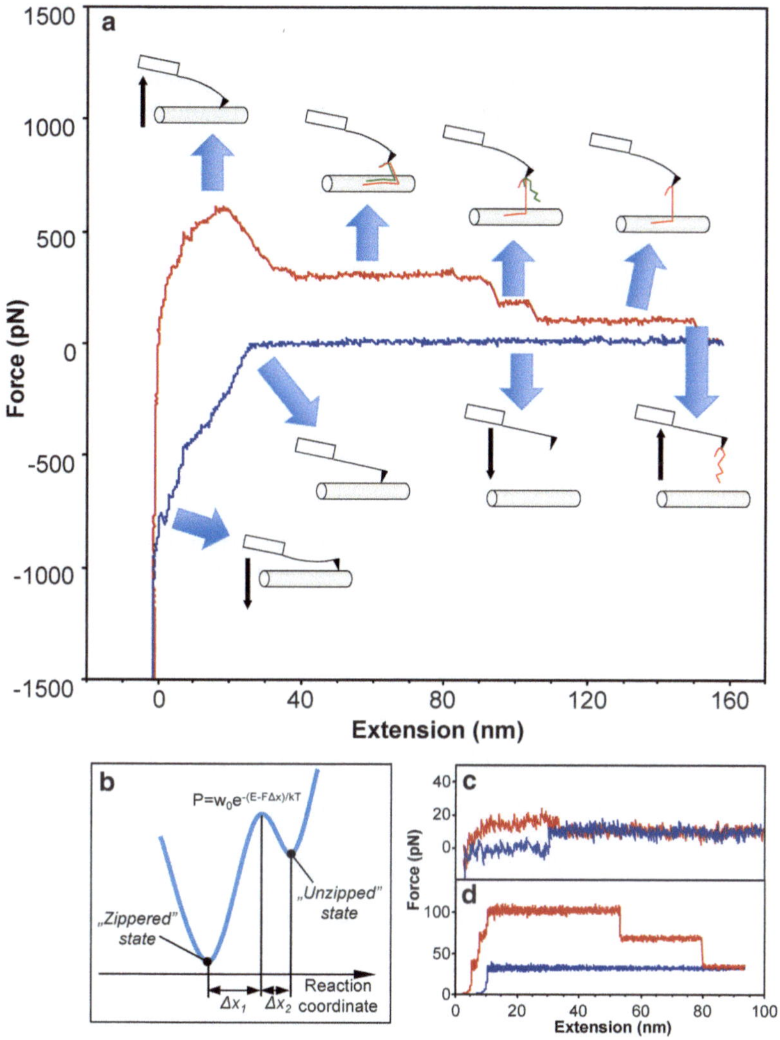

Fig. 7.11 Model and simulation of the unzipping nanomechanical behavior of amyloid fibrils. (**a**) Stretch-relaxation cycle displaying force steps. Schematics of the cantilever and fibril status are shown. (**b**) Simulation of amyloid fibril behavior with a two-state system, in which the fibril may transit between the zippered and unzipped states with probabilities (shown in the equation) that depend on the respective activation energy barriers modulated according to the external force and the shape of the potential. (**c**) Example of a reversible force plateau obtained on an Aβ25–35 fibril. *Red* and *blue* datapoints correspond to the departure and retraction of the AFM cantilever from and to the sample, respectively. (**d**) Force versus extension data obtained in Monte-Carlo simulation based on the two-state model. The force plateaus, force steps, and the reversible plateau are recovered by the simulation

between monomers along the subunit sheet are noncovalent, yet, because of parallel coupling, they stabilize the axial structure so much that continuity is maintained in spite of intramonomeric unfolding events. While such intramonomeric unfolding events may dominate the nanomechanical behavior in the oligomeric stages of the amyloid fibrillogenesis, they function as sacrificial bond breakage so as to maintain a global adhesive behavior in functional amyloids.

Acknowledgments This work was supported by grants from the Hungarian Science Foundation (OTKA K73256 and K84133) and the Hungarian National Development Agency (TAMOP-4.2.1. B-09/1/KMR-2010-0001).

References

Adlard PA, Cummings BJ (2004) Alzheimer's disease - a sum greater than its parts? Neurobiol Aging 25:725–733
Almeida MR, Macedo B et al (2004) Selective binding to transthyretin and tetramer stabilization in serum from patients with familial amyloidotic polyneuropathy by an iodinated diflunisal derivative. Biochem J 381(2):351–356
Alzheimer A (1907) Uber eine eigneartige Erkrankung der Hirnrinde. Centralblatt fur Nervenheilkunde und Psychiatrie 30:177–179
Ambrose WP, Moerner WE (1991) Fluorescence spectroscopy and spectral diffusion of single impurity molecules in a crystal. Nature 349:225–227
Ashkin A (1980) Application of laser radiation pressure. Science 210:1081–1088
Ashkin A, Dziedzic JM et al (1987) Optical trapping and manipulation of single cells using infrared laser beams. Nature 330:769–771
Ashkin A, Schutze K et al (1990) Force generation of organelle transport measured in vivo by an infrared laser trap. Nature 348(6299):346–348
Axelrod D (1989) Total internal reflection fluorescence microscopy. Methods Cell Biol 30:245–270
Axelrod D (2003) Total internal reflection fluorescence microscopy in cell biology. Methods Enzymol 361:1–33
Axelrod D, Burghardt TP et al (1984) Total internal reflection fluorescence. Annu Rev Biophys Bioeng 13:247–268
Ballard C, Gauthier S et al (2011) Alzheimer's disease. Lancet 377(9770):1019–1031
Ban T, Hamada D et al (2003) Direct observation of amyloid fibril growth monitored by thioflavin T fluorescence. J Biol Chem 278(19):16462–16465
Ban T, Hoshino M et al (2004) Direct observation of abeta amyloid fibril growth and inhibition. J Mol Biol 344(3):757–767
Bell GI (1978) Models for the specific adhesion of cells to cells. Science 200(4342):618–627
Bellotti V, Mangione P et al (2000) Review: immunoglobulin light chain amyloidosis–the archetype of structural and pathogenic variability. J Struct Biol 130(2–3):280–289
Bertoncini CW, Celej MS (2011) Small molecule fluorescent probes for the detection of amyloid self-assembly in vitro and in vivo. Curr Protein Pept Sci 12(3):205–220
Best RB, Brockwell DJ et al (2003) Force mode atomic force microscopy as a tool for protein folding studies. Anal Chim Acta 479:87–105
Betzig E, Chichester RJ (1993) Single molecules observed by near-field scanning optical microscopy. Science 262:1422–1425
Biancalana M, Koide S (2010) Molecular mechanism of Thioflavin-T binding to amyloid fibrils. Biochim Biophys Acta 1804(7):1405–1412

Biancalana M, Makabe K et al (2009) Molecular mechanism of thioflavin-T binding to the surface of beta-rich peptide self-assemblies. J Mol Biol 385(4):1052–1063
Bianco P, Nagy A et al (2007) Interaction forces between F-actin and titin PEVK domain measured with optical tweezers. Biophys J 93(6):2102–2109
Binnig G, Quate CF et al (1986) Atomic force microscope. Phys Rev Lett 56(9):930–933
Björkman PJ, Saper MA et al (1987) Structure of the human class I histocompatibility antigen HLA-A2. Nature 329:506–512
Blanco LP, Evans ML et al (2012) Diversity, biogenesis and function of microbial amyloids. Trends Microbiol 20(2):66–73
Brigatti MF, Guggenheim S et al (2003) Crystal chemistry of the 1M mica polytype: the octahedral sheet. Am Mineral 88:667–675
Bustamante C, Marko JF et al (1994) Entropic elasticity of lambda-phage DNA. Science 265 (5178):1599–1600
Bustamante C, Macosko JC et al (2000a) Grabbing the cat by the tail: manipulating molecules one by one. Nat Rev Mol Cell Biol 1(2):130–136
Bustamante C, Smith SB et al (2000b) Single-molecule studies of DNA mechanics. Curr Opin Struct Biol 10(3):279–285
Buxbaum JN (2003) The systemic amyloidoses. Curr Opin Rheumatol 16:67–75
Cardoso I, Saraiva MJ (2005) Doxycycline disrupts transthyretin amyloid: evidence from studies in a FAP transgenic mice model. FASEB J 20:234–239
Cardoso I, Goldsbury CS et al (2002) Transthyretin fibrillogenesis entails the assembly of monomers: a molecular model for in vitro assembled transthyretin amyloid-like fibrils. J Mol Biol 317(5):683–695
Carrion-Vazquez M, Oberhauser AF et al (2000) Mechanical design of proteins studied by single-molecule force spectroscopy and protein engineering. Prog Biophys Mol Biol 74(1–2):63–91
Castellani RJ, Rolston RK et al (2010) Alzheimer disease. Dis Mon 56(9):484–546
Castro CE, Dong J et al (2011) Physical properties of polymorphic yeast prion amyloid fibers. Biophys J 101(2):439–448
Cummings JL (2004) Alzheimer's disease. N Engl J Med 351(1):56–67
Damas AM, Saraiva MJ (2000) Review: TTR amyloidosis-structural features leading to protein aggregation and their implications on therapeutic strategies. J Struct Biol 130:290–299
Damas AM, Saraiva MJ (2005) Transthyretin. In: Sipe JD (ed) Amyloid proteins: the beta sheet conformation and disease. Wiley, New York, pp 571–583
del Mar Martinez-Senac M, Villalain J et al (1999) Structure of the Alzheimer β-amyloid peptide (25-35) and its interaction with negatively charged phospholipid vesicles. Eur J Biochem 265:744–753
del Mercato LL, Maruccio G et al (2008) Amyloid-like fibrils in elastin-related polypeptides: structural characterization and elastic properties. Biomacromolecules 9(3):796–803
Deniz AA, Dahan M et al (1999) Single-pair fluorescence resonance energy transfer on freely diffusing molecules: observation of Forster distance dependence and subpopulations. Proc Natl Acad Sci USA 96(7):3670–3675
Dickson RM, Cubitt AB et al (1997) On/off blinking and switching behaviour of single molecules of green fluorescent protein. Nature 388(6640):355–358
Dobson CM (1999) Protein misfolding, evolution and disease. Trends Biochem Sci 24(9):329–332
Dobson CM (2003) Protein folding and misfolding. Nature 426:884–889
Doi M, Edwards SF (1986) The theory of polymer dynamics. Clarendon, Oxford
Dong M, Hovgaard MB et al (2008) AFM-based force spectroscopy measurements of mature amyloid fibrils of the peptide glucagon. Nanotechnology 19(38):384013
Dong J, Castro CE et al (2010) Optical trapping with high forces reveals unexpected behaviors of prion fibrils. Nat Struct Mol Biol 17(12):1422–1430
Eakin CM, Miranker AD (2005) From chance to frequent encounters: origins of ß2-microglobulin fibrillogenesis. Biochim Biophys Acta 1753:92–99

Eakin CM, Berman AJ et al (2006) A native to amyloidogenic transition regulated by a backbone trigger. Nat Struct Mol Biol 13(3):202–208
Eichner T, Radford SE (2011) Understanding the complex mechanisms of beta2-microglobulin amyloid assembly. FEBS J 278(20):3868–3883
Eisenberg D, Jucker M (2012) The amyloid state of proteins in human diseases. Cell 148 (6):1188–1203
Evans E (2001) Probing the relation between force–lifetime–and chemistry in single molecular bonds. Annu Rev Biophys Biomol Struct 30:105–128
Evans E, Ritchie K (1997) Dynamic strength of molecular adhesion bonds. Biophys J 72 (4):1541–1555
Falk RH, Comenzo RL et al (1997) The systemic amyloidoses. New Eng J Med 337:898–909
Fandrich M (2012) Oligomeric intermediates in amyloid formation: structure determination and mechanisms of toxicity. J Mol Biol 421(4–5):427–40
Fandrich M, Fletcher MA et al (2001) Amyloid fibrils from muscle myoglobin. Nature 410:165–166
Fernandez JM, Li H (2004) Force-clamp spectroscopy monitors the folding trajectory of a single protein. Science 303(5664):1674–1678
Fisher TE, Oberhauser AF et al (1999) The study of protein mechanics with the atomic force microscope. Trends Biochem Sci 24(10):379–384
Florin EL, Moy VT et al (1994) Adhesion forces between individual ligand-receptor pairs. Science 264(5157):415–417
Forloni G, Chiesa R et al (1993) Apoptosis mediated neurotoxicity induced by chronic application of beta amyloid fragment 25-35. Neuroreport 4(5):523–526
Franzini M (1969) The A and B mica layers and the crystal structure of sheet silicates. Contr Mineral Petrol 21:203–224
Friedman R (2011) Aggregation of amyloids in a cellular context: modelling and experiment. Biochem J 438(3):415–426
Ganchev DN, Cobb NJ et al (2008) Nanomechanical properties of human prion protein amyloid as probed by force spectroscopy. Biophys J 95(6):2909–2915
Garcia-Saez AJ, Schwille P (2007) Single molecule techniques for the study of membrane proteins. Appl Microbiol Biotechnol 76(2):257–266
Giese RFJ (1979) Hydroxyl orientations in 2:1 phyllosilicates. Clays Clay Miner 27(3):213–223
Gimzewski JK, Joachim C (1999) Nanoscale science of single molecules using local probes. Science 283(5408):1683–1688
Gordon MP, Ha T et al (2004) Single-molecule high-resolution imaging with photobleaching. Proc Natl Acad Sci USA 101(17):6462–6465
Gorelik J, Shevchuk A et al (2002) Scanning surface confocal microscopy for simultaneous topographical and fluorescence imaging: application to single virus-like particle entry into a cell. Proc Natl Acad Sci USA 99(25):16018–16023
Greenwald J, Riek R (2010) Biology of amyloid: structure, function, and regulation. Structure 18 (10):1244–1260
Ha T, Enderle T et al (1996) Probing the interaction between two single molecules: fluorescence resonance energy transfer between a single donor and a single acceptor. Proc Natl Acad Sci USA 93(13):6264–6268
Ha T, Ting AY et al (1999) Single-molecule fluorescence spectroscopy of enzyme conformational dynamics and cleavage mechanism. Proc Natl Acad Sci USA 96(3):893–898
Hardy J, Selkoe DJ (2002) The amyloid hypothesis of Alzheimer's disease: progress and problems on the road to therapeutics. Science 297:353–3556
Harper JD, Wong SS et al (1997) Observation of metastable Abeta amyloid protofibrils by atomic force microscopy. Chem Biol 4(2):119–125
Harriman OL, Leake MC (2011) Single molecule experimentation in biological physics: exploring the living component of soft condensed matter one molecule at a time. J Phys Condens Matter 23(50):503101

Hawkins PN, Myers MJ et al (1988) Diagnostic radionuclide imaging of amyloid: biological targeting by circulating human serum amyloid P component. Lancet 1(8600):1413–1418
Huang B, Bates M et al (2009) Super-resolution fluorescence microscopy. Annu Rev Biochem 78:993–1016
Hutter JL, Bechhoefer J (1993) Calibration of atomic-force microscope tips. Rev Sci Instrum 64 (7):1868–1873
Ionescu-Zanetti C, Khurana R et al (1999) Monitoring the assembly of Ig light-chain amyloid fibrils by atomic force microscopy. Proc Natl Acad Sci USA 96(23):13175–13179
Ishii Y, Yanagida T (2000) Single molecule detection in life science. Single Mol 1:5–13
Ishii Y, Ishijima A et al (2001) Single molecule nanomanipulation of biomolecules. Trends Biotechnol 19(6):211–216
Jacobson DR, McFarlin DE et al (1992) Transthyretin Pro55, a variant associated with early-onset, aggressive, diffuse amyloidosis with cardiac and neurologic involvement. Hum Genet 89:353–356
Kad NM, Thomson NH et al (2001) Beta(2)-microglobulin and its deamidated variant, N17D form amyloid fibrils with a range of morphologies in vitro. J Mol Biol 313(3):559–571
Kad NM, Myers SL et al (2003) Hierarchical assembly of beta2-microglobulin amyloid in vitro revealed by atomic force microscopy. J Mol Biol 330(4):785–797
Kapanidis AN, Weiss S (2002) Fluorescent probes and bioconjugation chemistries for single-molecule fluorescence analysis of biomolecules. J Chem Phys 117:10953–10964
Kardos J, Yamamoto K et al (2004) Direct measurement of the thermodynamic parameters of amyloid formation by isothermal titration calorimetry. J Biol Chem 279(53):55308–55314
Kardos J, Okuno D et al (2005) Structural studies reveal that the diverse morphology of ß2-microglobulin aggregates is a reflection of different molecular architectures. Biochim Biophys Acta 1753:108–120
Karsai Á, Nagy A et al (2005) Effect of lysine-28 side-chain acetylation on the nanomechanical behavior of alzheimer amyloid beta25-35 fibrils. J Chem Inf Model 45(6):1641–1646
Karsai Á, Martonfalvi Z et al (2006) Mechanical manipulation of Alzheimer's amyloid beta1-42 fibrils. J Struct Biol 155(2):316–326
Karsai Á, Grama L et al (2007) Potassium-dependent oriented growth of amyloid ß25-35 fibrils on mica. Nanotechnology 18:345102
Kellermayer MSZ (2005) Visualizing and manipulating individual protein molecules. Physiol Meas 26:R119–R153
Kellermayer MS, Grama L (2002) Stretching and visualizing titin molecules: combining structure, dynamics and mechanics. J Muscle Res Cell Motil 23(5–6):499–511
Kellermayer MS, Smith SB et al (1997) Folding-unfolding transitions in single titin molecules characterized with laser tweezers. Science 276(5315):1112–1116
Kellermayer MS, Smith SB et al (2001) Mechanical fatigue in repetitively stretched single molecules of titin. Biophys J 80(2):852–863
Kellermayer MS, Bustamante C et al (2003) Mechanics and structure of titin oligomers explored with atomic force microscopy. Biochim Biophys Acta 1604(2):105–114
Kellermayer MS, Grama L et al (2005) Reversible mechanical unzipping of amyloid beta-fibrils. J Biol Chem 280(9):8464–8470
Kellermayer MS, Karsai Á et al (2006) Spatially and temporally synchronized atomic force and total internal reflection fluorescence microscopy for imaging and manipulating cells and biomolecules. Biophys J 91(7):2665–2677
Kellermayer MS, Karsai Á et al (2008) Stepwise dynamics of epitaxially growing single amyloid fibrils. Proc Natl Acad Sci USA 105(1):141–144
Khurana R, Ionescu-Zanetti C et al (2003) A general model for amyloid fibril assembly based on morphological studies using atomic force microscopy. Biophys J 85(2):1135–1144
Kimura Y, Bianco PR (2006) Single molecule studies of DNA binding proteins using optical tweezers. Analyst 131(8):868–874

Klar TA, Jakobs S et al (2000) Fluorescence microscopy with diffraction resolution broken by stimulated emission. Proc Natl Acad Sci USA 97:8206–8210

Kowalewski T, Holtzman DM (1999) In situ atomic force microscopy study of Alzheimer's beta-amyloid peptide on different substrates: new insights into mechanism of beta-sheet formation. Proc Natl Acad Sci USA 96(7):3688–3693

Lacoste TD, Michalet X et al (2000) Ultrahigh-resolution multicolor colocalization of single fluorescent probes. Proc Natl Acad Sci USA 97(17):9461–9466

Lakowicz JR (2006) Principles of fluorescence spectroscopy. Springer, New York, NY

LeVine H 3rd (1993) Thioflavine T interaction with synthetic Alzheimer's disease beta-amyloid peptides: detection of amyloid aggregation in solution. Protein Sci 2(3):404–410

Ling Y, Morgan K et al (2003) Amyloid precursor protein (APP) and the biology of proteolytic processing: relevance to Alzheimer's disease. Int J Biochem Cell Biol 35(11):1505–1535

Lomakin A, Chung DS et al (1996) On the nucleation and growth of amyloid ß-protein fibrils: detection of nuclei and quantitation of rate constants. Proc Natl Acad Sci USA 93:1125–1129

Lomakin A, Teplow DB et al (1997) Kinetic theory of fibrillogenesis of amyloid ß-protein. Proc Natl Acad Sci USA 94:7942–7947

Lührs T, Ritter C et al (2005) 3D structure of Alzheimer's amyloid-ß(1-42) fibrils. Proc Natl Acad Sci USA 102:17342–17347

Marko JF, Siggia ED (1995) Stretching DNA. Macromolecules 28:8759–8770

Mathur AB, Truskey GA et al (2000) Atomic force and total internal reflection fluorescence microscopy for the study of force transmission in endothelial cells. Biophys J 78(4):1725–1735

Mattson MP (2004) Pathways towards and away from Alzheimer's disease. Nature 430 (7000):631–639

Maury CP (2009) The emerging concept of functional amyloid. J Intern Med 265(3):329–334

Mehta AD, Finer JT et al (1997) Detection of single-molecule interactions using correlated thermal diffusion. Proc Natl Acad Sci USA 94(15):7927–7931

Mehta AD, Pullen KA et al (1998) Single molecule biochemistry using optical tweezers. FEBS Lett 430(1–2):23–27

Mehta AD, Rief M et al (1999a) Biomechanics, one molecule at a time. J Biol Chem 274 (21):14517–14520

Mehta AD, Rief M et al (1999b) Single-molecule biomechanics with optical methods. Science 283 (5408):1689–1695

Merkel R, Nassoy P et al (1999) Energy landscapes of receptor-ligand bonds explored with dynamic force spectroscopy. Nature 397(6714):50–53

Mesquida P, Riener CK et al (2007) Morphology and mechanical stability of amyloid-like peptide fibrils. J Mater Sci Mater Med 18(7):1325–1331

Michalet X, Weiss S (2002) Single-molecule spectroscopy and microscopy. C R Physique 3:619–644

Moerner WE, Orrit M (1999) Illuminating single molecules in condensed matter. Science 283 (5408):1670–1676

Morais-de-Sa E, Pereira PJ et al (2004) The crystal structure of transthyretin in complex with diethylstilbestrol: a promising template for the design of amyloid inhibitors. J Biol Chem 279 (51):53483–53490

Mostaert AS, Higgins MJ et al (2006) Nanoscale mechanical characterisation of amyloid fibrils discovered in a natural adhesive. J Biol Phys 32(5):393–401

Mostaert SA, Crockett R et al (2009) Mechanically functional amyloid fibrils in the adhesive of a marine invertebrate as revealed by Raman spectroscopy and atomic force microscopy. Arch Histol Cytol 72(4–5):199–207

Muller DJ, Janovjak H et al (2002) Observing structure, function and assembly of single proteins by AFM. Prog Biophys Mol Biol 79(1–3):1–43

Myers SL, Jones S et al (2006) A systematic study of the effect of physiological factors on ß2-microglobulin amyloid formation at neutral pH. Biochemistry 45:2311–2321

Neuman KC (2010) Single-molecule measurements of DNA topology and topoisomerases. J Biol Chem 285(25):18967–18971

Neuman KC, Nagy A (2008) Single-molecule force spectroscopy: optical tweezers, magnetic tweezers and atomic force microscopy. Nat Methods 5(6):491–505

Nichols MR, Moss MA et al (2002) Growth of beta-amyloid(1-40) protofibrils by monomer elongation and lateral association. Characterization of distinct products by light scattering and atomic force microscopy. Biochemistry 41(19):6115–6127

Nie S, Zare RN (1997) Optical detection of single molecules. Annu Rev Biophys Biomol Struct 26:567–596

Nie S, Chiu DT et al (1994) Probing individual molecules with confocal fluorescence microscopy. Science 266(5187):1018–1021

Nishida S, Funabashi Y et al (2002) Combination of AFM with an objective-type total internal reflection fluorescence microscope (TIRFM) for nanomanipulation of single cells. Ultramicroscopy 91(1–4):269–274

Nishizaka T, Miyata H et al (1995) Unbinding force of a single motor molecule of muscle measured using optical tweezers. Nature 377(6546):251–254

Noy A (2011) Force spectroscopy 101: how to design, perform, and analyze an AFM-based single molecule force spectroscopy experiment. Curr Opin Chem Biol 15:710–718

Orrit M, Bernard J (1990) Single pentacene molecules detected by fluorescence excitation in a p-terphenyl crystal. Phys Rev Lett 65(21):2716–2719

Orte A, Birkett NR et al (2008) Direct characterization of amyloidogenic oligomers by single-molecule fluorescence. Proc Natl Acad Sci USA 105(38):14424–14429

Orte A, Clarke R et al (2010) Single-molecule two-colour coincidence detection to probe biomolecular associations. Biochem Soc Trans 38(4):914–918

Orte A, Clarke RW et al (2011) Single-molecule fluorescence coincidence spectroscopy and its application to resonance energy transfer. Chemphyschem 12(3):491–499

Otzen D (2010) Functional amyloid: turning swords into plowshares. Prion 4(4):256–264

Paige MF, Bjerneld EJ et al (2001) A comparison of through-the-objective total internal reflection microscopy and epifluorescence microscopy for single-molecule fluorescence imaging. Single Mol 2:191–201

Pepys MB (2006) Amyloidosis. Annu Rev Med 57(8):1–19

Pepys MB, Herbert J et al (2002) Targeted pharmacological depletion of serum amyloid P component for treatment of human amyloidosis. Nature 417(6886):254–259

Peterman EJ, Sosa H et al (2004) Single-molecule fluorescence spectroscopy and microscopy of biomolecular motors. Annu Rev Phys Chem 55:79–96

Petkova AT, Ishii Y et al (2002) A structural model for Alzheimer's ß-amyloid fibrils based on experimental constraints from solid state NMR. Proc Natl Acad Sci USA 99(26):16742–16747

Petkova AT, Yau WM et al (2006) Experimental constraints on quaternary structure in Alzheimer's beta-amyloid fibrils. Biochemistry 45(2):498–512

Pike CJ, Walencewicz-Wasserman AJ et al (1995) Structure-activity analyses of beta-amyloid peptides: contributions of the beta 25-35 region to aggregation and neurotoxicity. J Neurochem 64(1):253–265

Pires RH, Saraiva MJ et al (2011) Structure and assembly-disassembly properties of wild-type transthyretin amyloid protofibrils observed with atomic force microscopy. J Mol Recognit 24 (3):467–476

Pope LH, Bennink ML et al (2002) Optical tweezers stretching of chromatin. J Muscle Res Cell Motil 23(5–6):397–407

Qu X, Wu D et al (2004) Nanometer-localized multiple single-molecule fluorescence microscopy. Proc Natl Acad Sci USA 101(31):11298–11303

Rief M, Grubmuller H (2002) Force spectroscopy of single biomolecules. Chemphyschem 3 (3):255–261

Rief M, Gautel M et al (1997a) Reversible unfolding of individual titin immunoglobulin domains by AFM. Science 276(5315):1109–1112

Rief M, Oesterhelt F et al (1997b) Single molecule force spectroscopy on polysaccharides by atomic force microscopy. Science 275(5304):1295–1297
Rief M, Fernandez JM et al (1998a) Elastically coupled two-level systems as a model for biopolymer extensibility. Phys Rev Lett 81(21):4764–4767
Rief M, Gautel M et al (1998b) The mechanical stability of immunoglobulin and fibronectin III domains in the muscle protein titin measured by atomic force microscopy. Biophys J 75 (6):3008–3014
Rochet JC, Lansbury PT (2000) Amyloid fibrillogenesis: themes and variations. Curr Opin Struct Biol 10:60–68
Sandal M, Valle F et al (2008) Conformational equilibria in monomeric alpha-synuclein at the single-molecule level. PLoS Biol 6(1):e6
Saraiva MJ (1995) Transthyretin mutations in health and disease. Hum Mutat 5(3):191–196
Saraiva MJ (2001) Transthyretin mutations in hyperthyroxinemia and amyloid diseases. Hum Mutat 17(6):493–503
Saraiva MJ (2002) Hereditary transthyretin amyloidosis: molecular basis and therapeutical strategies. Expert Rev Mol Med 14:1–11
Schwille P, Korlach J et al (1999) Fluorescence correlation spectroscopy with single-molecule sensitivity on cell and model membranes. Cytometry 36(3):176–182
Schwille P, Kummer S et al (2000) Fluorescence correlation spectroscopy reveals fast optical excitation-driven intramolecular dynamics of yellow fluorescent proteins. Proc Natl Acad Sci USA 97(1):151–156
Seidel R, Dekker C (2007) Single-molecule studies of nucleic acid motors. Curr Opin Struct Biol 17(1):80–86
Seisenberger G, Ried MU et al (2001) Real-time single-molecule imaging of the infection pathway of an adeno-associated virus. Science 294(5548):1929–1932
Selkoe DJ (1997) Images in neuroscience. Alzheimer's disease: from genes to pathogenesis. Am J Psychiatry 154(9):1198
Selkoe DJ (2001) Alzheimer's disease: genes, proteins and therapy. Physiol Rev 81(2):741–766
Selkoe DJ (2003) Folding proteins in fatal ways. Nature 426(6968):900–904
Serpell LC (2000) Alzheimer's amyloid fibrils: structure and assembly. Biochim Biophys Acta 1502(1):16–30
Serpell LC, Sunde M et al (1995) Examination of the structure of the transthyretin amyloid fibril by image reconstruction from electron micrographs. J Mol Biol 254(2):113–118
Serpell LC, Sunde M et al (2000) The protofilament substructure of amyloid fibrils. J Mol Biol 300 (5):1033–1039
Slayter EM, Slayter HS (1992) Light and electron microscopy. Cambridge University Press, Cambridge
Smith SB, Finzi L et al (1992) Direct mechanical measurements of the elasticity of single DNA molecules by using magnetic beads. Science 258(5085):1122–1126
Smith SB, Cui Y et al (1996) Overstretching B-DNA: the elastic response of individual double-stranded and single-stranded DNA molecules. Science 271(5250):795–799
Smith SB, Cui Y et al (2003) Optical-trap force transducer that operates by direct measurement of light momentum. Methods Enzymol 361:134–162
Smith JF, Knowles TP et al (2006) Characterization of the nanoscale properties of individual amyloid fibrils. Proc Natl Acad Sci USA 103(43):15806–15811
Snyder GE, Sakamoto T et al (2004) Nanometer localization of single green fluorescent proteins: evidence that myosin V walks hand-over-hand via telemark configuration. Biophys J 87:1776–1783
Sorce B, Sabella S et al (2009) Single-molecule mechanical unfolding of amyloidogenic beta2-microglobulin: the force-spectroscopy approach. Chemphyschem 10(9–10):1471–1477
Souillac PO, Uversky VN et al (2002a) Effect of association state and conformational stability on the kinetics of immunoglobulin light chain amyloid fibril formation at physiological pH. J Biol Chem 277(15):12657–12665

Souillac PO, Uversky VN et al (2002b) Elucidation of the molecular mechanism during the early events in immunoglobulin light chain amyloid fibrillation. Evidence for an off-pathway oligomer at acidic pH. J Biol Chem 277(15):12666–12679
Stevens PW, Raffen R et al (1995) Recombinant immunoglobulin variable domains generated from synthetic genes provide a system for in vitro characterization of light-chain amyloid proteins. Protein Sci 4(3):421–432
Stout AL (2001) Detection and characterization of individual intermolecular bonds using optical tweezers. Biophys J 80(6):2976–2986
Strunz T, Oroszlan K et al (2000) Model energy landscapes and the force-induced dissociation of ligand- receptor bonds. Biophys J 79(3):1206–1212
Sullan RM, Gunari N et al (2009) Nanoscale structures and mechanics of barnacle cement. Biofouling 25(3):263–275
Sunde M, Serpell LC et al (1997) Common core structure of amyloid fibrils by synchrotron X-ray diffraction. J Mol Biol 273(3):729–739
Svoboda K, Block S (1994) Biological applications of optical forces. Annu Rev Biophys Biomol Struct 23:247–285
Sweers K, van der Werf K et al (2011) Nanomechanical properties of alpha-synuclein amyloid fibrils: a comparative study by nanoindentation, harmonic force microscopy, and Peakforce QNM. Nanoscale Res Lett 6(1):270
Sweers KK, Bennink ML et al (2012) Nanomechanical properties of single amyloid fibrils. J Phys Condens Matter 24(24):243101
Tan SY, Pepys MB (1994) Amyloidosis. Histopathology 25:403–414
Tennent GA (1999) Isolation and characterization of amyloid fibrils from tissue. Methods Enzymol 309:26–47
Tennent GA, Lovat LB et al (1995) Serum amyloid P component prevents proteolysis of the amyloid fibrils of Alzheimer disease and systemic amyloidosis. Proc Natl Acad Sci USA 92 (10):4299–4303
Terzi E, Holzemann G et al (1994) Alzheimer beta-amyloid peptide 25-35: electrostatic interactions with phospholipid membranes. Biochemistry 33(23):7434–7441
Tjernberg LO, Pramanik A et al (1999) Amyloid β-peptide polymerization studied using fluorescence correlation spectroscopy. Chem Biol 6:53–62
Török M, Milton S et al (2002) Structural and dynamic features of Alzheimer's Abeta peptide in amyloid fibrils studied by site-directed spin labeling. J Biol Chem 277(43):40810–40815
Tskhovrebova L, Trinick J et al (1997) Elasticity and unfolding of single molecules of the giant muscle protein titin. Nature 387(6630):308–312
Tycko R (2004) Progress towards a molecular-level understanding of amyloid fibrils. Curr Op Struct Biol 14:96–103
Van Craenenbroeck E, Engelborghs Y (2000) Fluorescence correlation spectroscopy: molecular recognition at the single molecule level. J Mol Recognit 13(2):93–100
Walsh DM, Hartley DM et al (1999) Amyloid beta-protein fibrillogenesis. Structure and biological activity of protofibrillar intermediates. J Biol Chem 274(36):25945–25952
Wang MD (1999) Manipulation of single molecules in biology. Curr Opin Biotechnol 10(1):81–86
Wang MC, Uhlenbeck GE (1945) On the theory of Brownian motion II. Rev Mod Phys 17:323–341
Webb RH (1999) Theoretical basis of confocal microscopy. Methods Enzymol 307:3–20
Weiss S (1999) Fluorescence spectroscopy of single biomolecules. Science 283(5408):1676–1683
Westermark P, Sletten K et al (1990) Fibril in senile systemic amyloidosis is derived from normal transthyretin. Proc Natl Acad Sci USA 87(7):2843–2845
Westermark P, Benson MD et al (2005) Amyloid: toward terminology clarification. Amyloid 12 (1):1–4
Yanagida T, Iwaki M et al (2008) Single molecule measurements and molecular motors. Philos Trans R Soc Lond B Biol Sci 363(1500):2123–2134

Yang DS, Yip CM et al (1999) Manipulating the amyloid-beta aggregation pathway with chemical chaperones. J Biol Chem 274(46):32970–32974

Yildiz A, Forkey JN et al (2003) Myosin V walks hand-over-hand: single fluorophore imaging with 1.5-nm localization. Science 300:2061–2065

Zhu M, Souillac PO et al (2002) Surface-catalyzed amyloid fibril formation. J Biol Chem 277 (52):50914–50922

Zhuang X, Rief M (2003) Single-molecule folding. Curr Opin Struct Biol 13(1):88–97

Zlatanova J, Lindsay SM et al (2000) Single molecule force spectroscopy in biology using the atomic force microscope. Prog Biophys Mol Biol 74(1–2):37–61

Chapter 8
Stretching and Imaging Individual Proteins on Live Cells Using Atomic Force Microscopy

David Alsteens and Yves F. Dufrêne

8.1 Introduction

8.1.1 Atomic Force Microscopy

Atomic force microscopy (AFM) works by scanning a sharp tip over the surface of a sample while sensing the near-field physical interactions between the tip and the sample. This interaction is used to generate three-dimensional views of biological structures directly in aqueous solution. The sample is mounted on a piezoelectric scanner, which ensures three-dimensional positioning with high accuracy. While the tip (or sample) is being scanned in the (x, y) directions, the force interacting between tip and specimen is monitored with piconewton sensitivity. This force is measured by the deflection of a soft cantilever, which is detected by a laser beam focused on the free end of the cantilever and reflected into a photodiode.

Besides imaging, AFM is also widely used to detect and manipulate single biomolecules, a modality known as single-molecule force spectroscopy (SMFS). Here, the tip is continuously approached to and retracted from the biological sample, and the cantilever deflection measures the interaction force. The obtained force–distance curves provide key insight into the localization, conformational properties, elasticity, and binding strength of single molecules, either on isolated systems or on live cells. The detection and controlled manipulation of a single molecule in complex environments such as living cells often require labeling the tip with specific antibodies or ligands using appropriate protocols (Hinterdorfer and Dufrêne 2006). A common strategy is to covalently anchor proteins on tips via a polyethylene glycol (PEG) crosslinker, which provides motional freedom and prevents denaturation (Ebner et al. 2007). Tips are first modified with amino groups,

D. Alsteens (✉) • Y.F. Dufrêne (✉)
Institute of Condensed Matter and Nanosciences, Université catholique de Louvain, Louvain-La-Neuve, Belgium
e-mail: david.alsteens@uclouvain.be; yves.dufrene@uclouvain.be

A.F. Oberhauser (ed.), *Single-molecule Studies of Proteins*, Biophysics for the Life Sciences 2, DOI 10.1007/978-1-4614-4921-8_8,

further reacted with PEG linkers carrying benzaldehyde functions that are then directly attached to proteins through their lysine residues. Another option is to use self-assembled monolayers of alkanethiols on gold tips (Dupres et al. 2007). Both the thiol and silane approaches make it possible to orientate the attached biomolecules via their C-terminal or N-terminal domains by linking recombinant histidine-tagged proteins onto an AFM tip coated with nitrilotriacetate groups.

8.1.2 Describing the Elastic Behavior of Proteins and Polysaccharides

Pulling on single biomolecules with an AFM tip typically yields force curves with nonlinear elongation forces, reflecting the stretching of the flexible molecule, until the adhesion "pull-off" force is observed. At very small displacements, reduction in the number of conformations gives rise to entropic elasticity forces. This can be viewed as an ideal chain and described by the Gaussian Chain model. In the presence of an applied force F, the expression of the extension x can be written as

$$x = \frac{L_c}{3} \frac{F l_K}{k_b T}. \tag{8.1}$$

where k_b is the Boltzmann constant, T the absolute temperature, L_c the contour length of the molecule, and l_k the characteristic length scale known as the Kuhn length and defined as the mean square distance between two consecutive monomers. The limitation of this model is that the distance between monomers can be extended without any limit.

This limitation is circumvented in the freely jointed chain (FJC) (Rief et al. 1997a) model, where the bond length is kept fixed and the bond direction can be oriented in any direction. This correction is crucial for larger extension (the limit is fixed at $F \ll k_b T / l_k$), where tension in the molecular backbone may lead to enthalpic elasticity effects (bond deformation, rupture of intramolecular hydrogen bonds, and even conformational changes of the entire molecule). The polymer can be considered as a series of rigid, orientationally independent statistical (Kuhn) segments, connected through flexible joints. The l_k is related to the contour length by $L_c = n\, l_k$. The extension x can be expressed as a function of the pulling force F as

$$x(F) = L_c \left[\coth\left(\frac{F l_k}{k_b T} \right) - \frac{k_b T}{F l_k} \right] \tag{8.2}$$

An extended FJC (FJC+) model has been developed in which Kuhn segments can stretch and align under force. The polymer consists of a series of n elastic springs characterized by a segment elasticity k_s. The extension x versus the pulling force F is then given by

$$x(F) = L_c \left[\coth\left(\frac{Fl_k}{k_b T}\right) - \frac{k_b T}{Fl_k} \right] \left[1 + \frac{F}{k_s L_c} \right] \tag{8.3}$$

The elastic response of biopolymers is described by the worm-like chain (WLC) (Bustamante et al. 1994). The polymer is an irregular curved filament, which is linear on the scale of the persistence length l_p, a parameter that represents the stiffness of the molecule. Hence, molecules with low persistence length have a tendency to form coils. Extension is limited by the contour length L_c of the molecule, i.e., the length of the linearly extended molecule without stretching the molecular backbone. In this model, the force F versus extension x is approximately given by

$$x(F) = L_c \left[\coth\left(\frac{Fl_k}{k_b T}\right) - \frac{k_b T}{Fl_k} \right] \left[1 + \frac{F}{k_s L_c} \right] \tag{8.4}$$

In AFM studies, proteins that behave like continuous deformable rods are generally well described by the WLC model, while polysaccharides that behave more as series of loosely connected segments can be described by the FJC+ model.

8.1.3 Unfolding Single Proteins

The past years have witnessed tremendous progress in applying AFM-based SMFS to measure the mechanical properties of proteins in vitro (Rief et al. 1997b; Oberhauser et al. 1998; Bustamante et al. 2000; Oesterhelt et al. 2000; Lee et al. 2006; Sotomayor and Schulten 2007). Pioneering studies have demonstrated the feasibility to subject proteins to controlled forces to provide details of their unfolding pathways and the forces that anchor them into the cell membrane. For instance, AFM was used to investigate the mechanical properties of titin, the giant sarcomeric protein of striated muscle (Rief et al. 1997b). The unfolding of titin exhibited a sawtooth like pattern, with a periodicity of about 25 nm. The forces required to unfold individual domains ranged from 150 to 300 pN. Upon relaxation, refolding of immunoglobulin domains was observed. Similarly, individual bacteriorhodopsins were extracted from native purple membranes (Oesterhelt et al. 2000). Upon retracting the AFM tip, the helices unfolded and the force spectra revealed the pattern of the unfolding of the individual molecules. Since then, single molecule AFM, combined with molecular dynamics simulations and protein engineering, has enabled to understand the mechanical behavior of a great variety of proteins (Puchner and Gaub 2009; Galera-Prat et al. 2010; Li and Cao 2010).

The mechanical properties of proteins play an important role in defining key cellular functions. Therefore, understanding how cellular proteins respond to mechanical force in live cells and how this response is related to function is a hot topic in cell biology. Here we discuss how single-protein manipulation and imaging experiments can be applied to live cells to shed new light into the highly sophisticated functions of cell surfaces. While we focus on membrane sensors and cell adhesion proteins from yeasts, these experiments should be applicable to other cell surface proteins.

8.2 Nanomechanics and Nanoclustering of Cell Surface Sensors

In the baker's yeast *Saccharomyces cerevisiae*, surface stresses acting on the cell wall or plasma membrane are detected by a group of five membrane sensors or wall stress component (Wsc): Wsc1, Wsc2, Wsc3, Mid2, and Mtl2 (Levin 2005). The sensors carry a cysteine motif near the amino terminal end, followed by a serine/threonine-rich (STR) highly mannosylated extracellular region, a single transmembrane domain (TMD), and a relatively short cytoplasmic tail. Mannosylation of the STR region is crucial for the sensor function and has been proposed to provide them with a rod-like structure. Yet, direct pieces of evidence for the biophysical properties and supramolecular organization of the Wsc sensors were not available until now. Hereafter we provide key insights into the mechanics, clustering, and functions of yeast sensors, owing to the integration of single-molecule AFM with the modern tools of molecular genetics. While the focus is on the yeast *S. cerevisiae*, the unicellular model eukaryote, our findings should be useful to understand the behavior and function of other cellular sensors.

8.2.1 Atomic Force Microscopy of Living Cells

AFM analysis of single cells requires the firm attachment of the cells onto a surface, which is far from being a simple task. Unlike animal cells, microbes have a well-defined shape and usually have no tendency to spread on surfaces under experimental conditions. As a result, the contact area between a cell and a support is very small, often leading to cell detachment by the scanning tip. Therefore, several approaches have been developed to promote cell attachment (Dufrêne 2008; Dupres et al. 2010). A first approach is to attach negatively charged cells onto glass (or mica) supports functionalized with positively charged macromolecules, like poly-L-lysine or polyethylenimine. This method has been successfully applied to lactic bacteria, Gram-negative bacteria, and diatoms, providing novel information on their surface structure and elasticity. Cells can also be immobilized onto gelatin-coated supports. To achieve that, thin freshly cleaved mica disks are vertically dipped into gelatin solutions at 40 °C and allowed to air dry overnight by standing on edge on a paper towel. An aliquot of concentrated cell suspension is spread on the support, allowed to stand for 10 min and rinsed in deionized water. Another convenient method for cell immobilization is the use of porous polymer

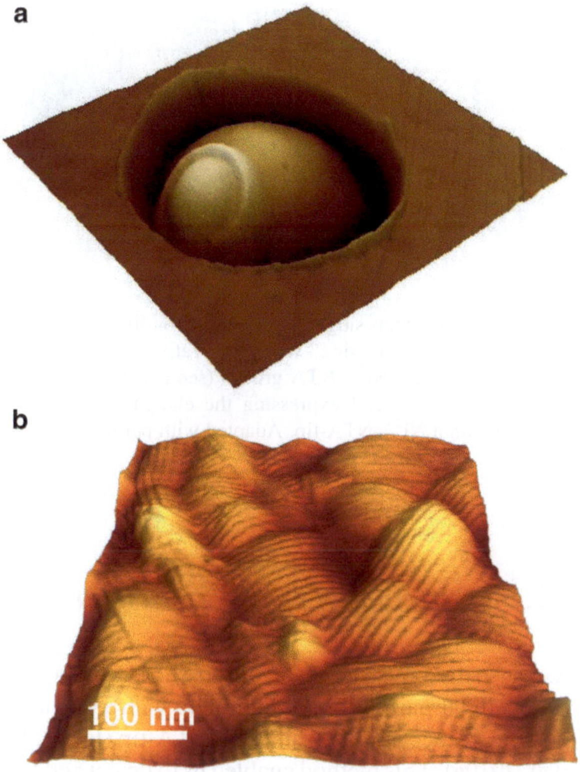

Fig. 8.1 High-resolution imaging of living microbial cells. (**a**) AFM image of a *Saccharomyces cerevisiae* yeast cell protruding from a porous polymer support. The cell shows a circular bud scar *left* after detachment of the daughter cell. (**b**) Higher-resolution AFM image of the surface of an *Aspergillus fumigatus* spore documenting the presence of ordered nanoscale rodlets. Reprinted with permission from (Dupres et al. 2010)

membranes. For this purpose, a concentrated cell suspension is gently sucked through a porous membrane with pore sizes similar to the size of the cells to be investigated. This procedure offers two advantages: it is fairly simple and does not involve a macromolecular support, thus preventing the risk of cell surface or tip contamination. However, the success of this method strongly depends on the cell geometry and on the cell surface properties. In particular, the method is generally not suited for rod-shaped bacteria.

Figure 8.1a shows a topographic image of a single cell of the baker's yeast *S. cerevisiae* trapped into a porous membrane. Using this approach, high-resolution images of the cell surface were obtained (Alsteens et al. 2008), revealing a very smooth surface that was consistent with the presence of the glucan/mannoprotein outer layer. By contrast, the surface of *Aspergillus fumigatus* spores is covered by a crystalline layer of regularly arranged rodlets (Fig. 8.1b) (Dague et al. 2008). These structures are composed of hydrophobins, a family of hydrophobic proteins of biological importance since they enhance spore dispersion by air currents and mediate the adherence to host cells. Notably, dramatic changes of the cell surface structure were observed after germination, i.e., the rodlet layer changed into a layer of amorphous material, presumably caused by the underlying polysaccharides.

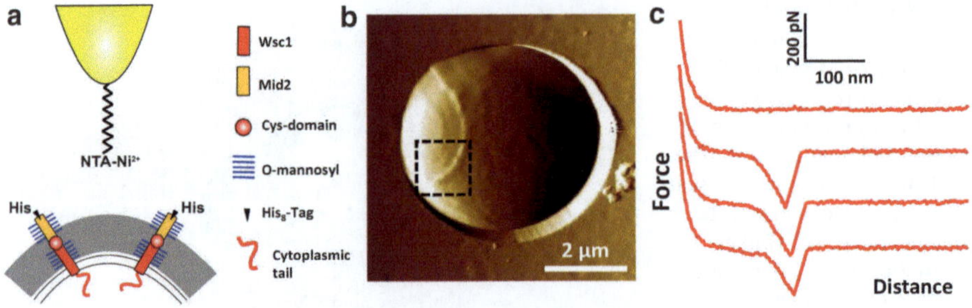

Fig. 8.2 AFM detects single Wsc1 sensors on live cells. (**a**) Force–distance *curves* were recorded on yeast cells expressing Wsc1 derivatives with an extended His-tag, using AFM tips functionalized with Ni^{2+}–NTA groups (see text for details). (**b**) AFM deflection image in buffer solution of a yeast cell expressing the elongated sensors and (**c**) representative force *curves* obtained with a Ni^{2+}–NTA-tip. Adapted with permission from (Dupres et al. 2009)

8.2.2 Detection of Single Sensors on Live Cells

As the native Wsc1 sensor is too short to reach the outermost cell surface, the extracellular part of the Mid2 protein with a His tag was added to render the sensor specifically accessible to an Ni^{2+}–NTA-tip (Fig. 8.2a) (Dupres et al. 2009; Heinisch et al. 2010a). This method enabled us to detect and force probe individual His-tagged sensors on living yeast cells (Fig. 8.2). *S. cerevisiae* cells expressing the elongated Wsc1 derivate were immobilized on a polycarbonate membrane (Fig. 8.2b). Force–distance curves recorded between the yeast surface and an NTA derivatized tip showed 162 $\pm$ 38 pN adhesion forces (Fig. 8.2c). A series of independent control experiments showed that these forces reflected the detection of individual His-tagged Wsc1 molecules. In addition to detecting individual His-tagged sensors, spatially resolved SMFS allowed us to localize them on the cell surface.

8.2.3 AFM Shows that Wsc1 Is a Nanospring

As we have seen before, force–extension curves obtained for proteins are usually well described by the WLC model. Notably, the force–extension curves recorded for Wsc1 (Fig. 8.3a) could not be fitted with the WLC model (Dupres et al. 2009). Instead, the curves showed a first nearly constant force extension, corresponding to the straightening of the extracellular polypeptide chain. Following this first regime, a linear region where the force is directly proportional to the extension was observed. This behavior is characteristic of a Hookean spring and may be interpreted as follows. Using the slope (*s*) of the linear portion of the curves, the sensor

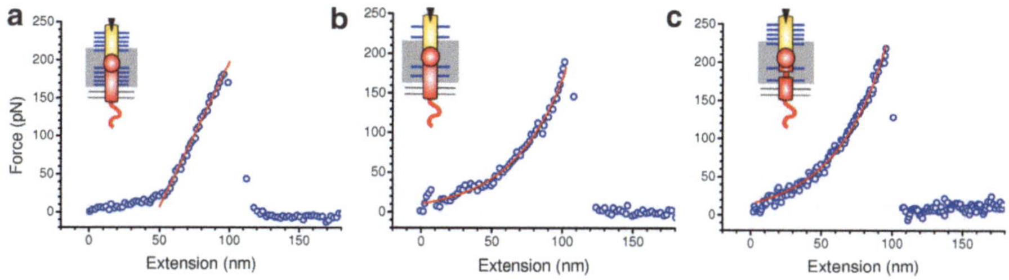

Fig. 8.3 Wsc1 sensors behave as linear springs while mutants altered in the extracellular STR region do not. (**a**) Representative force–extension *curve* obtained upon stretching a single Wsc1 molecule. Clearly visible are two extension regimes, reflecting elongation at nearly zero force, followed by a Hookean spring behavior. (**b**) Representative force–extension *curves* obtained upon stretching single Wsc1 sensors in a mutant in which mannosylation of the extracellular STR region is substantially reduced, and (**c**) in a mutant in which the STR region is interrupted by a run of glycines. Adapted with permission from (Dupres et al. 2009)

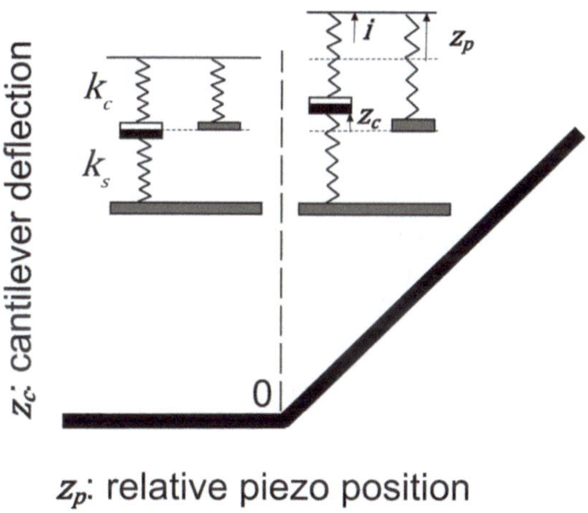

Fig. 8.4 Diagram representing the Hookean behavior of two springs in series. On the *left side* of the figure, no extension has occurred and the piezo position is defined as zero. As the piezo moves a distance Z_p, the cantilever spring (k_c) extends by a distance i and the sensor spring (k_s) extends by Z_c

spring constant (k_s) can be calculated assuming that the deflection of the cantilever is only due to the elastic deformation of the sensor and knowing the cantilever spring constant (k_c). Because both the sensor and the cantilever can be modeled as springs (Fig. 8.4), both the tip and the sensor can move during force experiment. As the piezo moves a distance Z_p, the cantilever spring (k_c) extends by a distance i and the sensor spring (k_s) extends by Z_c. The distance that the piezo has moved, Z_p, is defined as zero where the two springs first make contact and become positive as the springs extend.

The force F, at any piezo position, can be calculated as (Velegol and Logan 2002)

$$F(Z_p) = K_c Z_c = K_s i \tag{8.5}$$

For two linear springs in contact, the magnitude of the piezo movement is equal to the sum of the compression of the two springs, or

$$Z_p = Z_c + i \tag{8.6}$$

Since the value of the linear slope is the ratio of Z_c to Z_p, combining Eqs. (8.5) and (8.6) gives

$$k_s = \frac{k_c s}{1 - s} \tag{8.7}$$

Using Eq. (8.7), the spring constant of the Wsc1 sensor is found to be 4.6 $\pm$ 0.4 pN/nm, a value that is very close to the behavior of ankyrin repeats (Lee et al. 2006). We suggest that the sensor spring behavior reflects the straightening of the stiff, coil-like shape Wsc1 molecule and possibly of the whole protein-cell wall complex as well. Interestingly, we observed a substantial reduction of the sensor spring constant when lowering the salt concentration or increasing the temperature. This indicates that the mechanical behavior of Wsc1 is influenced by cell stress surface.

Glycosylation of the extracellular STR region of Wsc1 is believed to be responsible for its stiff, extended conformation (Straede and Heinisch 2007). Therefore, the relationship between the structure of Wsc1 and its mechanical properties was examined through mutagenesis. Two mutants affected in the extracellular STR-region were investigated (Dupres et al. 2009). The first mutant (Fig. 8.3b) lacking one of the protein *O*-mannosyltransferases resulted in an underglycosylation of the sensor. The second mutant (Fig. 8.3c) contained an insertion of ten glycine residues in the center of the STR region. As can been observed in Fig. 8.3b, c, both mutants no longer exhibited a spring behavior, indicating that glycosylation of Wsc1 substantially contributes to its linear spring properties. Could the measured spring properties be altered by the Mid2 elongation? To answer this question, we investigated a strain expressing His-tagged Wsc1 sensors lacking the Mid2 elongation (Dupres et al. 2009). As expected, these native sensors could only be localized on the bud scar where the cell wall is thinner. Importantly, native sensors behaved like linear nanosprings with a spring constant of 4.8 $\pm$ 0.2 pN/nm, very close to the elongated sensors.

These experiments demonstrate that AFM can be combined with advanced protein design to study the function and the mechanical properties of yeast sensors in living cells down to the single molecule level. SMFS is a valuable tool to force probe not only surface-associated proteins but also proteins that are embedded within the yeast cell wall.

8.2.4 Imaging Sensor Clustering in Live Cells

Understanding how membrane proteins assemble to form membrane micro- and nanodomains is an important question in cell biology. In higher eukaryotes, lateral microdomains (or "lipid rafts") enriched in sphingolipids and sterols have been speculated to favor segregation of specific membrane proteins like receptors or GPI-anchored proteins (Simons and Ikonen 1997; Lingwood and Simons 2010). Microdomains in higher eukaryotes are 20–100 nm in size and very transient, making their direct visualization in live cells very challenging. In contrast, stable microdomains have been observed within the plasma membrane of *S. cerevisiae* harboring specific GFP-labeled marker proteins (Stradalova et al. 2009). These microdomains are distributed into so-called MCPs (for "*m*embrane *c*ompartment with *P*ma1"), which form a network-like structure, defined by its constituent proton ATPase Pma1, and MCCs ("*m*embrane *c*ompartment with *C*an1"), which display 300 nm patches and house a number of proton symporters, as well as a component of the eisosomes (Straede and Heinisch 2007).

Using spatially resolved SMFS, we mapped the distribution of single Wsc1 sensors in living *S. cerevisiae* cells (Heinisch et al. 2010b). Adhesion maps of 1 μm × 1 μm size were recorded on cells expressing the above-mentioned His-tagged elongated sensors using Ni^{2+}–NTA tips (Fig. 8.5). Many sensor molecules appeared to cluster in areas of approximately 0.04 μm^2 and an equivalent diameter of 230 nm (Fig. 8.5b). This size of Wsc1 clusters matches the range of the 300 nm large patches reported for MCC marker proteins in *S. cerevisiae* and is clearly larger than the putative size of rafts in higher eukaryotes. Remarkably, both the total amount of wild-type Wsc1 sensors and their tendency to cluster increased when cells where stressed by either heat or in medium with low osmolarity (Figs. 8.5c, d). This correlates with the observation that stress conditions activate the cell wall integrity pathway and suggests that clustering is a stress-responsive process that is intimately connected to signaling. We propose that clustering is a means developed by yeast to locally concentrate Wsc1 sensors and their interacting downstream components. This process ultimately enhances the signal strength and the corresponding cellular response.

All Wsc-type sensors bear a cysteine-rich domain (CRD) barring eight conserved cysteine residues essential for sensor function. We therefore mapped the distribution of Wsc1 in CRD mutants, i.e., mutants in which some of these cysteines where replaced by alanine residues, leading to nonfunctional sensors (Heinisch et al. 2010b). Such mutants displayed adhesion frequencies, adhesion values, and spring behaviors similar to those of the wild-type protein. Thus, the sensors' mechanical properties are exclusively determined by the Ser/Thr-rich region (STR) as stated earlier and are not influenced by the CRD domain. However, adhesion maps of the CRD mutants demonstrated major differences in the lateral sensor organization at the surface as compared to wild-type cells, mutant sensors being evenly distributed rather than clustered.

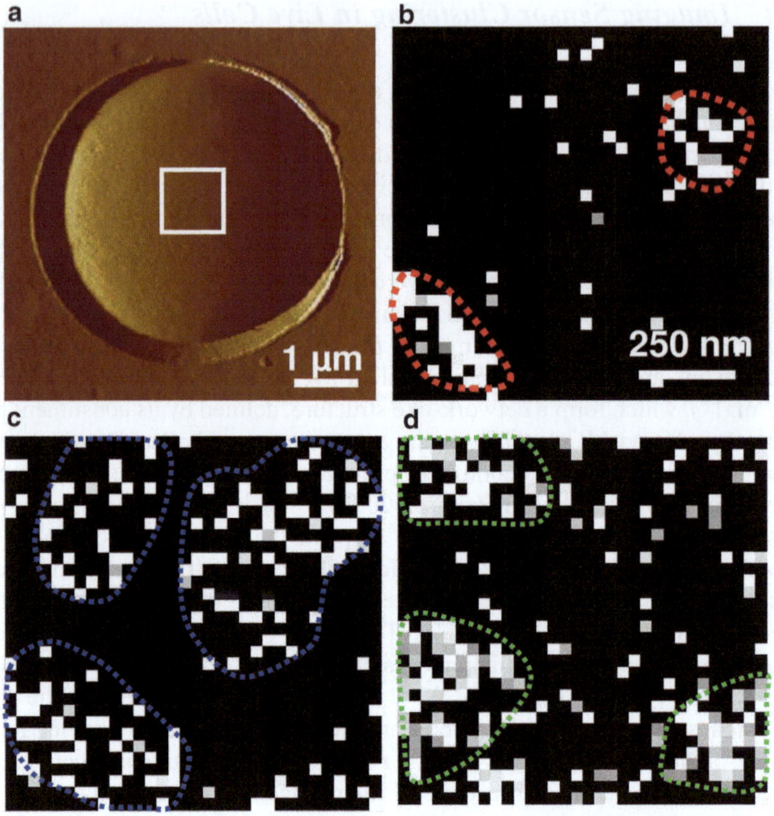

Fig. 8.5 Clustering of Wsc1 is stimulated under stressing conditions. (**a**) AFM deflection image of a yeast cell trapped into a porous polymer membrane, recorded in buffer solution at 25°C. The cells express elongated, fully functional Wsc1–Mid2 hybrid sensor bearing an His-tag. (**b**) Representative adhesion force map obtained by scanning a 1 µm × 1 µm area on the cell surface with a Ni^{++}–NTA-tip. The heterogeneous distribution of the bright pixels, which represent the detection of single sensors, clearly documents the formation of nanoscale clusters (highlighted by *dotted red lines*). (**c, d**) Adhesion force maps (1 µm × 1 µm) recorded with a Ni^{++}–NTA-tip either in buffer solution at 37°C (**c**, heat shock) or in deionized water at 25°C (**d**, hypoosmotic shock). Stressing conditions strongly enhance Wsc1 clustering (clusters are highlighted by *dotted blue* and *green lines*). Reprinted with permission from (Heinisch et al. 2010b)

We further demonstrated that Wsc1 clustering involves disulfide bridges of the CRD domain (Dupres et al. 2011). While Wsc1 formed nanoscale clusters on native cells, clustering was no longer observed after treatment with the reducing agent dithiothreitol (DTT), indicating that intra- or intermolecular disulfide bridges are required for clustering. Moreover, DTT treatment resulted in a significant increase in cell surface roughness, suggesting that disulfide bridges between other cell wall proteins are crucial for proper cell surface topology. These remarkable sensor properties may well apply to other sensors and receptors with CRDs throughout biology.

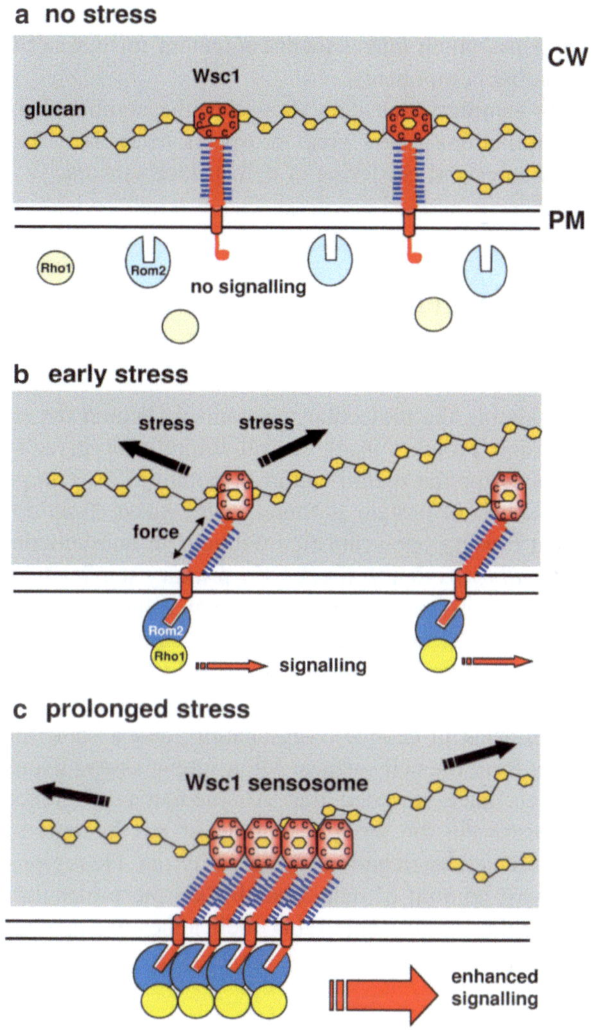

Fig. 8.6 AFM unravels the Wsc1 "sensosome." The schematic drawing presents our current working model of sensor function and signal enhancement by clustering. In the absence of stress (**a**) the sensors do not cluster, are in loose contact with the cell wall by their CRD domain, and their cytoplasmic tail does not interact with the downstream signaling components. Under stress conditions, the sensors are stretched by mechanical forces (**b**), leading to a conformational change in the CT and a recruitment of the interaction partners at the cytoplasmic face of the plasma membrane. Protein/protein interactions of the deformed CRD domains induce sensor clustering, including the downstream components, thus forming a "sensosome" with enhanced signaling capacity (**c**). Reprinted with permission from (Heinisch et al. 2010b)

These findings lead us to propose that in yeast, like in higher eukaryotes, sensor function is coupled to a localized enrichment of sensors/receptors in membrane patches. We expect our newly observed Wsc1 nanoclusters, for which we propose the term Wsc1 "sensosome" (Fig. 8.6), to promote the concomitant accumulation of downstream signaling components of the CWI signaling pathway at the inner leaflet of the plasma membrane. Regarding the mechanics of the signal reception, stretching of either the cell wall or the plasma membrane could alter the conformation of the CRD domain, thereby exposing interfaces to promote intermolecular protein–protein interactions. These would trigger the association of further sensor molecules within the plasma membrane. In addition to these changes in the CRD domain, all sensor molecules will also assume a conformation of their cytoplasmic domains, which makes them competent for interaction with the downstream CWI signaling components.

In summary, our data show that the combined method of SMFS with genetic manipulations offers great prospects to explore the mechanisms underlying the mechanics and clustering of cell surface sensors.

8.3 Nanomechanics and Nanoclustering of Cell Adhesion Proteins

Elucidating the molecular mechanisms behind the strength and mechanics of cell adhesion proteins is of central importance in cell biology and offers exciting avenues for the identification of potential drug targets. A prominent example of cell adhesion protein is found in the yeast *Candida albicans*, the most common agent causing opportunistic infections in immunocompromised patients. Adhesion of *C. albicans* to host tissues, the primary step leading to infection, is mediated by a family of cell surface proteins known as the agglutinin-like sequence (Als) proteins (Hoyer 2001; Dranginis et al. 2007). Als proteins possess four functional regions (Fig. 8.7a): an immunoglobulin (Ig)-like region near the N terminus that initiates cell adhesion, a threonine-rich region (T), a tandem repeat (TR) region that participates in cell–cell aggregation, and a stalk region projecting the molecule away from the cell surface. Microscopic assays using *S. cerevisiae* surface display models have revealed that Als-mediated adherence involves two steps: initial adhesion via the specific binding of the Ig region to peptide ligands, followed by cell–cell aggregation mediated by the TR regions (Lipke et al. 2012). Despite the vast amount of data available on the molecular biology of Als proteins, the molecular details underlying their adhesion, mechanics, and clustering have not been elucidated yet.

8.3.1 *Unfolding Single Als5p Proteins in Vitro*

To probe the mechanical unfolding of isolated Als5p proteins, soluble Ig-T-TR_6 fragments terminated with a His tag were grafted onto gold surfaces modified with NTA groups and stretched by their Ig domain using AFM tips modified with Ig-T fragments (Fig. 8.7b) (Alsteens et al. 2009). Force–extension curves showed a saw-tooth pattern with well-defined peaks (Fig. 8.7c). The first six peaks showed ascending forces ranging from 160 to 240 pN corresponding to the force-induced unfolding of the secondary structures of individual TR domains and could be fitted by the WLC model (Fig. 8.7c, red lines). Our measured forces (160–240 pN) were in the range of unfolding forces reported for β-fold domains, such as Ig and fibronectin domains in titin

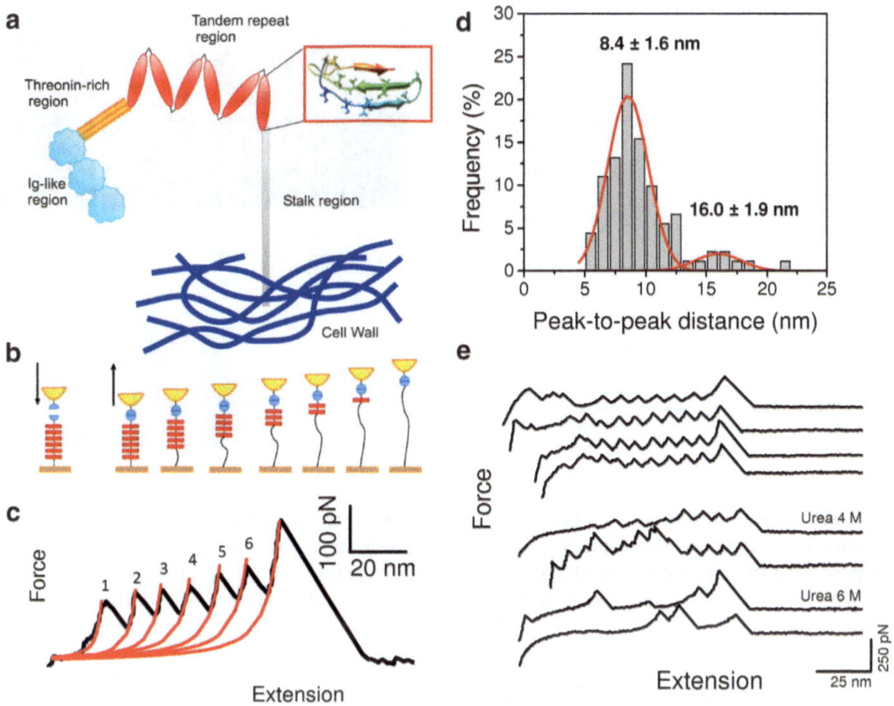

Fig. 8.7 Unfolding isolated Als5p proteins. (**a**) Representation of the Als5p from *C. albicans*. The tandem repeat (TR) region is made of multiple glycosylated 36-amino acid repeats that are arranged in antiparallel β-sheets (*inset*). (**b**) Principle of the SMFS experiments. Ig-T-TR_6 fragments were attached on a gold surface and stretched via their Ig domains using an Ig-T-tip. (**c**) Force–extension *curves* obtained by stretching single Ig-T-TR_6 showed periodic features reflecting the sequential unfolding of the TR domains. Force peaks were well described by the WLC model (*red lines*). (**d**) Histogram of the peak-to-peak distances for the first six peaks. (**e**) Addition of urea dramatically altered the unfolding peaks due to hydrogen bond disruption (*lower traces*). Adapted with permission from (Alsteens et al. 2009)

(Rief et al. 1997b) and tenascin (Oberhauser et al. 1998), and larger than the forces needed to unfold α-helical domains (Rief et al. 1999). The change in contour length between the first consecutive six peaks was constant, $\Delta L_c = 8.4 \pm 1.6$ nm, and very close to the peak-to-peak distance (Fig. 8.7d) revealing that the chain is fully extended at this force. This length corresponds to the expected 36 amino acids of a single TR repeat supporting our interpretation of the sawtooth pattern. The seventh peak showed a much higher force and larger spacing and could be attributed to the stretching of Ig domains, leading to the rupture of the Ig–Ig complex. Urea altered the shape of the unfolding peaks (Fig. 8.7e, lower traces), reflecting a loss of the mechanical stability of the TR domains due to hydrogen bond disruption. This observation correlated with cellular behavior since Als5p-mediated adhesion is known to be reversibly inhibited by urea (Rauceo et al. 2006).

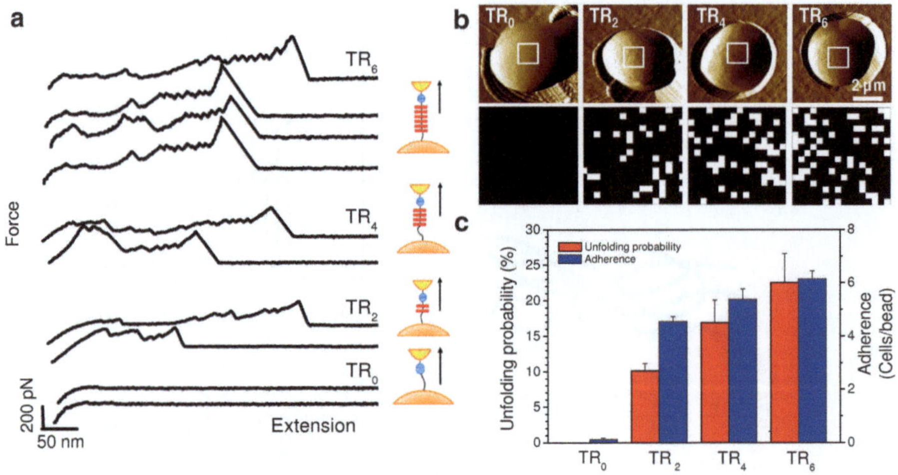

Fig. 8.8 Unfolding Als5p proteins on live cells. (**a**) Force–extension *curves* on *S. cerevisiae* expressing Als5p with six, four, two, and no TR. (**b**) Mapping unfolding forces on yeast cells expressing different numbers of TR. (**c**) Histogram showing the unfolding probability and the level of yeast adherence as a function of the number of TR. Reprinted with permission from (Alsteens et al. 2009)

8.3.2 Unfolding Single Als5p Proteins in Vivo

Notably, single Als proteins were localized and unfolded directly on live cells (Alsteens et al. 2009). Force–extension curves were recorded across the surface of *S. cerevisiae* cells expressing Als5p with different numbers of TR (Fig. 8.8a). The curves obtained on cells expressing six repeats (TR_6) were similar to those found on isolated proteins. Cells expressing a smaller number of repeats showed patterns with fewer force peaks. The distribution of unfolding forces over the cell surface (Fig. 8.8b) revealed an increasing number of exposed proteins with the number of TR. As shown in Fig. 8.8c, cells expressing a smaller number of repeats showed lower unfolding probability. The level of yeast adherence (number of cells per fibonectin-coated bead) was correlated with the unfolding probability (Fig. 8.8c), suggesting that the mechanical properties of the TR region are important for yeast adherence. These SMFS experiments demonstrate that stretching of Als5p leads to the unfolding of their individual TR domains. Presumably, unfolding of the domains leads to the exposure of hydrophobic groups that would promote fungal adhesion.

8.3.3 Imaging Force-Induced Amyloid Nanodomains

In animal cells, there is evidence that the early stage of adhesion involves the formation of adhesion domains composed of aggregated adhesion proteins (Vogel and Sheetz 2006; Geiger et al. 2009). Such adhesion domains have been shown to grow and strengthen under force (Bershadsky et al. 2006; Smith et al. 2008). Until now, however, such force-dependent clustering has never been reported for microbial adhesins. We used SMFS to demonstrate the force-induced formation and propagation of Als5p adhesin nanodomains in living yeast cells.

8.3.3.1 Dual Detection of Single Als5p Proteins

We mapped the distribution of Als5p on *S. cerevisiae* surface display models using SMFS with tips bearing specific antibodies (Alsteens et al. 2010). To this end, a V5 epitope tag was inserted at the N-terminal end of the protein (Fig. 8.9a), and AFM tips terminated with anti-V5 antibodies (Fig. 8.9b) were used to detect and map individual Als5p by recording spatially resolved force curves on 1 μm^2 areas. As shown in Fig. 8.9c, 16% of the force curves featured adhesion force peaks. The corresponding adhesion force histogram displayed four maxima centered at 45 ± 15 pN, 98 ± 32 pN, 180 ± 12 pN, and 220 ± 20 pN. The two first peaks were attributed to the relatively weak binding of the anti-V5 antibody to the V5 epitope. Strikingly, force curves with large adhesion peaks (~180 pN and ~220 pN) showed profiles characteristic of Als5p binding to protein ligands (Fig. 8.9d, red curves). These curves showed sawtooth patterns with multiple force peaks, which we reported earlier as the sequential unfolding of the Als5p TR domains. These observations lead us to conclude that the anti-V5 tip is capable of dual detection, i.e., molecular recognition of Als5p via the V5 epitope or strong multipoint attachment to ligands on the antibody (Fig. 8.9d, insets).

8.3.3.2 Amyloid Nanodomains Are Triggered by Force

What is the spatial arrangement of Als5p proteins on the cell surface? To answer this question, spatially resolved SMFS was used to localize individual Als5p proteins on live cells (Alsteens et al. 2010). Adhesion force maps on 1 μm^2 obtained from wild-type cells (Fig. 8.10a, first column) on which force was never applied revealed that the proteins were homogeneously distributed (Fig. 8.10b, first column). Recording a second map on the same area led to an increase in the total density from 172 ± 16 proteins/μm^2 to 268 ± 13 proteins/μm^2. In addition, proteins in the second force map formed clusters of 100–500 nm (Fig. 8.10c, first column). As shown in Fig. 8.10e, the fraction of clustered molecules increased from 8 to 61%. This dramatic change in spatial organization was reproducibly observed upon recording additional maps.

We observed that heat-killed cells (WT_k) showed the same behavior as live cells (Fig. 8.10a–c, second column), i.e., increased Als5p density (Fig. 8.10e) and nanodomains formation. These results show that the force-induced clustering of Als5p is independent of cellular metabolic activity. We also found that clustering is abolished with a single-site mutation in the conserved amyloid-forming sequence.

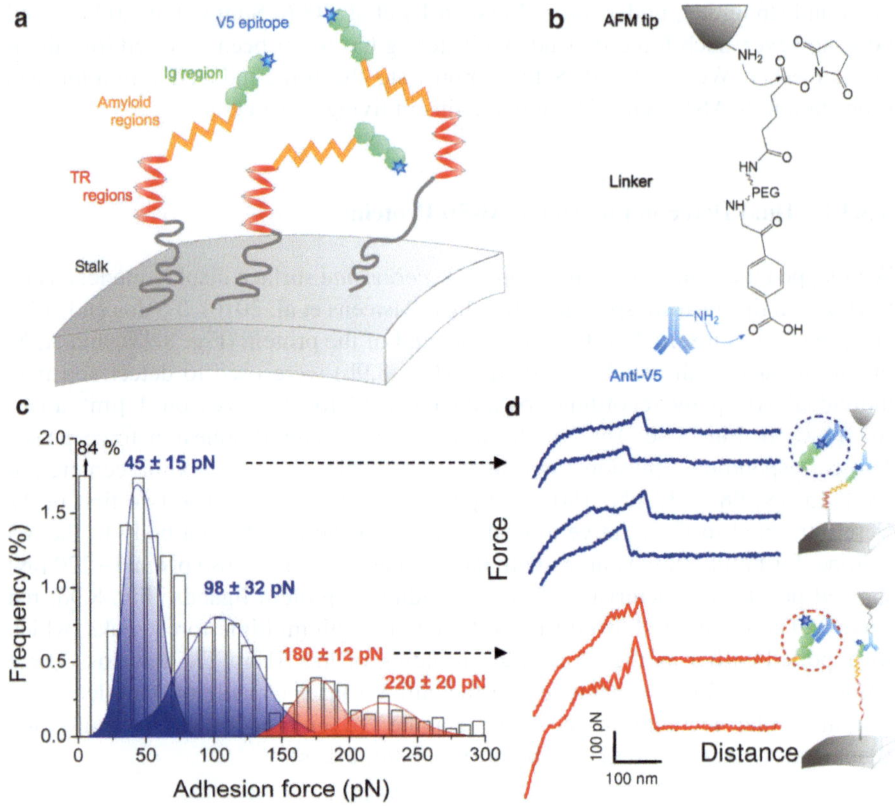

Fig. 8.9 Dual detection of single Als5p on live cells. (**a**, **b**) Principle of the SMFS experiment. *S. cerevisiae* cells expressing Als5p proteins tagged with a V5 epitope (**a**) were probed using AFM tips terminated with an anti-V5 antibody (**b**). Adhesion force histogram from four maps of 1,024 data points (**c**) and representative force *curves* (**d**) obtained by recording spatially resolved force *curves* over 1 × 1-μm areas of the cell surface using anti-V5 tips. The anti-V5 tip is capable of dual detection: the *blue curves* show single weak adhesion peaks reflecting recognition of the V5-tag, and the *red curves* feature sawtooth patterns with multiple force peaks documenting the unfolding of the entire protein via Ig binding. Reprinted with permission from (Alsteens et al. 2010)

A V326N mutation in the Als5p sequence reduces TANGO β-aggregation potential of the amyloid region from 93 to 4% (Otoo et al. 2008). To test whether the amyloid-forming sequence is critical to nanodomains formation, the V326N mutation was incorporated into full length Als5p. Remarkably, Als5p clustering properties were almost completely abolished in the V326N mutant (Fig. 8.10a–c, third column). Since mutant cells also showed reduced aggregation, our data suggest (1) that amyloid interactions represent the driving force underlying Als5p clustering and (2) that the resulting adhesion domains play a pivotal role in cellular aggregation.

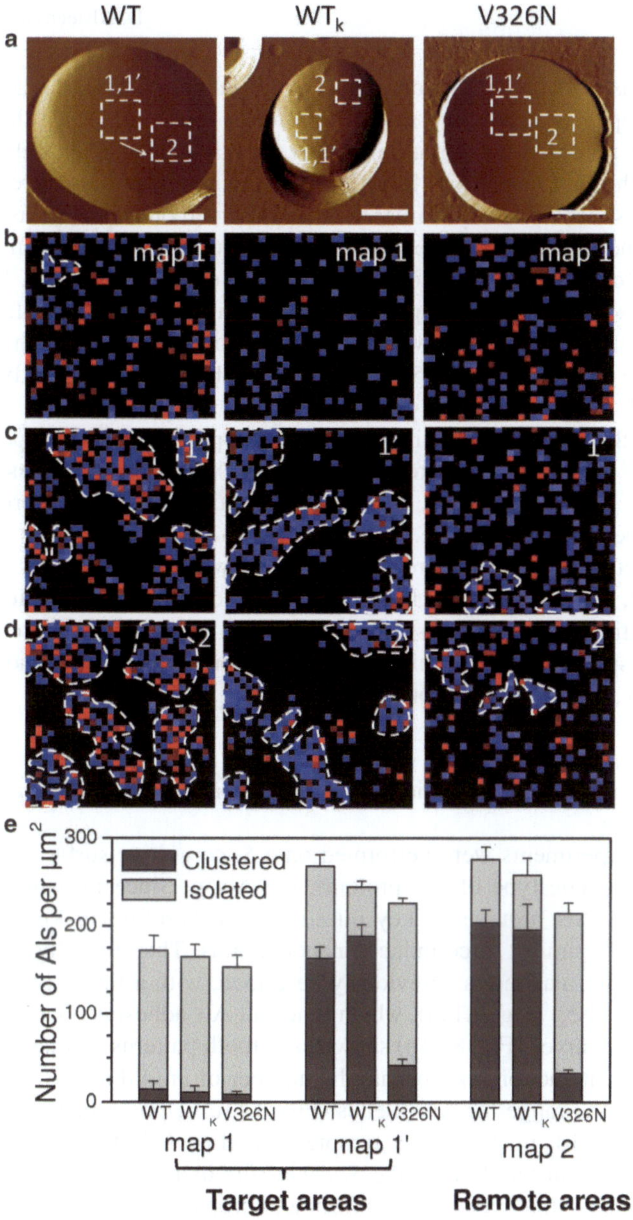

Fig. 8.10 Formation and propagation of Als5p nanodomains. (**a**) AFM topographic images of wild-type (WT, *first column*), heat-killed (WT$_k$, *second column*), and V326N mutant (V326N, *third column*) *S. cerevisiae* cells expressing V5-tagged proteins. (**b**) Adhesion force maps recorded with an anti-V5 tip on a given target area of the native cells, i.e., cells that were never subjected to force (maps #1; recorded on the *dashed squares* in (**a**). *Blue* and *red* pixels correspond to forces smaller and larger than 150 pN, respectively, thus to V5-tagged Als5p recognition and unfolding. (**c**) Second adhesion force maps recorded on the same target area (maps #1′). Formation of nanoscale clusters is highlighted by *dashed lines*. (**d**) Adhesion force maps recorded on remote areas (maps #2) localized several hundred nanometers away (see *dashed squares* in **a**). (**e**) Surface density histograms showing the number of proteins/µm^2 measured for WT, WT$_k$, and for V326N cells in three different conditions. Reprinted with permission from (Alsteens et al. 2010)

We then asked whether the force-induced nanodomains are localized or whether they can propagate across the surface (Alsteens et al. 2010). To this end, force curves were recorded on remote areas, i.e., areas localized several micrometers away from the first maps (Fig. 8.10d). Strikingly, nanodomains were observed on remotes areas that were very similar to those observed on targets areas for both WT and WT_k. These results indicate that delivering piconewton forces on a target area induces the formation of Als5p clusters across the entire cell surface. The extended length of Als proteins can be up to 500 nm, giving a radius of gyration of almost 1 μm across the cell surface. Owing to this radius of gyration, Als adhesins are able to reorganize on a scale of 100–500 nm even though they are covalently anchored to the cell wall.

Because clusters were essentially never seen in primary maps, we postulated that nanodomain formation and propagation are rather slow processes. To further analyze this time dependency, cells were preactivated by force probing during increasing periods of time, i.e., 12.5, 25, and 37.5 min (Alsteens et al. 2010). These preactivation steps were followed by two classical maps. From these experiments, we learned that (1) 25 min of preactivation induces formation, but not propagation of the clusters; (2) propagation occurs in a timescale of above 35 min at a speed of ~20 nm/min. Hence, nanodomain formation and propagation are two slow, time-dependent processes.

8.3.3.3 Formation of Amyloid Nanodomains in *Candida albicans*

The SMFS experiments were performed on a *S. cerevisiae* surface display model expressing a single type of Als proteins, the Als5p. Since in nature *C. albicans* displays eight Als proteins, a key question was therefore to know whether *C. albicans* shows similar force-induced nanodomains. Therefore, living *C. albicans* yeast cells were mapped, as previously described, with a tip derivatized with the Ig-T form of Als5p (Fig. 8.11a), which binds all Als adhesins (Garcia et al. 2011). The extension curves (Fig. 8.11b) showed sawtooth patterns similar to those of the Als5p–Als5p interaction, except that the number of unfolding TR domains varied from 3 to 33, the range for Als allelles (Fig. 8.11c). Upon remapping of the same region (Fig. 8.11d, map 1′) or a remote area (Fig. 8.11d, map 2), the adhesin molecules were clustered, as in the *S. cerevisiae* model. These observations lead us to conclude (1) that nanodomains form and propagate in response to force in *C. albicans*, just as they do in Als5p-expressing *S. cerevisiae* and (2) that our findings made on the surface display model are biologically relevant.

8.3.3.4 Biological Significance of Amyloid Nanodomains

Amyloids are protein aggregates formed by the ordered crystallization of identical short sequences in many molecules of the same protein. Although protein amyloids are characteristic of pathological conditions, including neurodegenerative diseases

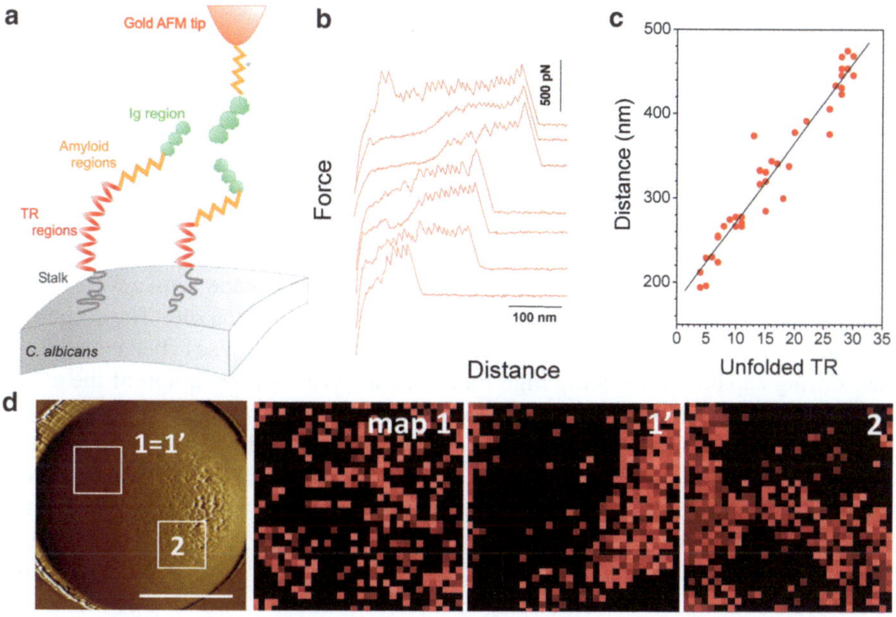

Fig. 8.11 AFM reveals Als adhesion nanodomains in *C. albicans*. (**a**) Principle of the single-molecule AFM experiment. *C. albicans* cells were probed, in buffer, using AFM tips terminated with Ig-T-tip. (**b**) Force–extension *curves* obtained by stretching Als proteins showed periodic features reflecting the sequential unfolding of the TR domains. (**c**) The rupture distance depended on the number of unfolded TR. (**d**) Imaging nanodomains: from *left* to *right*, AFM topographic image showing a *C. albicans* cell (scale bar: 2 μm), adhesion force map (1 μm × 1 μm) recorded with an Ig tip on a given target area of the native cell (map #1 recorded on the *dashed squares* in topographic image; *red scale*: 300 pN), second adhesion force map recorded on the same target area (map #1′), and third adhesion force map recorded on a remote area (map #2) localized several hundred nanometers away from the first map. Adapted with permission from (Garcia et al. 2011)

(Barten and Albright 2008), they have also been shown to have positive functional consequences. In microbes, functional amyloids have been shown to occur in various proteins, including adhesive bacterial curli (Barnhart and Chapman 2006) and fungal adhesins (Ramsook et al. 2010).

Our single-molecule experiments, combined with genomics, biochemistry, and cell biology approaches, have provided new insights into the fascinating mechanisms underlying Als-mediated adhesion (Lipke et al. 2012). The strength of adhesion results from force-activated amyloid-like clustering of hundreds of adhesin molecules to form arrays of ordered multimeric binding sites. Hence, AFM has demonstrated a novel function for amyloids in eukaryotes, i.e., the clustering and activation of cell adhesion proteins in response to force. Because bioinformatics shows similar amyloid sequences in many cell adhesion proteins, and force-induced activation is a common biological phenomenon, these observations imply that force-dependent amyloid formation may be a general mechanism for activating cell adhesion.

8.4 Conclusions

Understanding how single proteins localize, assemble, and interact on the surface of living cells is an important challenge. Of particular interest is to decipher how surface proteins respond to mechanical stimuli and how they dynamically assemble into nanodomains. This chapter demonstrates that AFM offers unprecedented possibilities for probing the mechanics, organization, assembly, and interactions of single proteins on live cells in relation with function (mechanosensing, adhesion).

AFM unraveled the mechanical properties of the membrane sensor Wsc1 from *S. cerevisiae* and of the cell adhesion protein Als5p from *C. albicans*. The combined use of SMFS with genetic manipulations revealed that Wsc1 behaves like a nanospring capable of resisting high mechanical force. Glycosylation of the extracellular Ser/Thr region of Wsc1 is responsible for its stiff, extended conformation and is thus required for its spring properties. Stressing conditions (deionized water or elevated temperature) were shown to substantially reduce the sensor spring constant, supporting the notion that sensor stretching plays a role in activating the CWI signaling pathway. By contrast, stretching single Als proteins on yeast cells with a biologically modified AFM tip revealed sawtooth patterns with multiple force peaks, each peak corresponding to the force-induced unfolding of individual tandem repeats engaged in cell adhesion. We suggest that the force-induced unfolding of Als proteins leads to extend conformations in which hydrophobic groups are freshly exposed, thus favoring hydrophobic interactions between opposing cells.

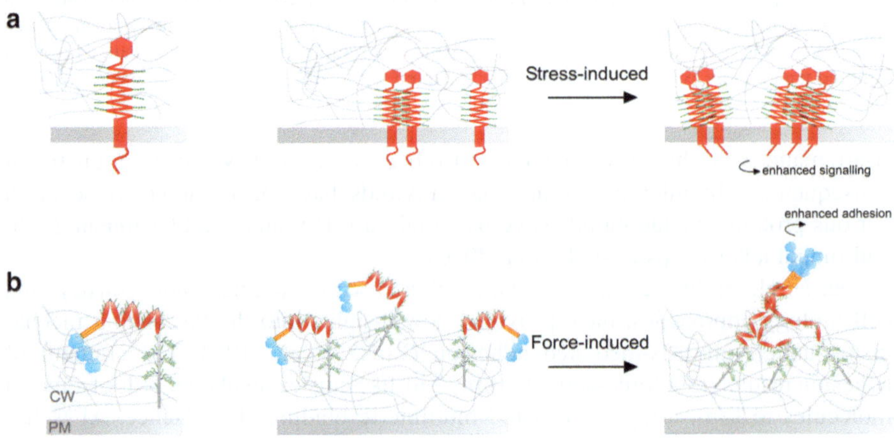

Fig. 8.12 AFM demonstrates that Wsc1 and Als5p form functional protein nanodomains on yeast cells in response to stress and that these domains are important for adhesion and mechanosensing. *Cartoon* illustrating the stress induced clustering of Wsc sensor (**a**) and the force induced clustering of Als adhesins (**b**). *Left panels*: the different regions of the proteins in their native, unstressed state. *Right panels*: clustered molecules resulting from stress conditions (**a**) or mechanical stimuli (**b**). *CW* cell wall, *PM* plasma membrane

Single-molecule imaging demonstrated that Wsc1 and Als5p form functional protein nanodomains on yeast cells in response to stress and that these domains are important for adhesion and mechanosensing. Wsc1 sensors were shown to form clusters of ~200 nm size. Clustering was strongly enhanced in stress conditions (osmotic stress, elevated temperature) (Fig. 8.12a), indicating that signaling is coupled to the localized enrichment of sensors within membrane patches (nanosensosomes). For Als5p, mechanical stimuli trigger the formation of nanodomains (Fig. 8.12b). Pulling on single Als5p proteins with AFM tips functionalized with specific antibodies induced the formation and propagation of amyloid adhesion domains of 100–500 nm size (nanoadhesomes). We suggest that clustering of adhesins in response to mechanical stimuli may be a general mechanism for activating cell adhesion in microbial pathogens.

Acknowledgments Work in our team was supported by the National Foundation for Scientific Research (FNRS), the Université catholique de Louvain (Fonds Spéciaux de Recherche), the Federal Office for Scientific, Technical and Cultural Affairs (Interuniversity Poles of Attraction Programme), and the Research Department of the Communauté française de Belgique (Concerted Research Action). Y.F.D. and D.A. are Senior Research Associate and Postdoctoral Researcher of the FRS-FNRS.

References

Alsteens D, Dupres V, Klotz SA, Gaur NK, Lipke PN, Dufrene YF (2009) Unfolding individual Als5p adhesion proteins on live cells. ACS Nano 3:1677–1682

Alsteens D, Dupres V, Mc Evoy K, Wildling L, Gruber HJ, Dufrêne YF (2008) Structure, cell wall elasticity and polysaccharide properties of living yeast cells, as probed by AFM. Nanotechnology 19:384005

Alsteens D, Garcia MC, Lipke PN, Dufrene YF (2010) Force-induced formation and propagation of adhesion nanodomains in living fungal cells. Proc Natl Acad Sci USA 107:20744–20749

Barnhart MM, Chapman MR (2006) Curli biogenesis and function. Annu Rev Microbiol 60:131–147

Barten DM, Albright CF (2008) Therapeutic strategies for Alzheimer's disease. Mol Neurobiol 37:171–186

Bershadsky AD, Kozlov M, Geiger B (2006) Adhesion-mediated mechanosensitivity: a time to experiment, and a time to theorize. Curr Opin Cell Biol 18:472–481

Bustamante C, Macosko JC, Wuite GJL (2000) Grabbing the cat by the tail: manipulating molecules one by one. Nat Rev Mol Cell Biol 1:130–136

Bustamante C, Marko JF, Siggia ED, Smith S (1994) Entropic elasticity of lambda-phage DNA. Science 265:1599–1600

Dague E, Alsteens D, Latgé JP, Dufrêne YF (2008) High-resolution cell surface dynamics of germinating *Aspergillus fumigatus* conidia. Biophys J 94:656–660

Dranginis AM, Rauceo JM, Coronado JE, Lipke PN (2007) A biochemical guide to yeast adhesins: glycoproteins for social and antisocial occasions. Microbiol Mol Biol Rev 71:282–294

Dufrêne YF (2008) Atomic force microscopy and chemical force microscopy of microbial cells. Nat Protoc 3:1132–1138

Dupres V, Alsteens D, Andre G, Dufrene YF (2010) Microbial nanoscopy: a closer look at microbial cell surfaces. Trends Microbiol 18:397–405

Dupres V, Alsteens D, Wilk S, Hansen B, Heinisch JJ, Dufrene YF (2009) The yeast Wsc1 cell surface sensor behaves like a nanospring in vivo. Nat Chem Biol 5:857–862

Dupres V, Heinisch JJ, Dufrêne YF (2011) Atomic force microscopy demonstrates that disulfide bridges are required for clustering of the yeast cell wall integrity sensor wsc1. Langmuir 27:15129–15134

Dupres V, Verbelen C, Dufrêne YF (2007) Probing molecular recognition sites on biosurfaces using AFM. Biomaterials 28:2393–2402

Ebner A, Wildling L, Kamruzzahan ASM, Rankl C, Wruss J, Hahn CD, Holzl M, Zhu R, Kienberger F, Blaas D, Hinterdorfer P, Gruber HJ (2007) A new, simple method for linking of antibodies to atomic force microscopy tips. Bioconjug Chem 18:1176–1184

Galera-Prat A, Gomez-Sicilia A, Oberhauser AF, Cieplak M, Carrion-Vazquez M (2010) Understanding biology by stretching proteins: recent progress. Curr Opin Struct Biol 20:63–69

Garcia MC, Lee JT, Ramsook CB, Alsteens D, Dufrene YF, Lipke PN (2011) A role for amyloid in cell aggregation and biofilm formation. PLoS One 6:e17632

Geiger B, Spatz JP, Bershadsky AD (2009) Environmental sensing through focal adhesions. Nat Rev Mol Cell Biol 10:21–33

Heinisch JJ, Dupres V, Alsteens D, Dufrêne YF (2010a) Measurement of the mechanical behavior of yeast membrane sensors using single-molecule atomic force microscopy. Nat Protoc 5:670–677

Heinisch JJ, Dupres V, Wilk S, Jendretzki A, Dufrêne YF (2010b) Single-molecule atomic force microscopy reveals clustering of the yeast plasma-membrane sensor Wsc1. PLoS One 5:9

Hinterdorfer P, Dufrêne YF (2006) Detection and localization of single molecular recognition events using atomic force microscopy. Nat Methods 3:347–355

Hoyer LL (2001) The ALS gene family of *Candida albicans*. Trends Microbiol 9:176–180

Lee G, Abdi K, Jiang Y, Michaely P, Bennett V, Marszalek PE (2006) Nanospring behaviour of ankyrin repeats. Nature 440:246–249

Levin DE (2005) Cell wall integrity signaling in *Saccharomyces cerevisiae*. Microbiol Mol Biol Rev 69:262–291

Li HB, Cao Y (2010) Protein mechanics: from single molecules to functional biomaterials. Acc Chem Res 43:1331–1341

Lingwood D, Simons K (2010) Lipid rafts as a membrane-organizing principle. Science 327:46–50

Lipke PN, Garcia MC, Alsteens D, Ramsook CB, Klotz SA, Dufrene YF (2012) Strengthening relationships: amyloids create adhesion nanodomains in yeasts. Trends Microbiol 20(2):59–65

Oberhauser AF, Marszalek PE, Erickson HP, Fernandez JM (1998) The molecular elasticity of the extracellular matrix protein tenascin. Nature 393:181–185

Oesterhelt F, Oesterhelt D, Pfeiffer M, Engel A, Gaub HE, Muller DJ (2000) Unfolding pathways of individual bacteriorhodopsins. Science 288:143–146

Otoo HN, Lee KG, Qiu WG, Lipke PN (2008) *Candida albicans* Als adhesins have conserved amyloid-forming sequences. Eukaryot Cell 7:768–782

Puchner EM, Gaub HE (2009) Force and function: probing proteins with AFM-based force spectroscopy. Curr Opin Struct Biol 19:605–614

Ramsook CB, Tan C, Garcia MC, Fung R, Soybelman G, Henry R, Litewka A, O'Meally S, Otoo HN, Khalaf RA, Dranginis AM, Gaur NK, Klotz SA, Rauceo JM, Jue CK, Lipke PN (2010) Yeast cell adhesion molecules have functional amyloid-forming sequences. Eukaryot Cell 9:393–404

Rauceo JM, De Armond R, Otoo H, Kahn PC, Klotz SA, Gaur NK, Lipke PN (2006) Threonine-rich repeats increase fibronectin binding in the *Candida albicans* adhesin Als5p. Eukaryot Cell 5:1664–1673

Rief M, Oesterhelt F, Heymann B, Gaub HE (1997a) Single molecule force spectroscopy on polysaccharides by atomic force microscopy. Science 275:1295–1297

Rief M, Gautel M, Oesterhelt F, Fernandez JM, Gaub HE (1997b) Reversible unfolding of individual titin immunoglobulin domains by AFM. Science 276:1109–1112

Rief M, Pascual J, Saraste M, Gaub HE (1999) Single molecule force spectroscopy of spectrin repeats: low unfolding forces in helix bundles. J Mol Biol 286:553–561

Simons K, Ikonen E (1997) Functional rafts in cell membranes. Nature 387:569–572

Smith AS, Sengupta K, Goennenwein S, Seifert U, Sackmann E (2008) Force-induced growth of adhesion domains is controlled by receptor mobility. Proc Natl Acad Sci USA 105:6906–6911

Sotomayor M, Schulten K (2007) Single-molecule experiments *in vitro* and *in silico*. Science 316:1144–1148

Stradalova V, Stahlschmidt W, Grossmann G, Blazikova M, Rachel R, Tanner W, Malinsky J (2009) Furrow-like invaginations of the yeast plasma membrane correspond to membrane compartment of Can1. J Cell Sci 122:2887–2894

Straede A, Heinisch JJ (2007) Functional analyses of the extra- and intracellular domains of the yeast cell wall integrity sensors Mid2 and Wsc1. FEBS Lett 581:4495–4500

Velegol SB, Logan BE (2002) Contributions of bacterial surface polymers, electrostatics, and cell elasticity to the shape of AFM force curves. Langmuir 18:5256–5262

Vogel V, Sheetz M (2006) Local force and geometry sensing regulate cell functions. Nat Rev Mol Cell Biol 7:265–275

Chapter 9
Individual Proteins Under Mechanical Stress: Lessons from Theory and Computer Simulations

Dmitrii E. Makarov

9.1 Introduction

This chapter focuses on theoretical ideas required to explain how proteins respond to mechanical forces. Particular emphasis will be placed on the scenario where the ends of the polypeptide chain are pulled apart, which eventually results in the unfolding of the protein (Fig. 9.1). This scenario commonly arises in single-molecule experiments that employ either the atomic force microscope (AFM) or laser tweezers to pull on biomolecules. Mechanically induced protein unfolding also occurs in living organisms. For example, it is believed to account for viscoelasticity of muscle fiber (Minajeva et al. 2001) and for high toughness and strength of natural materials such as spider silk (Becker et al. 2003; Fantner et al. 2006; Smith et al. 1999). Mechanically driven protein unfolding has also been implicated as a necessary step in protein degradation (Lee et al. 2001; Makarov 2009b; Matouschek 2003; Matouschek and Bustamante 2003; Prakash and Matouschek 2004; Shariff et al. 2004). Furthermore, mechanical forces can be used to control enzymatic activity through mechanically induced conformational changes (Alegre-Cebollada et al. 2010). Conversely, conformational transitions can be used to generate mechanical forces, both in biological, protein-based molecular motors (Fisher and Kolomeisky 1999, 2001) and in synthetic nanomachines (McCullagh et al. 2011). More broadly, the above phenomena, all exhibiting coupling between chemical and mechanical processes, are embraced by the quickly growing field of mechanochemistry (Beyer and Clausen-Schaumann 2005; Boulatov and Kucharski 2011; Brantley et al. 2011; Kochhar et al. 2010; Ribas-Arino et al. 2009). Central to mechanochemistry is the concept of a force dependent transition rate, whether it is the rate of unfolding or of any other conformational

D.E. Makarov (✉)
Department of Chemistry and Biochemistry and Institute for Computational Engineering and Sciences, University of Texas at Austin, Austin, TX, USA
e-mail: makarov@cm.utexas.edu

A.F. Oberhauser (ed.), *Single-molecule Studies of Proteins*, Biophysics for the Life Sciences 2, DOI 10.1007/978-1-4614-4921-8_9,

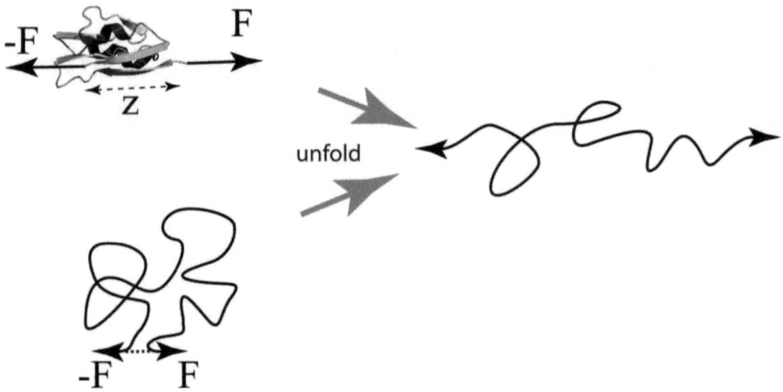

Fig. 9.1 Mechanically induced protein unfolding and mechanically induced cross-link rupture. The latter system is used here to elucidate some of the properties of the mechanical unfolding process. The force effect on the thermodynamics of the stretched chain is described in terms of two conjugate variables, the pulling force F and the end-to-end extension z in the direction of the force. The actual protein structure shown is that of ubiquitin

change, and so a discussion of this rate and its properties will also be central to this chapter. The outline of this chapter is as follows. After a discussion of thermodynamics of molecules subjected to mechanical stress (Sect. 9.2), I will discuss current theoretical views regarding the computation and properties of force-dependent rates (Sects. 9.3 and 9.4) and, finally, focus on the mechanical response of proteins in the context of single-molecule pulling experiments (Sect. 9.5). Section 9.6 concludes with a discussion of challenges and future prospects.

9.2 Mechanical Extension as an Unfolding Coordinate

In this section, I will discuss how mechanical stress affects equilibrium thermodynamic properties of molecules (e.g., proteins and RNA) that can undergo a folding–unfolding transition. A typical single-molecule pulling experiment provides the relationship between molecular extension and the force acting on the molecule.[1] In the case where the molecule is stretched via the application of a force $\mathbf{F}$ acting, e.g., between the two ends of the molecular chain (see Fig. 9.1), the effect of the force amounts to an additional contribution to the molecule's energy, equal to $-\mathbf{FR}$, where $\mathbf{R}$ is the end-to-end vector. If a coordinate system is used, whose z-axis

[1] Note, however, that the force is usually measured through the *displacement* of the measuring device such as the AFM tip. A more accurate description of single-molecule pulling should therefore include the elastic properties of the pulling device itself. As further discussed in Sect. 9.5, the idealized description adopted here is adequate when the force is transmitted to the molecule via soft handles, whose compliance is higher than that of the molecule itself.

is aligned with the direction of the force vector, then this contribution becomes simply $-Fz$, where z is the molecule's extension measured along the direction of the force. This term is similar to the potential energy mgh of a mass m held at an elevation h in the Earth' gravitational field. From a thermodynamics standpoint, the force F and the extension z constitute a pair of conjugate variables that are analogous, respectively, to pressure and volume in the standard description of gasses or liquids. The thermodynamic properties of a molecule held at a constant force F are then described in terms of its free energy $G_F(z)$, as a function of z and temperature T.

The free energy $G_F(z)$ is also related to the probability distribution of the molecule's extension, $p_F(z)$, by

$$G_F(z) = -k_B T \ln p_F(z). \tag{9.1}$$

If, for example, two distinct conformations exist (say, folded and unfolded) that differ in the values of z, one expects two peaks in the distribution $p_F(z)$ and, correspondingly, two minima in the free energy $G_F(z)$. Such double-well structure is thus the signature of thermodynamic equilibrium between distinct molecular states.

To further understand the shape and the properties of $G_F(z)$, it is instructive to start with the exactly solvable toy model of a random polymer chain, whose ends are "sticky" (Barsegov et al. 2008). In the simplest model of a randomly coiled polymer (De Gennes 1979), the chain is composed of $N \gg 1$ statistically independent segments, each of length b, so that the mean squared end-to-end distance is given by the standard random-walk relation, $\langle R^2 \rangle = Nb^2$. In the absence of intermonomer interaction and of a mechanical force, the probability distribution of the end-to-end vector then obeys Gaussian statistics:

$$p_0(\mathbf{R}) = \left(\frac{3}{2\pi N b^2}\right)^{3/2} \exp\left(-\frac{3|\mathbf{R}|^2}{2Nb^2}\right). \tag{9.2}$$

To mimic folding, we now modify this model by adding an attractive potential dependent on the distance $R = |\mathbf{R}|$ between the chain ends:

$$U(R) = \begin{cases} -U_0, & R \leq R_0 \\ 0, & R > R_0 \end{cases}. \tag{9.3}$$

The attraction is felt only when the chain ends happen to be within a short distance $R_0 \ll b\sqrt{N}$. The chain can now be in two states: In the "folded" state the ends stick to one another and the chain forms a ring conformation. In the "unfolded" state, the chain ends have escaped their attractive interaction and are free; in this state, the chain statistics is essentially the same as that of a random coil.

Despite its simplicity, our model captures many features expected of protein folding. In particular, there is an energy barrier U_0 to unfolding. Likewise, there is a free energy barrier to folding, caused by the entropic cost of closing the loop

(Szabo et al. 1980). Furthermore, properties of the unfolded state are comparable to those of a denatured protein and, although our folded state lacks structural order characteristic of proteins, it captures essential features of the folded state as manifested in mechanical pulling experiments, which probe the chain extension rather than its finer structural details. In particular, just as in our toy model, protein folding often brings the ends of a polypeptide closer together, relative to the unfolded state.

Now let us calculate the free energy $G_F(z)$ for this model. Because the effect of the force amounts to subtracting Fz from the zero-force free energy profile,

$$G_F(z) = G_0(z) - fz, \tag{9.4}$$

it is sufficient to find $G_0(z)$, or, equivalently, the probability distribution $p_0(z)$ in the absence of the force. This is given by

$$\begin{aligned} p_0(z) &= \langle \delta(R\cos\theta - z)\rangle = Q^{-1}\int \mathrm{d}^3\mathbf{R}p_0(\mathbf{R})\mathrm{e}^{-\beta U(R)}\delta(R\cos\theta - z) \\ &= \left(\frac{3}{2\pi Nb^2}\right)^{3/2} Q^{-1}\int_0^\infty \mathrm{d}R\int_0^{2\pi}\mathrm{d}\varphi\int_0^\pi \mathrm{d}\theta R^2\sin\theta \mathrm{e}^{-\frac{3R^2}{2Nb^2}}\mathrm{e}^{-\beta U(R)}\delta(R\cos\theta - z). \end{aligned} \tag{9.5}$$

Here $\beta = (k_\mathrm{B}T)^{-1}$ and $Q = \int \mathrm{d}^3\mathbf{R}p_0(\mathbf{R})\exp[-\beta U(R)]$ is the partition function of the system. The integral in (9.5) is conveniently written by representing the end-to-end distance vector in terms of polar coordinates (R, θ, φ) such that the Cartesian components of $\mathbf{R}$ are given by $(x, y, z) = (R\sin\theta\cos\varphi, R\sin\theta\sin\varphi, R\cos\theta)$.

Performing integration over the angular variables, we obtain

$$p_0(z) = 2\pi\left(\frac{3}{2\pi Nb^2}\right)^{3/2} Q^{-1}\int_z^\infty R\mathrm{d}R\mathrm{e}^{-\beta U(R)}\mathrm{e}^{-\frac{3R^2}{2Nb^2}}. \tag{9.6}$$

Note that, because the integrand in the above equation is positive, $p_0(z)$ is a monotonically decreasing function of z. This is true for any interaction potential $U(R)$. For the specific potential defined by (9.3), we can further evaluate this integral to obtain

$$p_0(z) = A\begin{cases} \mathrm{e}^{-\frac{3z^2}{2Nb^2}}, & z > R_0 \\ \mathrm{e}^{-\frac{3R_0^2}{2Nb^2}} + \mathrm{e}^{\beta U_0}\left(\mathrm{e}^{-\frac{3z^2}{2Nb^2}} - \mathrm{e}^{-\frac{3R_0^2}{2Nb^2}}\right), & z \le R_0 \end{cases} \tag{9.7}$$

where A is a normalization factor that is unimportant for our purposes. A typical free energy profile resulting from (9.1) and (9.7) is plotted in Fig. 9.2. One feature we immediately observe is that $G_0(z)$ lacks a double-well structure anticipated for a system that has two distinct states. The reason for such behavior is simple: In the absence of a force, the most likely value of the extension z is zero for both

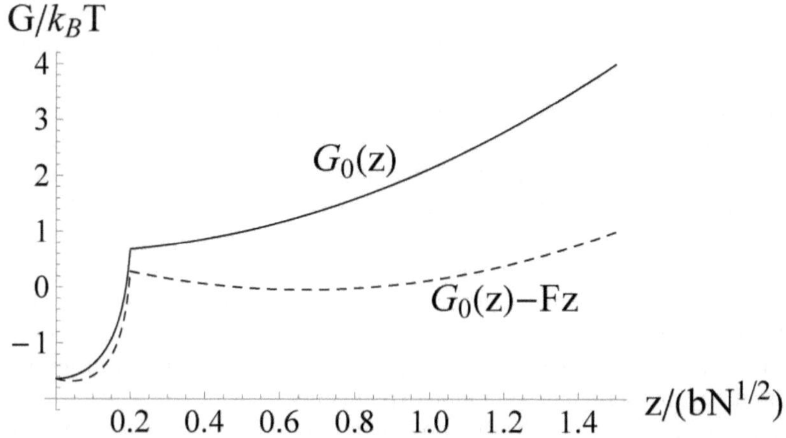

Fig. 9.2 Free energy profile, as a function of the end-to-end extension along the direction of the force, for a polymer chain with its ends interacting via the potential of (9.2), where $R_0 = 0.2bN^{1/2}$ and $\beta U_0 = 5$

the folded and the unfolded states, and so an observation of z does not yield any information as to whether the protein is folded or unfolded.[2] The failure of z to distinguish between the folded and the unfolded states is quite general and is not limited to our simple model. Indeed, in the absence of a force all configurations of the molecule that differ by a rotation are equally likely, negative values of z are as likely as positive ones, and the mean value of z is zero, regardless of the internal state of the protein. Note that Fig. 9.2 only shows the shape of $G_0(z)$ for positive values of z; For negative values of z we have $G_0(z) = G_0(-z)$.

Application of a force makes the system anisotropic and establishes a preferred direction in space. Consequently, we expect that a measurement of z made in the presence of a force may differentiate between the folded and the unfolded states. This is, indeed, observed in Fig. 9.2, where, at nonzero force, the free energy $G_F(z)$ now exhibits two minima. The minimum with the smaller value of extension corresponds to a molecule that is still folded (i.e., with the two ends stuck together). The minimum with the larger value of z corresponds to the two ends being pulled apart; Since the distance between the ends is greater than R_0, the chain is in its unfolded state.

One concludes, therefore, that the folding–unfolding transition will be manifested, at an appropriately chosen force, as an abrupt change in the molecule's extension. Because the mechanical compliance is different in the folded and the unfolded states (i.e., the folded molecule is stiff and the unfolded molecule is soft), the unfolding transition will be seen as extension of the molecule. Moreover, the relative population

[2] In contrast, an observation of the absolute value R of the end-to-end distance may distinguish between the folded and the unfolded states. Here, however, R is not a proper choice of the thermodynamic variable that is conjugate to the pulling force.

of the folded and unfolded states can be manipulated with a force: an increase in the force makes the extended (i.e., unfolded) state of the molecule more thermodynamically favorable relative to the low-extension, folded state. It is then possible to pick the value of the force F such that the two states are equally populated. Observation of the molecule subjected to such a force over a sufficient period of time will reveal jumps between the low- and the high-extension states. Such two-state behavior has, indeed, been observed in stretched RNA hairpins (Liphardt et al. 2001; Ritort et al. 2002). It is important to note that, while the high-extension state corresponds to an unfolded molecule, its properties are generally different from the zero-force unfolded state. Indeed, the stretched polymer is generally aligned along the force. Such an alignment entails loss of entropy. The free energy cost of mechanical unfolding includes the entropic cost of stretching and is not simply equal to the free energy of unfolding as measured, say, in thermal denaturation experiments.

More realistic models of the thermodynamics of mechanically stretched proteins confirm the picture developed above. Because the structural order within the folded conformation cannot be fully characterized by specifying the extension z, it is expedient to introduce a second, structural coordinate Q that is specifically designed to differentiate between the folded and the unfolded conformational ensembles. This can, for example, be the root mean square deviation of the structure from a reference structure representing the folded conformation: with such a choice, Q is close to zero in the native state and is a large positive number in the unfolded state. We can then define the free energy

$$G_0(z,Q) = -k_B T \ln p_0(z,Q) \tag{9.8}$$

in terms of the equilibrium joint probability distribution $p_0(z,Q)$ of the two variables. Application of a force tilts this free energy landscape, which becomes

$$G_F(z,Q) = G_0(z,Q) - zF. \tag{9.9}$$

An example of such a free energy landscape computed using a coarse-grained model of ubiquitin (Kirmizialtin et al. 2005) (see Fig. 9.1 for the structure of ubiquitin) is shown in Fig. 9.3. At zero force, the chain is not stretched and can be either in the native state N or in a "compact denatured" (CD) state. The extension z is close to zero in both states. Observation of z will thus fail to detect the folding transition at zero force. In contrast, the folding coordinate Q distinguishes between the two states: If z is close to zero while Q is varied, two minima of $G_0(z,Q)$ are observed, one corresponding to the native state and the other to the unfolded state CD. As explained above, the situation changes when a force is applied: because of the difference in the mechanical compliance of the unfolded and folded states, two states will be observed in the probability distribution of the extension, corresponding to the folded and the unfolded states (Kirmizialtin et al. 2005).

Figure 9.3 also demonstrates the coupling between the mechanical and the folding coordinates: Stretching the molecule readily eliminates native-like structures from the conformational ensemble. For sufficiently large values of

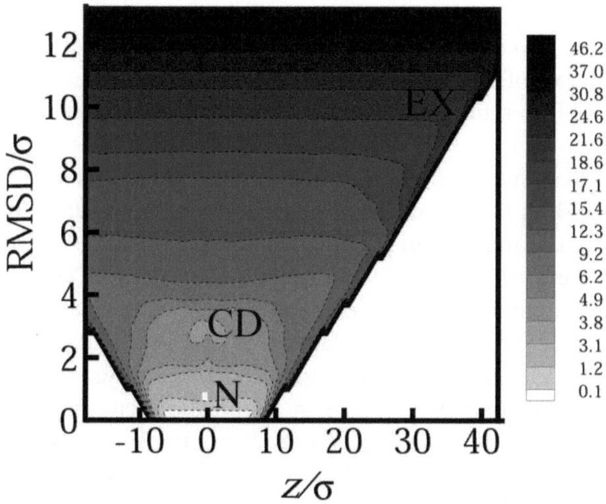

Fig. 9.3 (Courtesy of Dr. Serdal Kirmizialtin). Contour plot of the free energy of ubiquitin, as a function of its extension z and the root mean square deviation Q from the native structure (both are measured in units of peptide bond length σ). Native (N), compact denatured (CD), and extended (EX) conformations are shown. The energy scale is set by the typical magnitude of hydrophobic interactions in the model (Kirmizialtin et al. 2005; Sorenson and Head-Gordon 2000, 2002; Veitshans et al. 1996)

extension z, native-like structures are so improbable (or impossible!) that they do not appear in the plot because they could not be sampled in the simulations. This, of course, is to be expected since extended configurations of the chain have little structural overlap with the native structure shown in Fig. 9.1.

The lack of the second well in $G_0(z)$ and the failure of the extension to distinguish between the unfolded and folded states at zero force has been discussed in Kirmizialtin et al. (2005), demonstrated in multiple computer simulation studies (Best et al. 2008; Li and Makarov 2004a, b; Li et al. 2006), and has received renewed attention more recently (Berkovich et al. 2010; Dudko et al. 2011). It should be emphasized here that in no way does the absence of the second well at zero force mean that folding/unfolding transition cannot be observed in a pulling experiment performed at a finite force.

9.3 The Effect of Mechanical Force on the Rate of a Conformational Transition: The Zhurkov–Bell Formula and Its Extensions

The discussion of the preceding section was limited to equilibrium thermodynamic properties of molecules under mechanical loading. We have found that application of a force can shift the thermodynamic equilibrium between the unfolded and folded conformations of a protein. As a result, it is possible to observe the system jump between the two states, as a function of time. It is natural to expect that the frequency or the rate of the jumps from one of those states to another would also be force-dependent. This section introduces theoretical ideas used to understand such dependences.

It is useful to somewhat broaden the scope of our discussion, beyond the mechanical unfolding scenario. Conformational transitions other than folding and unfolding can also be manipulated by mechanical stress. For example, force can be used to modulate enzymatic activity, as reviewed in (Alegre-Cebollada et al. 2010). Likewise, forced unbinding of biomolecular complexes has attracted considerable attention (see, e.g., Florin et al. 1994; Heymann and Grubmuller 1999, 2001; Yuan et al. 2000; Izrailev et al. 1997; Guo et al. 2010). Finally, mechanical pulling is believed to be an essential step in protein degradation, causing proteins to unfold and thread through the proteasome pore and providing access to the proteolytic chamber (Matouschek and Bustamante 2003). The goal of the following discussion is, therefore, to establish a general framework for the description of the force effect on the rate of a (broadly defined) conformational transition.

Models of various levels of sophistication can be used to describe conformational rearrangements in molecular systems. At one end of the spectrum, there are all-atom models while at the other, there are simple, one-dimensional models that postulate that the molecule's conformation can be described in terms of a single degree of freedom, referred to as "reaction coordinate" or "order parameter." All-atom models are often too costly computationally and thus unsuitable for interpreting experimental data. In fact, until recently, most single-molecule pulling experiments have been interpreted in terms of the least sophisticated model consisting of a single reaction coordinate, the extension z.

On the other hand, the general problem of choosing a "good" reaction coordinate has received much attention in the context of chemical kinetics (Truhlar et al. 1996). Good reaction coordinates are typically ones associated with slow motions of the system, relative to the remaining degrees of freedom. If the timescales of the slow and the fast motions are well separated, it is often possible to treat the fast degrees of freedom in an approximate fashion. One particularly useful approximation (Szabo et al. 1980; Zwanzig 2001) results from the assumption that the system's motion can be described as (biased) diffusion in the potential of (9.4). This approximation assumes that z is a good reaction coordinate while the effect of the remaining degrees of freedom is lumped into a single quantity, the effective diffusion coefficient along z. This diffusion coefficient, together with the zero-force free energy $G_0(z)$, then completely specifies the system's response to a mechanical force.

By construction (Zwanzig 2001), the one-dimensional diffusion model automatically guarantees to reproduce the correct equilibrium distribution of the mechanical extension z [cf. Eq. (9.1)], and so it properly accounts for the *equilibrium properties*, as observed in a pulling experiment. Except by pure chance, however, there is no reason for the motion along z to be slow (relative to other degrees of freedom), and so z cannot be guaranteed to be a good reaction coordinate for the purposes of describing the system's *dynamics*. On the other hand, if a one-dimensional model is desired, one does not have the luxury of choosing a better (i.e., slower) reaction coordinate since the extension z must be included in the model explicitly to describe the system's coupling to the force. We therefore conclude that the minimal model for mechanochemical phenomena must contain at least two degrees of freedom: the mechanical extension z and some other slow coordinate Q. Two dimensional models have thus been recently pursued by several workers (Dudko et al. 2008; Hyeon and Thirumalai 2008; Suzuki and Dudko 2010, 2011). In the context of folding, for example, Q can be some quantity that is suitable for the description of folding in the absence of the force (such as the rmsd from the native conformation used in Fig. 9.3). The resulting minimalist model may then involve diffusion on a two-dimensional free energy landscape such as the one shown in Fig. 9.3.

In what follows, I will adopt a more general view, where the conformation of the molecule of interest can be defined by specifying the values of its n degrees of freedom. If $n = 1$ or 2, the above one- and two-dimensional models are recovered. For sufficiently small molecules, a fully atomistic description of mechanochemistry is possible, where n equals the total number of degrees of freedom (Boulatov and Kucharski 2011; Kochhar et al. 2010; Lenhardt et al. 2010; Huang and Boulatov 2011; Konda et al. 2011; Bailey and Mosey 2012). For larger, biologically relevant molecules, a coarser description formulated in terms of a smaller subset of relevant variables is often necessary. Depending on the nature of the problem, those variables can either be collective coordinates, as in Fig. 9.3, or the Cartesian coordinates of a subset of atoms.[3] For example, instead of explicitly treating each water molecule surrounding a protein, one can use an implicit solvent model, which includes only the atoms belonging to the protein. Likewise, instead of explicitly considering every atom of the protein, one can further use a coarser model, where each amino acid residue is represented as a single atom (Veitshans et al. 1996; Camacho and Thirumalai 1993; Klimov and Thirumalai 1997, 2000).

To be specific, let us assume that our problem is formulated in terms of the Cartesian coordinates

$$\mathbf{r} = (x_1, y_1, z_1, x_2, y_2, z_2, \ldots, x_N, y_N, z_N) \tag{9.10}$$

of a subset of N atoms belonging to the molecule. The case of collective coordinates is analogous. The equilibrium statistics of the molecule's conformations is governed by the probability distribution

$$p(\mathbf{r}) \propto \exp[-\beta G(\mathbf{r})], \tag{9.11}$$

[3] Note that Cartesian atomic coordinates are usually unsuitable as proper degrees of freedom for the description of protein folding because the unfolded state does not correspond to a particular structure and so is not close to any specific point in the Cartesian space. Instead, as exemplified in Fig. 9.3, it is often beneficial to use a collective "folding" coordinate Q that is a nonlinear function of the atomic coordinates.

where $G(\mathbf{r})$ defines the molecule's free energy surface in the 3N-dimensional space. In an all-atom model, of course, $G(\mathbf{r})$ would coincide with the underlying potential energy of the molecular system. Using, instead, a relatively small subset of relevant variables has the added benefit that coarse graining, through elimination of certain degrees of freedom, tends to smoothen the free energy landscape $G(\mathbf{r})$. That is, while the underlying energy of a molecule, as a function of all of its atomic coordinates, is relatively rough and may include numerous minima, maxima, and saddle points, low-dimensional models produced by integrating out multiple degrees of freedom tend to give rise to much simpler, smooth free energy landscapes (Li and Makarov 2003) such as the two-dimensional free energy surface shown in Fig. 9.3. According to the standard view of chemical kinetics, stable molecular conformations are associated with minima on the free energy landscape $G(\mathbf{r})$, such as the native and the condensed denatured minima appearing in Fig. 9.3. Most of the time, the molecule is found in the vicinity of one of the minima. A conformational transition then involves jumping from one minimum to another. This process involves the crossing of a free energy barrier separating the two stable states. Suppose the transition starts from the vicinity of a particular minimum $\mathbf{r} = \mathbf{r}^{(0)}$, which we call the reactant conformation. According to standard rate theories (Hanggi et al. 1990), the rate of the transition k (i.e., the probability of making the transition per unit time) is controlled by the free energy difference between the reactant conformation and the transition-state conformation $\mathbf{r} = \mathbf{r}^{(\mathrm{TS})}$. The latter conformation corresponds to the saddle-point of $G(\mathbf{r})$ that separates the reactant minimum from the final, or "product," minimum. To simplify the discussion, I will assume that only one such saddle exists. Then the rate can be estimated as

$$k = v \exp\left\{ -\frac{G[\mathbf{r}^{(\mathrm{TS})}] - G[\mathbf{r}^{(0)}]}{k_{\mathrm{B}}T} \right\}, \tag{9.12}$$

where v is a preexponential factor. This factor generally depends on the properties of the free energy surface $G(\mathbf{r})$ and on coupling of the system to the degrees of freedom that are not explicitly included in the model. This coupling is commonly lumped into an effective diffusion coefficient, or, more generally, a diffusion matrix, which then can be used to estimate the rate using Kramers' theory and its extensions (Szabo et al. 1980; Hanggi et al. 1990). Importantly, v is relatively weakly dependent on the force, in contrast to the *exponential* force dependence that results from the force-induced change in the barrier. For this reason, the preexponential factor will be treated here as a constant. To estimate the effect of the force on the transition rate, it then remains to calculate how the free energy of the reactant and transition-state conformations changes upon application of the force.

Although a typical experimental scenario involves application of a force to a pair of atoms, it is instructive to consider a more general situation, where each atom is subjected to a force. We thus define a 3N-dimensional force vector

$$\mathbf{f} = (f_{x1}, f_{y1}, f_{z1}, \ldots, f_{xN}, f_{yN}, f_{zN}), \tag{9.13}$$

where f_{xi}, f_{yi}, and f_{zi} are the Cartesian components of the force acting on the ith atom. When such a generalized force is applied to the molecule, its free energy becomes $G(\mathbf{r}) - \mathbf{fr}$ [cf. (Eq. 9.4)] and the positions of the reactant minimum and the transition-state saddle point become shifted. The new locations of those stationary points are determined by the condition $\nabla[G(\mathbf{r}) - \mathbf{fr}] = 0$ or $\nabla G(\mathbf{r}) = \mathbf{f}$, where ∇ denotes gradient in the $3N$-dimensional space. For a sufficiently low force, the shifts in the minimum and saddle positions can be calculated using Taylor series:

$$\mathbf{r}^{(0),(\mathrm{TS})}(\mathbf{f}) \approx \mathbf{r}^{(0),(\mathrm{TS})}(0) + (\mathbf{h}^{(0),(\mathrm{TS})})^{-1}\mathbf{f}, \tag{9.14}$$

where

$$\mathbf{h}^{(0),(\mathrm{TS})} = \left.\frac{\partial^2 G}{\partial r_\alpha \partial r_\beta}\right|_{\mathbf{r}=\mathbf{r}^{(0),(\mathrm{TS})}(0)} \tag{9.15}$$

is the Hessian matrix computed for the reactant/transition state at zero force. Using this result, we can calculate the free energy at the reactant and the transition states in the presence of a force:

$$\begin{aligned} G^{(0),(\mathrm{TS})}[\mathbf{r}^{(0)(\mathrm{TS})}(\mathbf{f})] &= G^{(0)(\mathrm{TS})}[\mathbf{r}^{(0)(\mathrm{TS})}(\mathbf{f})] - \mathbf{f}\mathbf{r}^{(0)(\mathrm{TS})}(\mathbf{f}) \\ &\approx G^{(0),(\mathrm{TS})}[\mathbf{r}^{(0)(\mathrm{TS})}(0)] - \mathbf{f}\mathbf{r}^{(0)(\mathrm{TS})}(0) \\ &\quad -\frac{1}{2}\mathbf{f}^T\left(\mathbf{h}^{(0)(\mathrm{TS})}\right)^{-1}\mathbf{f}. \end{aligned} \tag{9.16}$$

Here $\mathbf{f}^T$ denotes the transposed force vector. Using (9.12), we now find

$$\ln k(\mathbf{f})/k(0) \approx -\frac{\Delta\Delta G_1(\mathbf{f}) + \Delta\Delta G_2(\mathbf{f})}{k_\mathrm{B}T}. \tag{9.17}$$

Here

$$\Delta\Delta G_1(\mathbf{f}) = -\mathbf{f}\left[\mathbf{r}^{(\mathrm{TS})}(0) - \mathbf{r}^{(0)}(0)\right] \tag{9.18}$$

is linear in force and

$$\Delta\Delta G_2(\mathbf{f}) = -\frac{1}{2}\mathbf{f}^T\left[(\mathbf{h}^{(\mathrm{TS})})^{-1} - (\mathbf{h}^{(0)})^{-1}\right]\mathbf{f} \tag{9.19}$$

is quadratic. Note that in a case where the molecule is free to rotate and translate in space, its Hessian matrix $\mathbf{h}$ is singular (with six zero eigenvalues corresponding to rigid-body motion) and therefore not invertible (Konda et al. 2011). Equations (9.16) and (9.19), then, appear to be ill defined. This predicament is easily surmounted, however, by noting that the only physically meaningful force vectors $\mathbf{f}$ are those orthogonal to the null space of the Hessian matrix. Indeed, only such vectors can ensure that the system is in mechanical equilibrium such that the total force and the torque acting on the molecule are both zero. As a consequence, all the inverses in (9.16) and (9.19) should be understood as generalized inverses (pseudoinverses), with the six zero eigenmodes suppressed in their spectral expansion.

The treatment so far is general and applies to a molecule subjected to arbitrary mechanical stress. Now let us focus on the case where pulling forces are applied to a pair of atoms, *i* and *j*, as in Fig. 9.1. Let **F** be the three-dimensional vector representing such a force. From here on, upper-case bold symbols will be used for *three-dimensional* vectors representing positions of and forces on individual atoms. I will continue to use lower-case boldface symbols to represent *3N-dimensional* vectors and (3*N*) $\times$ (3*N*) matrices corresponding to the configuration of the entire molecule. The mechanical equilibrium of the system requires that **F** act along the line connecting the atoms. Let $\mathbf{E} = \mathbf{F}/|\mathbf{F}|$ be the unit vector along the force direction. Then we must have

$$\frac{\mathbf{R}_j - \mathbf{R}_i}{|\mathbf{R}_j - \mathbf{R}_i|} = \mathbf{E}, \tag{9.20}$$

where $\mathbf{R}_i$ is the three-dimensional vector describing the position of atom *i*. To evaluate (9.17), we note that the 3*N*-dimensional force vector has the following components:

$$\begin{aligned} f_{xi} = -FE_x,\ f_{yi} = -FE_y,\ f_{zi} = -FE_z, \\ f_{xj} = FE_x,\ f_{yj} = FE_y,\ f_{zj} = FE_z, \\ f_{xk} = f_{yk} = f_{zk} = 0, \quad \text{for } k \neq i, j. \end{aligned} \tag{9.21}$$

In particular, the expression for the first-order correction to the barrier height can be simplified to give:

$$\begin{aligned} \Delta\Delta G_1(F) &= -F\left\{\mathbf{E}\left[\mathbf{R}_j^{(\mathrm{TS})}(0) - \mathbf{R}_j^{(0)}(0)\right] - \mathbf{E}\left[\mathbf{R}_i^{(\mathrm{TS})}(0) - \mathbf{R}_i^{(0)}(0)\right]\right\} \\ &= -F\left[\left|\mathbf{R}_j^{(\mathrm{TS})}(0) - \mathbf{R}_i^{(\mathrm{TS})}(0)\right| - \left|\mathbf{R}_j^{(0)}(0) - \mathbf{R}_i^{(0)}(0)\right|\right] \equiv -F(R_{ij}^{(\mathrm{TS})} - R_{ij}^{(0)}). \end{aligned} \tag{9.22}$$

The second-order term is clearly quadratic in *F*. We can thus write it as follows:

$$\Delta\Delta G_2(F) = -\frac{1}{2}\chi_{ij}^{(\mathrm{TS})}F^2 + \frac{1}{2}\chi_{ij}^{(0)}F^2. \tag{9.23}$$

The parameters $\chi_{ij}^{(0)}$ and $\chi_{ij}^{(\mathrm{TS})}$ introduced above describe the mechanical compliance of the molecule (the inverse of stiffness) in its reactant and transition states. The explicit expression for these compliances, in terms of the underlying Hessian matrices, is given in the Appendix. Note that we always have $\chi_{ij}^{(0)} \geq 0$.

Indeed, the second term in (9.19) and (9.23) must be non-negative because the Hessian matrix $\mathbf{h}^{(0)}$, calculated at a minimum of G, is non-negative definite. In contrast, nothing can be said about the sign of $\chi^{(\mathrm{TS})}$ because the Hessian matrix $\mathbf{h}^{(\mathrm{TS})}$, corresponding to a saddle point of G, has both negative and positive eigenvalues.

To summarize our findings, the effect of a sufficiently weak force on the rate of a conformational rearrangement is given by

$$k(F) \approx k(0) \exp\left[\frac{F\Delta R_{ij}}{k_{\mathrm{B}}T}\right] \exp\left[\frac{F^2 \Delta\chi_{ij}}{2k_{\mathrm{B}}T}\right], \tag{9.24}$$

where $\Delta R_{ij} = R_{ij}^{(\mathrm{TS})} - R_{ij}^{(0)}$ is the change in the distance between the atoms when going from the reactant state to the transition state and $\Delta\chi_{ij} = \chi_{ij}^{(\mathrm{TS})} - \chi_{ij}^{(0)}$ is the corresponding change in the compliance. If the force is low enough that it satisfies the condition $F \ll \left|\Delta R_{ij}/\Delta\chi_{ij}\right|$, then the term quadratic in the force can be neglected, resulting in a simple exponential force dependence of the rate. This result, commonly referred to as Bell's formula, was originally introduced by Zhurkov (1965) and later popularized by Bell (1978).

In the regime where the Zhurkov–Bell formula is valid, pulling on a pair of atoms, i and j, will change the transition barrier by an amount that is proportional to the difference between the (zero-force) interatomic distances R_{ij} in the transition and the reactant states. If the distance R_{ij} between the two atoms is longer in the transition state than it is in the stable configuration, then application of a force between those two atoms will accelerate the transition. In the opposite case, application of the force will slow the reaction down.

The quadratic term, derived in various forms (Boulatov and Kucharski 2011; Huang and Boulatov 2011; Konda et al. 2011; Bailey and Mosey 2012), accounts for the difference between the elastic energies stored by the mechanically stretched molecule in the transition and reactant states and contributes to the overall activation barrier.

The harmonic transition-state picture developed above is summarized in the free energy diagram shown in Fig. 9.4. When in its stable, reactant state, the molecule's response to pulling on its atoms i and j is that of a Hookean spring whose energy is given by

$$G(\mathbf{R}_{ij}) = G(R_{ij}^{(0)}) + (R_{ij} - R_{ij}^{(0)})^2/(2\chi_{ij}^{(0)}). \tag{9.25}$$

Application of a force F stretches the spring such that the new equilibrium distance is

$$R_{ij}(F) = R_{ij}^{(0)} + F\chi_{ij}^{(0)} \tag{9.26}$$

while the new equilibrium energy, relative to that at zero force, is

$$\Delta G^{(0)}(F) = (R_{ij} - R_{ij}^{(0)})^2/(2\chi_{ij}^{(0)}) - FR_{ij} = -FR_{ij}^{(0)} - \frac{1}{2}\chi_{ij}^{(0)}F^2. \tag{9.27}$$

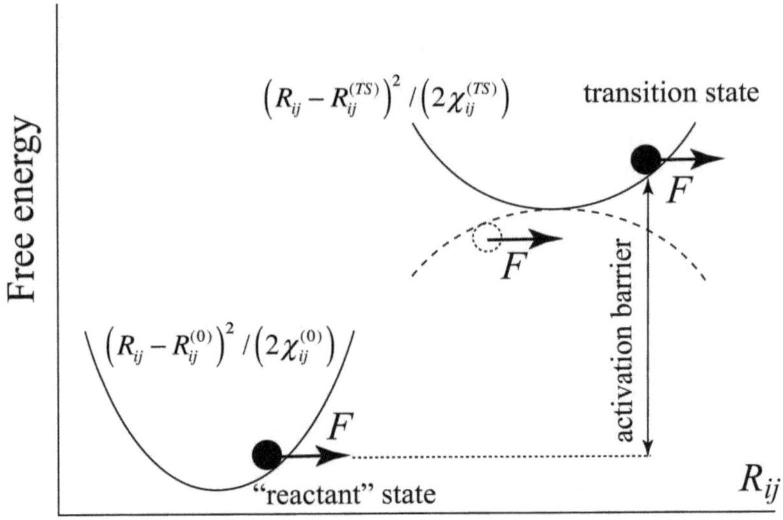

Fig. 9.4 Free energies corresponding to a molecule's reactant and transition-state configurations plotted as a function of the distance between a pair of atoms, i and j. While the reactant free energy always exhibits a minimum corresponding to the equilibrium distance between the atoms, the transition-state energy may exhibit either a minimum or a maximum. A stretching force F applied between these atoms lengthens the distance R_{ij} in the reactant state. In contrast, this force may either reduce or increase the value of R_{ij} corresponding to the transition state

A similar expression can be written for the force-induced change in the transition-state energy, $\Delta G^{(\mathrm{TS})}(F)$. The force-induced change in the activation barrier, $\Delta\Delta G(F) = \Delta G^{(\mathrm{TS})}(F) - \Delta G^{(0)}(F)$, then leads to (9.22) and (9.23). It should be emphasized, however, that despite the one-dimensional appearance of the free energy diagram in Fig. 9.4, our theory is essentially multidimensional. This is particularly evident from the shape of the transition-state free energy curve, which can exhibit either a maximum (i.e., $\chi_{ij}^{(\mathrm{TS})}<0$) or a minimum (i.e., $\chi_{ij}^{(\mathrm{TS})}>0$). In one dimension, the transition state must correspond to a maximum (i.e., a barrier). For more than one degree of freedom, however, the transition state corresponds to a saddle point. That is, a small excursion away from the transition state may either increase or decrease the free energy, depending on the direction of this excursion in the 3N-dimensional space. Accordingly, pulling on an arbitrary pair of atoms may either increase or decrease the transition-state energy depending on the geometry and the energetics of the molecule. Both possibilities are illustrated in Fig. 9.4. If $\chi_{ij}^{(\mathrm{TS})}<0$, then application of a stretching force shortens the distance between the transition state and the initial state, similarly to the prediction of one-dimensional theories (Li and Makarov 2003; Dudko et al. 2006, 2007; Makarov 2007; Evans and Ritchie 1997, 1999; Hyeon and Thirumalai 2005). If $\chi_{ij}^{(\mathrm{TS})}>0$, then this distance can either decrease or increase, depending on whether the transition state or the initial state is more compliant. The sign and the magnitude of the susceptibilities qualitatively affect the shape of the "Arrhenius plot" of $\ln k(F)$ vs.

F [cf. (9.24)]. If $\chi_{ij}^{(\mathrm{TS})}<0$, this results in a concave function, while if $\chi_{ij}^{(\mathrm{TS})}>0$ then it is either concave or convex, depending on the sign of $\Delta\chi_{ij}$. In contrast, one-dimensional theories cannot account for convex behavior of $\ln k(F)$.

The deficiencies of one-dimensional models have been brought to the spotlight in a number of recent studies (Suzuki and Dudko 2010, 2011; Yew et al. 2010). In practice, both convex and concave behavior has been predicted for reaction rates in small molecules (Kochhar et al. 2010; Lenhardt et al. 2010; Konda et al. 2011). For mechanically stretched biomolecules, however, a limited range of experimentally accessible forces often prevents one from detecting the curvature in $\ln k(F)$ (Li and Makarov 2003). As a result, it is quite common to describe experimental data in terms of the Zhurkov–Bell formula [Eq. (9.24), without the term quadratic in force]. Nevertheless, more sophisticated analyses of experimental data often reveal the existence of curvature (Dudko et al. 2008, 2007; Makarov 2007; Schlierf and Rief 2006).

The harmonic theory developed here assumes that the applied force is low enough that the Taylor expansion of the free energy barrier can be truncated to omit cubic and higher-order terms. A more qualitative limitation of the theory is that it assumes that the force does not change the topography of the free energy landscape, and, therefore, does not alter the pathway of the rearrangement. However, force-induced changes in the free energy landscapes may involve complete disappearance of a barrier or the emergence of an entirely different pathway (Dudko et al. 2003; Lacks 2005; Lacks et al. 2010; Maloney and Lacks 2006). Such phenomena cannot be treated using the present approach. Instead, the global shape of the force-modified $G(\mathbf{r}) - \mathbf{fr}$ free energy landscape has to be explicitly considered.

Another assumption implicit in the above derivation is that only one saddle point and, therefore, only one transition pathway exists. Our theory is, however, trivially generalized to the case of multiple pathways, as the force dependence of the rate for each pathway can still be approximated by (9.24). Multiple pathway scenarios can lead to a nontrivial force dependence of the rate (Graham and Best 2011). If, for example, two pathways exist, exhibiting different force dependences of the transition rate, one pathway may dominate the overall transition rate at low forces and the other at high forces. Such a switch between two dominant mechanisms of mechanically induced protein unfolding has, indeed, been observed in computer simulations (Graham and Best 2011).

Finally, note that the above discussion assumed a force-controlled scenario, where a constant force is imposed on the molecule. One can also envisage a displacement-controlled experiment, where the distance R_{ij} between a pair of atoms is manipulated directly. The change in the reaction barrier, as a result of constraining the distance R_{ij} to a specified value (relative to the no-strain situation), is also easily computed using our theory:

$$\Delta\Delta G(R_{ij}) = \frac{\left(R_{ij} - R_{ij}^{(\mathrm{TS})}\right)^2}{2\chi_{ij}^{(\mathrm{TS})}} - \frac{\left(R_{ij} - R_{ij}^{(0)}\right)^2}{2\chi_{ij}^{(0)}}.$$

Typically, single-molecule pulling experiments exert forces on molecules by controlling the displacement of "handles" attached to the molecules (e.g., the AFM cantilever). That is, strictly speaking, they correspond neither to the force-controlled nor to the displacement-controlled scenarios. The displacement-controlled scenario is recovered when the handle is much stiffer than the molecule so that the displacement of the handle is equal to the molecule's extension. The much more common force-controlled scenario occurs in the opposite limit of a handle that is much softer than the molecule (Franco et al. 2009). If the compliances of the handle and of the molecule are comparable in magnitude, then the elastic response of the handles must be considered explicitly.

9.4 Molecular Jamming

The theory developed in the preceding section predicts the possibility of an interesting and counterintuitive effect: Application of a stretching force to a "bond" between a pair of atoms, *i* and *j*, may, in fact, suppress the dissociation of this bond.[4] Indeed, according to (9.24), the transition rate will decrease with the increasing force whenever the distance between the atoms, to which the force is applied, is shorter in the transition state than in the initial state.

One can think of this as the "jamming" of a molecular bond. Such jamming is ubiquitous in the macroscopic world: The common use of doorstop wedges and sailing knots, for example, relies on this effect. The molecular jamming is commonly referred to as the catch-bond phenomenon, where the lifetime of the bond increases upon application of a stretching force [see Barsegov and Thirumalai (2005); Prezhdo and Pereverzev (2009) for a review of relevant literature]. In contrast to their macroscopic analogs, however, molecular catch bonds are considerably less common. Experimental examples of catch bonds mostly involve molecular adhesion bonds (Marshall et al. 2003). Computer simulations of mechanically assisted protein unfolding also predict catch-bond behavior in certain cases (Graham and Best 2011; Peplowski et al. 2011), although, to the author's knowledge, such behavior has not been observed experimentally. Another example of catch-bond behavior is provided by computer simulations of friction knots tied between a pair of polymer strands, which showed that, when a force is applied to separate the strands joined with a knot, this force may, in fact, increase the time it takes for the knot to unravel (Kirmizialtin and Makarov 2008). A similar molecular jamming effect has been found in computer-simulated mechanical pulling of proteins that exhibit "slip-knot" structural motifs (Peplowski et al. 2011). This case will be further discussed in Sect. 9.5.

[4] Here, by dissociation, we simply mean a conformational transition through which the two atoms come apart. This definition thus includes unfolding of a protein stretched by a mechanical force.

Molecular knot jamming may be of biological significance in a different context: The existence of a rather complex knot in the native structure of human ubiquitin hydrolase has been surmised to render this protein resistant to degradation (Virnau et al. 2006). Specifically, the degradation process requires that the protein pass through a narrow constriction within the proteasome, a process that is widely believed to be mechanically driven (Matouschek and Bustamante 2003). It was then conjectured that the bulky knot may jam within the constriction, thus preventing the translocation and the subsequent degradation of the protein. Computer simulations of this process (Huang and Makarov 2008) confirmed that the presence of a knot can considerably suppress the translocation rate but did not find the catch-bond effect in this case, possibly because of inadequacy of the force field used or because sufficiently low forces could not be accessed by the simulations.

The molecular jamming or catch-bond behavior is fundamentally a multidimensional effect. It results from the force pushing the system along the "wrong" path in its multidimensional conformational space, in much the same way as trying to take one's coat off by pulling hard on one of its sleeves fails by missing the correct procedure. In contrast, simple one-dimensional theories of force-induced conformational rearrangements in biomolecules inevitably miss the catch-bond behavior. This is a critical limitation that should be kept in mind.

One may wonder why the catch-bond behavior is not more commonly found in single-molecule pulling studies. Pulling at the ends of a polypeptide chain in an AFM setup, for example, is consistently found to speed up protein unfolding, in contrast to theory and simulations predicting, in certain cases, the opposite trend (Graham and Best 2011). A possible answer is that typical experimental forces may be too high. Indeed, whenever an alternative transition pathway exists, with a rate that increases with increasing force, such a pathway will eventually become dominant at a high enough force. This results in a rollover in the force dependence of the transition rate, where it first decreases and then increases with increasing force (Graham and Best 2011; Prezhdo and Pereverzev 2009). It is then possible that AFM pulling studies, which usually involve forces of order of tens to hundreds of piconewtons, miss the more complex low-force behavior. Given that biologically relevant forces are often considerably lower and often in the range of a few piconewtons or less, it is not inconceivable that the catch-bond behavior would be a common phenomenon in vivo.

9.5 Exploring and Optimizing the Mechanical Stability of Proteins

9.5.1 Sacrificial Bond Mechanism as the Molecular Basis of High Strength and Toughness of Protein-Based Materials

A number of nature-made protein-based materials, such as spider capture silk or bone, exhibit a combination of high strength and toughness unsurpassed by any artificial material (Becker et al. 2003; Smith et al. 1999). Strength measures the

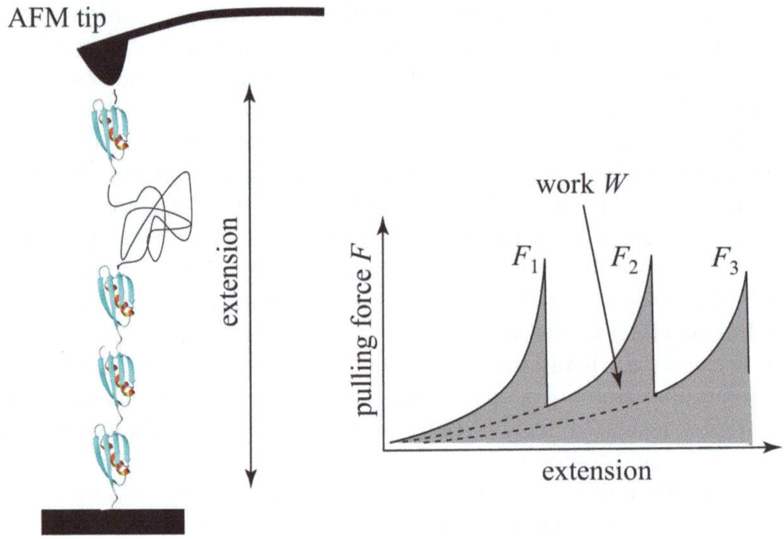

Fig. 9.5 An AFM pulling experiment (schematically depicted on the *left*) demonstrates the "sacrificial bond" mechanism. When the molecular chain, containing multiple protein domains and attached to an AFM tip and a substrate, is extended, the resulting force vs. extension curve exhibits a series of peaks. Each abrupt drop in the force results from the unfolding of an individual domain, which releases slack and increases the overall compliance of the chain. The overall work exerted in such an experiment (equal to the *shaded area* under the curve) may be significantly greater than the thermodynamic cost of unfolding the domains because of the nonequilibrium character of this experiment

maximum stress a material can withstand before it breaks. Toughness quantifies the ability of a material to absorb energy, when loaded, before it fails. High strength and toughness are often mutually exclusive. That is, many strong materials are brittle and therefore have low toughness.

The reconciliation of high toughness and high strength found in natural biomaterials is believed to be achieved through the *sacrificial bond* mechanism (Fantner et al. 2006; Smith et al. 1999). This mechanism is readily exposed by single-molecule AFM pulling studies of polyprotein chains, which consist of multiple linked protein domains. Such chains are common structural units in many load-bearing natural materials, as exemplified by the giant muscle protein titin (Erickson 1997). Polyproteins with specified domain sequences can also be engineered. Schematics of an AFM pulling experiment is illustrated in Fig 9.5, where a polyprotein chain is being stretched between a substrate and an AFM tip. The overall extension of the chain is controlled by moving the substrate relative to the tip, while the stretching force F is measured by observing the deflection of the AFM cantilever. It is often the case that the observed force vs. extension curve displays the following features:

- The observed stretching force exhibits a succession of peaks followed by sudden drops in the force F.

- The average height of the peaks depends on the rate at which the extension is increased. The faster the pulling, the higher the peaks.
- The work done in such a process, estimated as the area under the curve, is considerably higher than the thermodynamic cost of unraveling all the domains within the chain.

The above observations are readily explained within the picture, where protein domains unfold one at a time. Unfolding of a domain results in the release of a "hidden length" thereby reducing the overall compliance of the chain, as described in Sect. 9.1. This results in an abrupt drop in the force, followed by a gradual rise as work is being done against entropic elasticity of the unfolded chain, until the next domain unfolds. As a result, a sawtooth pattern is observed, as shown in Fig. 9.5. It should be noted that insufficient time resolution often precludes one to see finer details of the unfolding event taking place during a single force drop, although unfolding intermediates can be observed in certain cases (Marszalek et al. 1999). In contrast, computer simulations of mechanical unfolding often reveal further details of the unfolding pathways (see, e.g., Irback et al. 2005; Lu et al. 1998; Lu and Schulten 1999; Kleiner and Shakhnovich 2007; Li and Kouza 2009; Li et al. 2007). Unfortunately, direct experimental verification of such computational predictions remains a challenge. See Sect. 9.5.2 for further discussion.

The magnitude of the unfolding force peaks can be estimated using a model where the domain unfolding is an all-or-none process characterized by a force-dependent unfolding rate $k_u(F)$. This model is often validated by experimental data (Rief et al. 1998), although, as noted above, finer details of the unfolding process may be masked by insufficient time resolution. Let $F(t)$ be the time dependence of the force starting from a particular unfolding event. Without loss of generality, we can set the time t of that event to be equal to zero. We would like to estimate the magnitude of the force peak when the next domain unfolds. If we assume that all the folded domains in the chain are identical and that there are n such domains, then the probability $S(t)$ that none of them unfold over a period of time t is described by a first-order kinetics equation

$$dS(t)/dt = -nk_u[F(t)]S(t). \tag{9.28}$$

Equation (9.28) neglects the possibility for an unfolded domain to refold, an adequate assumption in an out-of-equilibrium case where the force is ramped up quickly enough that refolding becomes thermodynamically unfavorable before any domain has a chance to refold. Experimentally, the validity of this assumption can be readily verified, as domain refolding would be signified as a sharp increase in the force. The probability $p(t)dt$ that one of the domains unfolds between t and $t + dt$ is equal to the decrement $-dS = S(t) - S(t + dt)$ in the survival probability S. Combining this with (9.28) gives the probability distribution of the unfolding time

$$p(t) = k_u[F(t)]e^{-\int_0^t k_u[F(t)]dt}. \tag{9.29}$$

To estimate the magnitude of the force peak, let us suppose that the force increases linearly with time, $F = \varphi t$. While not necessarily realistic, this assumption will not qualitatively affect any of our conclusions. Let us further assume, for simplicity, that the force dependence of the unfolding rate is described by the Zhurkov–Bell formula, i.e., by (9.24) with the term quadratic in the force neglected,

$$k(F) = k_0 e^{F/F_c}. \tag{9.30}$$

Here $F_c = k_B T/\Delta R$ defines a characteristic force scale that quantifies the sensitivity of the unfolding rate to the force and ΔR is the change in the domain's end-to-end distance when going from the folded state to its mechanical transition state. Using (9.29) and (9.30), we find

$$p(t) = nk_0 \exp\left(\frac{\varphi t}{F_c}\right) \exp\left[-\frac{nk_0 F_c}{\varphi}\left(e^{\varphi t/F_c} - 1\right)\right]. \tag{9.31}$$

The most likely unfolding time t^* corresponds to the maximum of this function, where $dp/dt = 0$. Solving this equation we, finally, find the most probable unfolding force

$$F^* = \varphi t^* = F_c \ln \frac{\varphi}{nk_0 F_c}. \tag{9.32}$$

Of course, since the unfolding of a domain is a probabilistic event, the actual heights $F_1, F_2, \ldots$ of the unfolding peaks will not be exactly equal to the prediction of (9.32). Nevertheless, this equation offers insight into the experimental observations listed above. In particular, it predicts that the typical height of the peaks will increase logarithmically with the pulling rate (Evans and Ritchie 1997, 1999) and will decrease with the number of folded domains present in the chain (Best et al. 2003a; Makarov et al. 2001), the latter observation simply being a consequence of the fact that, the more domains are present in the chain the more likely it is for one of them to unfold.

Equation (9.32) further shows that the key quantity that controls the strength (i.e., the peak unfolding force) is the force scale F_c, which, in turn, is inversely proportional to the shift ΔR in the end-to-end distance of a domain between its transition-state and native structures.

To understand what physical parameters control toughness (i.e., the area under the force vs. extension curve), consider the simple model, in which the entropic elasticity of the unfolded polypeptide chain is described by Hook's law,

$$F = \gamma R, \tag{9.33}$$

where γ is a spring constant. This spring constant is readily estimated for the Gaussian chain model introduced in Sect. 9.1. Indeed, the probability distribution of the end-to-end distance in a chain of N statistically independent segments, (9.2), is identical to that of just two beads connected by a Hookean spring, whose stiffness is

$$\gamma = \frac{3k_\mathrm{B}T}{Nb^2}. \tag{9.34}$$

Every time a domain unfolds, the overall length N of the unfolded chain increases and the spring constant drops. If we further assume that folded domains are much stiffer than the unfolded parts of the polypeptide chain and that the end-to-end distance of a folded domain is much shorter than the overall polyprotein chain extension, then the total work done between two subsequent peaks, say F_1 and F_2, is given by the integral:

$$W_{12} = \int_{R_1}^{R_2} F(R)\mathrm{d}R = \int_{F_1/\gamma_1}^{F_2/\gamma_2} \gamma_2 R\mathrm{d}R = \gamma_2\left(\frac{F_2^2}{2\gamma_2^2} - \frac{F_1^2}{2\gamma_1^2}\right),$$

where γ_i denotes the spring constant prior to the ith unfolding event (where the force F_i was observed). To further simplify this expression, let us assume that the two peaks have roughly the same height F. Then

$$\begin{aligned} W_{12} &\approx \gamma_2\left(\frac{F^2}{2\gamma_2^2} - \frac{F^2}{2\gamma_1^2}\right) = \frac{F^2}{2}\left(\frac{1}{\gamma_2} - \frac{1}{\gamma_1}\right)\left(1 + \frac{\gamma_2}{\gamma_1}\right) \\ &= \frac{F^2}{6k_\mathrm{B}T} b^2 \Delta N\left(1 + \frac{N_1}{N_2}\right). \end{aligned} \tag{9.35}$$

Here, N_i is the total length of all the unfolded parts of the chain within the polyprotein, and $\Delta N = N_2 - N_1$ is the length of the chain released upon unfolding of one domain. Since $1 + N_1/N_2$ is simply a number between 1 and 2, Eq. (9.35) shows that high toughness of a polyprotein chain can be achieved by increasing either the hidden length ΔN or the force F (or both). Therefore, a polyprotein can be both strong and tough.

Note that, although Hook's law is often a reasonable approximation for low chain extensions (De Gennes 1979), it is often inadequate at higher forces relevant for AFM studies (Marszalek et al. 1999; Rief et al. 1997, 1998; Oberhauser et al. 1998; Hugel et al. 2005), where more realistic polymer models have to be used. This model is adequate in the present context, as the goal of the above exercise was not to develop a realistic model for polyprotein stretching but rather to establish the key factors that control the mechanical response of polyprotein chains. If, however, a more quantitative description of experimental data is desired, the reader should consult additional literature; see, e.g., Rief et al. (1998).

I will conclude this subsection with a general remark on the fundamental origin of high toughness and strength of naturally occurring biomaterials. As the above discussion indicates, those properties hinge on the ability of the system to dissipate large amounts of energy, in a highly nonequilibrium pulling process. Should the pulling process be performed slowly, in a quasiequilibrium fashion, the work done to stretch the polyprotein would be equal to the free energy cost of unraveling all of its domains, $W = \Delta G$. In contrast, a nonequilibrium pulling process may result in substantial energy dissipation so that $W \gg \Delta G$, a property that is commonly exhibited by molecular chains of high toughness. These considerations imply that the notion of mechanical stability of a chain is only meaningful when the rate at which the chain is loaded is specified. The same system that exhibits high strength and toughness at a high pulling rate may be quite labile when loaded slowly. Likewise, not every protein, when stretched under typical AFM pulling conditions, will be far away from equilibrium and exhibit the sawtooth pattern characteristic of "strong" proteins. Instead, many proteins readily unfold without showing a significant force peak (Li et al. 2000; Oberhauser et al. 2002).

9.5.2 The Mechanical Stability of an Individual Protein Domain: A Computational Perspective

Let us focus now on the mechanical response of an individual globular protein domain. It follows from the discussion of the preceding subsection that, under the conditions that produce a sawtooth force-extension curve, it is commonly subjected to a force that is much higher than the force required to destabilize the protein thermodynamically. It is not surprising then that, under such nonequilibrium pulling conditions, there is no significant correlation between mechanical and thermodynamic stability of a protein. Moreover, the mechanism of mechanical unfolding can be quite different from that of thermal or chemical denaturation (Li and Makarov 2003; Graham and Best 2011). For example, while thermal unfolding of repeat proteins (i.e., proteins that consist of interacting repeat units) is a cooperative process, the same proteins unravel sequentially when loaded mechanically (Makarov 2009a).

In view of a large disparity existing between the unfolding timescales probed experimentally (or relevant biologically) and those accessible by computer simulations, computational exploration of the mechanical response of proteins presents a rather formidable challenge. Steered molecular dynamics (SMD) simulations (Marszalek et al. 1999; Lu et al. 1998; Lu and Schulten 1999; Israelewitz et al. 2001; Paci and Karplus 1999) are brute force computer experiments, where a molecule is subjected to a mechanical force. While providing significant mechanistic insight, in order to accomplish unfolding within a computationally feasible timescale, they often have to be carried out in a regime that involves much higher forces and/or pulling rates than those employed experimentally. Because of this difference and because, as remarked above, mechanical stability is strongly dependent on the rate of loading, conclusions regarding unfolding mechanisms and mechanical stability drawn directly from SMD data should be taken with a grain of salt. To overcome this difficulty, various approaches have been proposed in order to extrapolate simulation data to lower forces/pulling rates; see, e.g., (Li and Makarov 2003; Balsera et al. 1997; Park et al. 2003; Xiong et al. 2006; Nummela and Andricioaei 2007; West et al. 2006a). Unfortunately such extrapolations, even if exact in principle, can be unreliable in practice (West et al. 2006a). Moreover, in cases where different unfolding mechanisms are dominant at different forces, extrapolation methods may even miss the correct unfolding mechanism altogether (Makarov et al. 2001; Makarov 2009b).

Less costly, coarse-grained models have also been employed (Kirmizialtin et al. 2005; Klimov and Thirumalai 2000; Kleiner and Shakhnovich 2007; West et al. 2006b; Sulkowska and Cieplak 2008) in order to bridge the timescales of simulations and experiments. They provide access to longer timescales, albeit, at the expense of force field accuracy and molecular detail.

Another approach avoids the problem of pulling too fast by, instead, employing efficient sampling techniques to reconstruct the *equilibrium* free energy profile $G_0(z)$ (cf. Sect.9.1) along the pulling coordinate z. The force-modified free energy profile $G_0(z) - Fz$ is then used to estimate the force-dependent free energy barrier and the force-dependent unfolding rate using Kramers' expression for the rate (Li and Makarov 2003, 2004a, b; Li et al. 2006). The underlying assumption of this approach is that the dynamics along the mechanical coordinate z can be described as a one-dimensional diffusion process in the effective potential $G_0(z)$ $-Fz$. The only other input parameter of the theory is the effective diffusion constant, which can also be estimated from molecular dynamics simulations (Izrailev et al. 1997; Li and Makarov 2003; Isralewitz et al. 2001; Balsera et al. 1997; Straub et al. 1987). For a number of proteins, this approach has yielded predictions that are in remarkably good agreement with experimental force vs. extension curves (Li and Makarov 2003, 2004a, b). However, as argued in Sect. 9.3, a one-dimensional free energy landscape is generally insufficient for reliable predictions regarding the dynamics of the processes induced by a force because z is not necessarily a "good" coordinate. In this regard, two-dimensional minimalist models may offer a sensible tradeoff between computational feasibility and realism (Suzuki and Dudko 2010, 2011).

Yet another class of approaches employs much simpler, ad hoc models to describe the energetics and kinetics of proteins subjected to forces (Makarov 2009a; Barsegov et al. 2008; Makarov et al. 2001; Imparato et al. 2007a, b; Hyeon et al. 2008; Dietz and Rief 2008; Staple et al. 2008). Given the limitations of all of the above methods, reliable modeling of the mechanically induced phenomena in proteins remains an open issue.

It is then clear that identifying new proteins that exhibit desirable mechanical behavior, such as high resistance to mechanical unfolding, is even a more difficult task. Indeed, it requires an analysis of the mechanical response of numerous possible structures, whereas detailed simulations of even a single protein are costly and not necessarily reliable. Two strategies have been put forth to systematically identify highly mechanically resistant proteins. The first one, proposed by us (Eom et al. 2003, 2005), was motivated by the observation that the mechanical response of globular proteins is primarily controlled by their native topology

(Klimov and Thirumalai 2000). We then reasoned that much could be learned about the relationship between topology and mechanical response by studying toy models that capture the key topological features of proteins but otherwise are extremely simplistic and—as a consequence—easily solvable. Specifically, we used a model in which the polypeptide chain was mimicked by a Gaussian polymer and where native contacts were mimicked by breakable cross-links. In support of this approach, similar models have been used to study various aspects of protein dynamics and folding (Barsegov et al. 2008; Dietz and Rief 2008; Erman and Dill 2000; Haliloglu et al. 1997; Tirion 1996). Furthermore, the cross-linked Gaussian chain model has been successful in explaining the relationship between native topology and folding rate (Makarov et al. 2002; Makarov and Metiu 2002; Makarov and Plaxco 2003).

By performing optimization over possible cross-link arrangements, we have found that conformations optimal with respect to either strength or toughness all involve a "mechanical clamp" formed by a pair of parallel strands, as highlighted in Fig. 9.6. The main activation barrier to unfolding is then associated with a cooperative rupture of this clamp via shearing. Given the robustness of this mechanical clamp, it is not surprising that Nature took advantage of this arrangement in proteins that have mechanical function. Perhaps the most known and studied example featuring such a clamp is the I27 domain of the muscle protein titin (shown in Fig. 9.6), which is known to sustain high mechanical forces when stretched between the ends in single-molecule pulling experiments (Marszalek et al. 1999; Rief et al. 1997, 1998; Fisher et al. 1999; Oberhauser et al. 2001; Kellermayer et al. 1997). Indeed, the unfolding mechanism that proceeds via the shearing of parallel strands has been found in computer simulations (Marszalek et al. 1999; Lu et al. 1998; Lu and Schulten 1999; Makarov et al. 2001; Best et al. 2003b). In contrast, another well-studied protein, the FNIII domain (also shown in Fig. 9.6), lacks this mechanical clamp and has been found to unfold under relatively weak loads (Oberhauser et al. 2002).

More importantly, studies of simple toy models have suggested that proteins that have no mechanical function but feature the same parallel strand structural motif may be good candidates as mechanically resistant proteins. This has led us to study the mechanical response of the GB1 domain (Fig. 9.6), ubiquitin, and related protein domains (Li and Makarov 2004a, b; Li et al. 2006), which, indeed, were found to exhibit high mechanical stability. The high mechanical stability of those domains has also been demonstrated experimentally (Cao et al. 2006, 2008; Carrion-Vazquez et al. 2003; Li et al. 2008; Sharma et al. 2007), and, in an exciting recent development (Cao and Li 2008), was employed to design, from bottom up, a material that mimics the mechanical properties of muscle.

A more recent systematic search for mechanically resistant proteins was undertaken by the Cieplak group (Sulkowska and Cieplak 2008; Sikora et al. 2009, 2011), which has used computer simulations to unravel, mechanically, proteins in the entire protein databank. Using native-centered, coarse-grained models to reduce computational cost, their study has confirmed that the mechanical clamp produced by terminal parallel strands is commonly associated with high

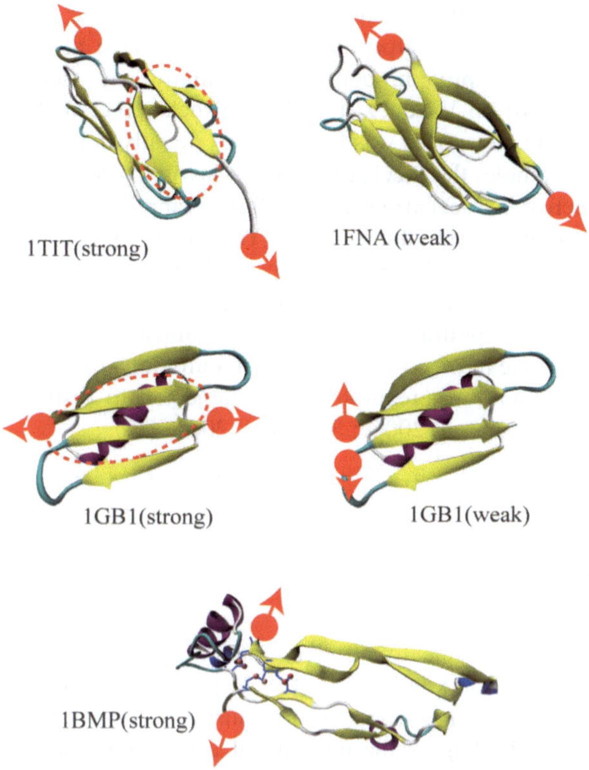

Fig. 9.6 Mechanical stability of globular proteins depends both on their structure and the points at which the forces are applied. Domains such as the I27 domain of the muscle protein titin (pdb id 1TIT) and the GB1 domain (1GB1) have a mechanical clamp formed by their terminal parallel strands, as indicated by *red ovals*). When pulled by the ends, they exhibit high mechanical stability. In contrast, the FNIII domain (1FNA), which lacks such a clamp, exhibits low mechanical stability despite its structural similarity to 1TIT. The parallel strand clamp is only effective in certain pulling scenarios; when the force is applied so as to unzip the same strands, the mechanical stability is low (Li and Makarov 2004b), as shown here for the GB1 domain. A protein-databank-wide computational survey of the mechanical stability of various proteins (Sulkowska and Cieplak 2008) has revealed other mechanically resistant motifs involving multiple disulfide bonds, e.g., in the bone morphogenetic protein 7 (1BMP)

mechanical stability. In addition, they have also discovered another family of mechanically resistant proteins that exhibit high unfolding forces produced by a "slipknot" mechanism involving disulfide bonds (Peplowski et al. 2011). Force-induced tightening of the slipknots in those proteins may result in molecular jamming (Peplowski et al. 2011) discussed in Sect. 9.4. One of the most mechanically stable proteins, as predicted by their study, is shown in Fig. 9.6. It should be noted that the slipknot mechanical clamp could not be predicted by our studies of cross-linked chains (Eom et al. 2003, 2005) because the "phantom" polymer model employed therein allowed chain self-crossings.

Whereas the focus of this section was on the case where the force is applied between the ends of the polypeptide chain, forces have also been exerted on different parts of the protein, both experimentally and theoretically (Li and Makarov 2004b; Carrion-Vazquez et al. 2003; Brockwell et al. 2003). Not surprisingly, the mechanical response was found to be critically dependent on the pulling geometry. For example, the mechanical clamp involving terminal parallel strands becomes ineffective in a scenario where the force is applied such that the strands are unzipped, rather than sheared (Fig. 9.6). The unzipping of the strands in this case resembles the tearing of a sheet of paper, which—as our everyday experience suggests—is the easiest way to break the material. The insights from the studies that probe the effect of the pulling geometry on mechanical unfolding are particularly relevant in the context of co-translocational unfolding of proteins by the cell degradation machinery, where proteins are subjected to distributed rather than localized forces (West et al. 2006c; Sato et al. 2005).

9.6 Outlook

Early theories of mechanically driven protein unfolding relied either on the Zhurkov–Bell model (Zhurkov 1965; Bell 1978) predicting exponential force dependence of the unfolding rate or on more sophisticated, but still one-dimensional, theories viewing unfolding as the crossing of a one-dimensional barrier (Dudko et al. 2007, 2008; Evans and Ritchie 1997, 1999; Hummer and Szabo 2001, 2003). While often successful in fitting experimental data, it is argued here that such models are generally insufficient if quantitative, first principles predictions of proteins' mechanical behavior are desired. Higher-dimensional models have therefore been emphasized in this chapter. Therein, however, lies the challenge. The choice of the "good" coordinates that need to be explicitly considered in such models is an open issue. First principles computation of multidimensional free energy landscapes is costly. Likewise, extraction of multidimensional landscapes from experimental data is difficult as such landscapes require more fitting parameters. Nevertheless, the essential multidimensionality of the underlying energy landscape is sometimes evident (Yew et al. 2010). An alternative to those relatively low-dimensional models is offered by brute force computer simulations. Despite growing CPU power, however, we are still far from directly accessing experimentally and/or biologically relevant regimes. Simulation methods such as transition path sampling (Bolhuis et al. 2002), forward flux (Allen et al. 2006), milestoning (Kirmizialtin and Elber 2011; Elber 2011), and transition interface methods (Van Erp and Bolhuis 2005) extend the accessible timescales and thus show significant promise. Likewise, reweighting (Nummela and Andricioaei 2007) and other extrapolation techniques that infer low-force behavior from high-force data are likely to see applications in the future. Despite significant difficulties involved in the simulations of even relatively small globular proteins, molecular simulation studies of larger protein assemblies, often combined with continuum mechanics approaches or coarse-grained models (Cetinkaya et al. 2011; Keten et al. 2010; Nova et al. 2010; Xiao et al. 2009; Yoon et al. 2009), begin to offer insights into the structure–function relationships of many fascinating biological materials.

Acknowledgments I am indebted to Christopher W. Bielawski, Johnathan N. Brantley, Kilho Eom, Helen Hansma, Paul Hansma, Lei Huang, Serdal Kirmizialtin, Sai Sriharsha M. Konda, Horia Metiu, and Gregory J. Rodin, who have collaborated with me on many of the topics described in this chapter. Financial support from the Robert A. Welch Foundation (grant no. F-1514), the National Science Foundation (grant no. CHE 0848571), and from W. A. "Tex" Moncrief, Jr. Endowment In Simulation-Based Engineering Sciences through a Grand Challenge Faculty Fellowship is gratefully acknowledged.

Appendix

In the Appendix, explicit expressions for the mechanical compliance of a molecule pulled between two chosen atoms i and j are provided, assuming that the molecule's free energy $G(\mathbf{r})$ can be approximated as a Taylor expansion, to second order, in the vicinity of its minimum or saddle point. For example, near the minimum free energy configuration $\mathbf{r}^{(0)}$, we have

$$G(\mathbf{r}) \approx G(\mathbf{r}^{(0)}) + \frac{1}{2}(\mathbf{r} - \mathbf{r}^{(0)})^T \mathbf{h}^{(0)} (\mathbf{r} - \mathbf{r}^{(0)}),$$

where $\mathbf{h}^{(0)}$ is the molecule's Hessian matrix. When the distance R_{ij} between the two selected atoms is increased, other atoms, of course, also become displaced. Since no external forces act on those atoms, their positions are determined from the condition that they are in mechanical equilibrium (i.e., the total force exerted on each of them by other atoms is zero). Given the assumed linearity of the system, its response to an increase in the distance R_{ij} is that of a Hookean spring, with a compliance (inverse stiffness) $\chi_{ij}^{(0)}$. Finding $\chi_{ij}^{(0)}$ can be viewed as a coarse-graining procedure, in which all of the atomic coordinates, except for the coordinates of the atoms i and j, are eliminated based on the above mechanical equilibrium condition. As the atoms of the molecule can be arbitrarily relabeled, it is convenient to assume that one always pulls on the first two atoms. We then write the molecule's Hessian matrix in the block-diagonal form:

$$\mathbf{h}^{(0)} = \begin{pmatrix} \mathbf{h}_{11}^{(0)} & \mathbf{h}_{12}^{(0)} \\ \mathbf{h}_{21}^{(0)} & \mathbf{h}_{22}^{(0)} \end{pmatrix}.$$

Here $\mathbf{h}_{11}^{(0)}, \mathbf{h}_{12}^{(0)}, \mathbf{h}_{21}^{(0)}$, and $\mathbf{h}_{22}^{(0)}$ are, respectively, 6×6, $6 \times (3N-6)$, $(3N-6) \times 6$, and $(3N-6) \times (3N-6)$ matrices. The $(3N-6)$ degrees of freedom of all the atoms other than the first two are eliminated through the standard coarse-graining procedure, to obtain an effective 6 × 6 Hessian matrix that describes the mechanical response of the pair of atoms one is pulling on. This matrix is given by the Schur complement (Konda et al. 2011; Eom et al. 2007; Soheilifard et al. 2011):

$$\bar{\mathbf{h}}_{11}^{(0)} = \mathbf{h}_{11}^{(0)} - \mathbf{h}_{12}^{(0)}(\mathbf{h}_{22}^{(0)})^{-1}\mathbf{h}_{21}^{(0)}.$$

This matrix should, of course, coincide with the Hessian matrix computed from the assumption that the free energy of the system is that of a simple Hookean spring given by (9.25). This, in particular, means that it has five zero eigenvalues and one nonzero eigenvalue equal to $2/\chi_{ij}^{(0)}$. Thus diagonalization of the 6 × 6 matrix $\mathbf{h}_{11}^{(0)}$ readily solves the problem of finding the effective compliance $\chi_{ij}^{(0)}$ in terms of the full Hessian matrix of the molecule.

The effective compliance $\chi_{ij}^{(TS)}$ of the molecule corresponding to its transition state (as well as to any critical point of the molecule's potential energy surface) can be computed in an analogous manner, using the Hessian matrix corresponding to the transition state. Of course, stretching the molecule while maintaining its transition-state configuration does not correspond to any experimental scenario. Nevertheless, as discussed in Sect. 9.3, this quantity is expedient in calculations of force-dependent rates.

References

Alegre-Cebollada J, Perez-Jimenez R, Kosuri P, Fernandez JM (2010) Single-molecule force spectroscopy approach to enzyme catalysis. J Biol Chem 285:18961

Allen RJ, Frenkel D, ten Wolde PR (2006) Forward flux sampling-type schemes for simulating rare events: efficiency analysis. J Chem Phys 124:194111

Bailey A, Mosey NJ (2012) Prediction of reaction barriers and force-induced instabilities under mechanochemical conditions with an approximate model: a case study of the ring opening of 1,3-cyclohexadiene. J Chem Phys 136:044102

Balsera M, Stepaniants S, Izrailev S, Oono Y, Schulten K (1997) Reconstructing potential energy functions from simulated force-induced unbinding processes. Biophys J 73:1281

Barsegov V, Thirumalai D (2005) Dynamics of unbinding of cell adhesion molecules: transition from catch to slip bonds. Proc Natl Acad Sci USA 102:1835

Barsegov V, Morrison G, Thirumalai D (2008) Role of internal chain dynamics on the rupture kinetic of adhesive contacts. Phys Rev Lett 100:248102

Becker N et al (2003) Molecular nanosprings in spider capture-silk threads. Nat Mater 2:278

Bell GI (1978) Models for specific adhesion of cells to cells. Science 200:618

Berkovich R, Garcia-Manyes S, Klafter J, Urbakh M, Fernandez JM (2010) Hopping around an entropic barrier created by force. Biochem Biophys Res Commun 403:133

Best RB et al (2003a) Force mode atomic force microscopy as a tool for protein folding studies. Anal Chim Acta 479:87

Best RB et al (2003b) Mechanical unfolding of a titin Ig domain: structure of transition state revealed by combining atomic force microscopy, protein engineering and molecular dynamics simulations. J Mol Biol 330:867

Best RB, Paci E, Hummer G, Dudko OK (2008) Pulling direction as a reaction coordinate for the mechanical unfolding of single molecules. J Phys Chem B 112:5968

Beyer MK, Clausen-Schaumann H (2005) Mechanochemistry: the mechanical activation of covalent bonds. Chem Rev 105:2921

Bolhuis PG, Chandler D, Dellago C, Geissler PL (2002) Transition path sampling: throwing ropes over rough mountain passes, in the dark. Annu Rev Phys Chem 53:291

Boulatov R, Kucharski TJ (2011) The physical chemistry of mechanoresponsive polymers. J Mater Chem 21:8237

Brantley JN, Wiggins KM, Bielawski CW (2011) Unclicking the click: mechanically facilitated 1,3-Dipolar cycloreversions. Science 333:1606

Brockwell DJ et al (2003) Pulling geometry defines the mechanical resistance of a beta-sheet protein. Nat Struct Biol 10:731

Camacho CJ, Thirumalai D (1993) Kinetics and thermodynamics of folding in model proteins. Proc Natl Acad Sci USA 90:6369

Cao Y, Li H (2008) Engineered elastomeric proteins with dual elasticity can be controlled by a molecular regulator. Nat Nanotechnol 3:512

Cao Y, Lam C, Wang M, Li H (2006) Nonmechanical protein can have significant mechanical stability. Angew Chem Int Ed Engl 45:642

Cao Y, Yoo T, Li H (2008) Single molecule force spectroscopy reveals engineered metal chelation is a general approach to enhance mechanical stability of proteins. Proc Natl Acad Sci USA 105:11152

Carrion-Vazquez M et al (2003) The mechanical stability of ubiquitin is linkage dependent. Nat Struct Biol 10:738

Cetinkaya M, Xiao S, Markert B, Stacklies W, Grater F (2011) Silk fiber mechanics from multiscale force distribution analysis. Biophys J 100:1298

De Gennes PG (1979) Scaling concepts in polymer physics. Cornell University Press, Ithaca, NY

Dietz H, Rief M (2008) Elastic bond network model for protein unfolding mechanics. Phys Rev Lett 100:098101

Dudko OK, Filippov AE, Klafter J, Urbakh M (2003) Beyond the conventional description of dynamic force spectroscopy of adhesion bonds. Proc Natl Acad Sci USA 100:11378

Dudko OK, Hummer G, Szabo A (2006) Intrinsic rates and activation free energies from single-molecule pulling experiments. Phys Rev Lett 96:108101

Dudko OK, Mathe J, Szabo A, Meller A, Hummer G (2007) Extracting kinetics from single-molecule force spectroscopy: nanopore unzipping of DNA hairpins. Biophys J 92:4188

Dudko OK, Hummer G, Szabo A (2008) Theory, analysis, and interpretation of single-molecule force spectroscopy experiments. Proc Natl Acad Sci USA 105:15755

Dudko OK, Graham TG, Best RB (2011) Locating the barrier for folding of single molecules under an external force. Phys Rev Lett 107:208301

Elber R (2011) Simulations of allosteric transitions. Curr Opin Struct Biol 21:167

Eom K, Li P-C, Makarov DE, Rodin GJ (2003) Relationship between the mechanical properties and topology of cross-linked polymer molecules: Parallel strands maximize the strength of model polymers and protein domains. J Phys Chem B 107:8730

Eom K, Makarov DE, Rodin GJ (2005) Theoretical studies of the kinetics of mechanical unfolding of cross-linked polymer chains and their implications for single molecule pulling experiments. Phys RevE 71:021904

Eom K, Baek SC, Ahn JH, Na S (2007) Coarse-graining of protein structures for the normal mode studies. J Comput Chem 28:1400

Erickson HP (1997) Stretching single protein molecules: titin is a weird spring. Science 276:1090

Erman B, Dill KA (2000) Gaussian model of protein folding. J Chem Phys 112:1050

Evans E, Ritchie K (1997) Dynamic strength of molecular adhesion bonds. Biophys J 72:1541

Evans E, Ritchie K (1999) Strength of a weak bond connecting flexible polymer chains. Biophys J 76:2439

Fantner GE et al (2006) Sacrificial bonds and hidden length: unraveling molecular mesostructures in tough materials. Biophys J 90:1411

Fisher ME, Kolomeisky AB (1999) The force exerted by a molecular motor. Proc Natl Acad Sci USA 96:6597
Fisher ME, Kolomeisky AB (2001) Simple mechanochemistry describes the dynamics of kinesin molecules. Proc Natl Acad Sci USA 98:7748
Fisher TE, Oberhauser AF, Vezquez MC, Marszalek PE, Fernandez J (1999) The study of protein mechanics with the atomic force microscope. TIBS 24:379
Florin E, Moy V, Gaub H (1994) Adhesion forces between individual ligand-receptor pairs. Science 264:415
Franco I, Schatz GC, Ratner MA (2009) Single-molecule pulling and the folding of donor-acceptor oligorotaxanes: phenomenology and interpretation. J Chem Phys 131:124902
Graham TG, Best RB (2011) Force-induced change in protein unfolding mechanism: discrete or continuous switch? J Phys Chem B 115:1546
Guo S, Li N, Lad N, Ray C, Akhremitchev BB (2010) Mechanical distortion of protein receptor decreases the lifetime of a receptor-ligand bond. J Am Chem Soc 132:9681
Haliloglu T, Bahar I, Erman B (1997) Gaussian dynamics of folded proteins. Phys Rev Lett 79:3090
Hanggi P, Talkner P, Borkovec M (1990) 50 years after Kramers. Rev Mod Phys 62:251
Heymann B, Grubmuller H (1999) AN02/DNP-hapten unbinding forces studied by molecular dynamics atomic force microscopy simulations. Chem Phys Lett 303:1
Heymann B, Grubmuller H (2001) Molecular dynamics force probe simulations of antibody/antigen unbinding: entropic control and nonadditivity of unbinding forces. Biophys J 81:1295
Huang Z, Boulatov R (2011) Chemomechanics: chemical kinetics for multiscale phenomena. Chem Soc Rev 40:2359
Huang L, Makarov DE (2008) Translocation of a knotted polypeptide through a pore. J Chem Phys 129:121107
Hugel T, Rief M, Seitz M, Gaub HE, Netz RR (2005) Highly stretched single polymers: atomic-force-microscope experiments versus ab-initio theory. Phys Rev Lett 94:048301
Hummer G, Szabo A (2001) Free energy reconstruction from nonequilibrium single-molecule pulling experiments. Proc Natl Acad Sci USA 98:3658
Hummer G, Szabo A (2003) Kinetics from nonequilibrium single-molecule pulling experiments. Biophys J 85:5
Hyeon C, Thirumalai D (2005) Mechanical unfolding of RNA hairpins. Proc Natl Acad Sci USA 102:6789
Hyeon C, Thirumalai D (2008) Multiple probes are required to explore and control the rugged energy landscape of RNA hairpins. J Am Chem Soc 130:1538
Hyeon C, Morrison G, Thirumalai D (2008) Force-dependent hopping rates of RNA hairpins can be estimated from accurate measurement of the folding landscapes. Proc Natl Acad Sci USA 105:9604
Imparato A, Pelizzola A, Zamparo M (2007a) Protein mechanical unfolding: a model with binary variables. J Chem Phys 127:145105
Imparato A, Pelizzola A, Zamparo M (2007b) Ising-like model for protein mechanical unfolding. Phys Rev Lett 98:148102
Irback A, Mitternacht S, Mohanty S (2005) Dissecting the mechanical unfolding of ubiquitin. Proc Natl Acad Sci USA 102:13427
Isralewitz B, Gao M, Schulten K (2001) Steered molecular dynamics and mechanical functions of proteins. Curr Opin Struct Biol 11:224
Izrailev S, Stepaniants S, Balsera M, Oono Y, Schulten K (1997) Molecular dynamics study of unbinding of the avidin-biotin complex. Biophys J 72:1568
Kellermayer MSZ, Smith SB, Granzier HL, Bustamante C (1997) Folding-unfolding transitions in single titin molecules characterized with laser tweezers. Science 276:1112
Keten S, Xu Z, Ihle B, Buehler MJ (2010) Nanoconfinement controls stiffness, strength and mechanical toughness of beta-sheet crystals in silk. Nat Mater 9:359

Kirmizialtin S, Elber R (2011) Revisiting and computing reaction coordinates with Directional Milestoning. J Phys Chem A 115:6137
Kirmizialtin S, Makarov DE (2008) Simulations of the untying of molecular friction knots between individual polymer strands. J Chem Phys 128:094901
Kirmizialtin S, Huang L, Makarov DE (2005) Topography of the free energy landscape probed via mechanical unfolding of proteins. J Chem Phys 122:234915
Kleiner A, Shakhnovich E (2007) The mechanical unfolding of ubiquitin through all-atom Monte Carlo simulation with a Go-type potential. Biophys J 92:2054
Klimov D, Thirumalai D (1997) Viscosity Dependence of Folding Rates of Protein. Phys Rev Lett 79: 317
Klimov DK, Thirumalai D (2000) Native topology determines force-induced unfolding pathways in globular proteins. Proc Natl Acad Sci USA 97:7254
Kochhar GS, Bailey A, Mosey NJ (2010) Competition between orbitals and stress in mechanochemistry. Angew Chem Int Ed 49:7452
Konda SSM, Brantley JN, Bielawski CW, Makarov DE (2011) Chemical reactions modulated by mechanical stress: extended Bell theory. J Chem Phys 135:164103
Lacks DJ (2005) Energy landscape distortions and the mechanical unfolding of proteins. Biophys J 88:3493
Lacks DJ, Willis J, Robinson M-P (2010) Fold catastrophes and the dependence of free-energy barriers to conformational transitions on applied force. J Phys Chem B 114:10821
Lee C, Schwartz MP, Prakash S, Iwakura M, Matouschek A (2001) ATP-dependent proteases degrade their substrates by processively unraveling them from the degradation signal. Mol Cell 7:627
Lenhardt JM et al (2010) Trapping a diradical transition state by mechanochemical polymer extension. Science 329:1057
Li MS, Kouza M (2009) Dependence of protein mechanical unfolding pathways on pulling speeds. J Chem Phys 130:145102
Li P-C, Makarov DE (2003) Theoretical studies of the mechanical unfolding of the muscle protein titin: bridging the time-scale gap between simulation and experiment. J Chem Phys 119:9260
Li P-C, Makarov DE (2004a) Ubiquitin-like protein domains show high resistance to mechanical unfolding similar to that of the I27 domain in titin: evidence from simulations. J Phys Chem B 108:745
Li P-C, Makarov DE (2004b) Simulation of the mechanical unfolding of ubiquitin: probing different unfolding reaction coordinates by changing the pulling geometry. J Chem Phys 121:4826
Li H, Oberhauser AF, Fowler SB, Clarke J, Fernandez JM (2000) Atomic force microscopy reveals the mechanical design of a modular protein. Proc Natl Acad Sci USA 97:6527
Li PC, Huang L, Makarov DE (2006) Mechanical unfolding of segment-swapped protein G dimer: results from replica exchange molecular dynamics simulations. J Phys Chem B 110:14469
Li MS, Kouza M, Hu CK (2007) Refolding upon force quench and pathways of mechanical and thermal unfolding of ubiquitin. Biophys J 92:547
Li H, Wang H-C, Cao Y, Sharma D, Wang M (2008) Configurational entropy modulates the mechanical stability of protein GB1. J Mol Biol 379:871
Liphardt J, Onoa B, Smith SB, Tinoco IJ, Bustamante C (2001) Reversible unfolding of single RNA molecules by mechanical force. Science 292:733
Lu H, Schulten K (1999) Steered molecular dynamics simulation of conformational changes of immunoglobulin domain I27 interpret atomic force microscopy observations. Chem Phys 247:141
Lu H, Isralewitz B, Krammer A, Vogel V, Schulten K (1998) Unfolding of titin immunoglobulin domains by steered molecular dynamics. Biophys J 75:662
Makarov DE (2007) Unraveling individual molecules by mechanical forces: theory meets experiment. Biophys J 92:4135
Makarov DE (2009a) A theoretical model for the mechanical unfolding of repeat proteins. Biophys J 96:2160

Makarov DE (2009b) Computer simulations and theory of protein translocation. Acc Chem Res 42:281
Makarov DE, Metiu H (2002) A model for the kinetics of protein folding: Kinetic Monte Carlo simulations and analytical results. J Chem Phys 116:5205
Makarov DE, Plaxco KW (2003) The topomer search model: a quantitive, fisrt principles description of two-state protein folding kinetics. Protein Sci 12:17
Makarov DE, Hansma PK, Metiu H (2001) Kinetic Monte Carlo simulation of titin unfolding. J Chem Phys 114:9663
Makarov DE, Keller C, Plaxco KW, Metiu H (2002) How the folding rate constant of simple, single-domain proteins depends on the number of native contacts. Proc Natl Acad Sci USA 99:3535
Maloney CE, Lacks DJ (2006) Energy barrier scalings in driven systems. Phys Rev E Stat Nonlin Soft Matter Phys 73:061106
Marszalek PE et al (1999) Mechanical unfolding intermediates in titin modules. Nature 402:100
Marshall BT et al (2003) Direct observation of catch bonds involving cell-adhesion molecules. Nature 423:190
Matouschek A (2003) Protein Unfolding - an important process in vivo? Curr Opin Struct Biol 13:98
Matouschek A, Bustamante C (2003) Finding a protein's Achilles heel. Nat Struct Biol 10:674
McCullagh M, Franco I, Ratner MA, Schatz GC (2011) DNA-based optomechanical molecular motor. J Am Chem Soc 133:3452
Minajeva A, Kulke M, Fernandez JM, Linke WA (2001) Unfolding of titin domains explains the viscoelastic behavior of skeletal myofibrils. Biophys J 80:1442
Nova A, Keten S, Pugno NM, Redaelli A, Buehler MJ (2010) Molecular and nanostructural mechanisms of deformation, strength and toughness of spider silk fibrils. Nano Lett 10:2626
Nummela J, Andricioaei I (2007) Exact low-force kinetics from high-force single-molecule unfolding events. Biophys J 93:3373
Oberhauser AF, Marszalek PE, Erickson H, Fernandez JM (1998) The molecular elasticity of the extracellular matrix protein tenascin. Nature 393:181
Oberhauser AF, Hansma PK, Carrion-Vazquez M, Fernandez JM (2001) Stepwise unfolding of titin under force-clamp atomic force microscopy. Proc Natl Acad Sci USA 98:468
Oberhauser AF, Badilla-Fernandez C, Carrion-Vazquez M, Fernandez JM (2002) The mechanical hierarchies of fibronectin observed with single-molecule AFM. J Mol Biol 319:433
Paci E, Karplus M (1999) Forced unfolding of fibronectin Type 3 modules: an analysis by biased molecular dynamics simulations. J Mol Biol 288:441
Park S, Khalili-Araghi F, Tajkhorshid E, Schulten K (2003) Free energy calculation from steered molecular dynamics simulations using Jarzynski's equality. J Chem Phys 119:3559
Peplowski L, Sikora M, Nowak W, Cieplak M (2011) Molecular jamming–the cystine slipknot mechanical clamp in all-atom simulations. J Chem Phys 134:085102
Prakash S, Matouschek A (2004) Protein unfolding in the cell. TIBS 29:593
Prezhdo OV, Pereverzev YV (2009) Theoretical aspects of the biological catch bond. Acc Chem Res 42:693
Ribas-Arino J, Shiga M, Marx D (2009) Understanding Covalent Mechanochemistry. Angew Chem Int Ed Engl 48:4190
Rief M, Gautel M, Oesterhelt F, Fernandez JM, Gaub HE (1997) Reversible unfolding of individual titin immunoglobulin domains by AFM. Science 276:1109
Rief M, Fernandez JM, Gaub HE (1998) Elastically coupled two-level systems as a model for biopolymer extensibility. Phys Rev Lett 81:4764
Ritort F, Bustamante C, Tinoco I (2002) A two-state kinetic model for the unfolding of single molecules by mechanical force. Proc Natl Acad Sci USA 99:13544
Sato T, Esaki M, Fernandez JM, Endo T (2005) Comparison of the protein-unfolding pathways between mitochondrial protein import and atomic-force microscopy measurements. Proc Natl Acad Sci USA 102:17999

Schlierf M, Rief M (2006) Single-molecule unfolding force distributions reveal a funnel-shaped energy landscape. Biophys J 90:L33
Shariff K, Ghosal S, Matouschek A (2004) The force exerted by the membrane potential during protein import into the mitochondrial matrix. Biophys J 86:3647
Sharma D et al (2007) Single-molecule force spectroscopy reveals a mechanically stable protein fold and the rational tuning of its mechanical stability. Proc Natl Acad Sci USA 104:9278
Sikora M, Sulkowska JI, Cieplak M (2009) Mechanical strength of 17,134 model proteins and cysteine slipknots. PLoS Comput Biol 5:e1000547
Sikora M, Sulkowska JI, Witkowski BS, Cieplak M (2011) BSDB: the biomolecule stretching database. Nucleic Acids Res 39:D443
Smith BL et al (1999) Molecular mechanistic origin of the toughness of natural adhesives, fibres and composites. Nature 399:761
Soheilifard R, Makarov DE, Rodin GJ (2011) Rigorous coarse-graining for the dynamics of linear systems with applications to relaxation dynamics in proteins. J Chem Phys 135:054107
Sorenson JM, Head-Gordon T (2000) Matching simulation and experiment: a new simplified model for Simulating protein folding. J Comput Biol 7:469
Sorenson JM, Head-Gordon T (2002) Towards minimalist models of larger proteins: a ubiquitin-like protein. Proteins Struct Funct Genet 46:368
Staple DB, Payne SH, Reddin AL, Kreuzer HJ (2008) Model for stretching and unfolding the giant multidomain muscle protein using single-molecule force spectroscopy. Phys Rev Lett 101:248301
Straub JE, Borkovec M, Berne BJ (1987) Calculation of dynamic friction on intramolecular degrees of freedom. J Phys Chem 91:4995
Sulkowska JI, Cieplak M (2008) Stretching to understand proteins - a survey of the protein data bank. Biophys J 94:6
Suzuki Y, Dudko OK (2010) Single-molecule rupture dynamics on multidimensional landscapes. Phys Rev Lett 104:048101
Suzuki Y, Dudko OK (2011) Biomolecules under mechanical stress: a simple mechanism of complex behavior. J Chem Phys 134:065102
Szabo A, Schulten K, Schulten Z (1980) First passage time approach to diffusion controlled reactions. J Chem Phys 72:4350
Tirion MM (1996) Large amplitude elastic motions in proteins from a single-arameter, atomic analysis. Phys Rev Lett 77:1905
Truhlar DG, Garrett BC, Klippenstein SJ (1996) Current status of transition-state theory. J Phys Chem 100:12771
Van Erp T, Bolhuis P (2005) Elaborating transition interface sampling methods. J Comput Phys 205:157–181
Veitshans T, Klimov D, Thirumalai D (1996) Protein folding kinetics: timescales, pathways and energy landscapes in terms of sequence-dependent properties. Fold Des 2:1
Virnau P, Mirny LA, Kardar M (2006) Intricate knots in proteins: function and evolution. PLoS Comput Biol 2:e122
West DK, Olmsted PD, Paci E (2006a) Free energy for protein folding from nonequilibrium simulations using the Jarzynski equality. J Chem Phys 125:204910
West DK, Brockwell DJ, Olmsted PD, Radford SE, Paci E (2006b) Mechanical resistance of proteins explained using simple molecular models. Biophys J 90:287
West DK, Brockwell DJ, Paci E (2006c) Prediction of the translocation kinetics of a protein from its mechanical properties. Biophys J 91:L51
Xiao S, Stacklies W, Cetinkaya M, Markert B, Grater F (2009) Mechanical response of silk crystalline units from force-distribution analysis. Biophys J 96:3997
Xiong H, Crespo A, Marti M, Estrin D, Roitberg AE (2006) Free energy calculations with non-equilibrium methods: applications of the Jarzynski relationship. Theor Chem Acc 116:338

Yew ZT, Schlierf M, Rief M, Paci E (2010) Direct evidence of the multidimensionality of the free-energy landscapes of proteins revealed by mechanical probes. Phys Rev E Stat Nonlin Soft Matter Phys 81:031923

Yoon G, Park HJ, Na S, Eom K (2009) Mesoscopic model for mechanical characterization of biological protein materials. J Comput Chem 30:873

Yuan C, Chen A, Kolb P, Moy VT (2000) Energy landscape of streptavidin-biotin complexes measured by atomic force microscopy. Biochemistry 39:10219

Zhurkov SN (1965) Kinetic concept of strenght of solids. Int J Fract Mech 1:311

Zwanzig R (2001) Nonequilibrium statistical mechanics. Oxford University Press, Oxford

Index

A

AL amyloidosis, 196
Alpha-galactosylceramide (α-GalCer), 53
American Type Culture Collection (ATCC), 51
Amyloid fibrillogenesis, 171
Amyloidogenic proteins
  amyloid β1–40, 187, 189–190
  amyloid β1–42, 186–188
  amyloid β25–35, 190–192
  amyloid fibrillogenesis, 171
  amyloidosis, 169, 170
  architectural feature, 170
  etiological factor, 170
  ex vivo amyloid
    AFM topography, 199
    AL amyloidosis, 196
    extraction, 196, 198
    fibrillogenesis, 197
  fluorescence
    position and distance, 185–186
    signal quality, 185
  force measurement
    elasticity, 181–182
    spectroscopy, 182–184
  kymography, 184
  β_2-microglobulin, 194, 196
  natural adhesives, 199–200
  past decade, 172–173
  prion fibrils, 199
  protofibrils, 171
  SAP, 172
  single-molecule method
    AFM, 174–176
    fluorescence, 177–179
    optical tweezer, 176–177
  stiffness calibration, 179–181
  structural techniques, 172
  transthyretin
    clinical forms, 192
    prevalence of, 192–193
    protofibrils, 194
    wild-type, 195
  unzipping nanomechanical, 201
Atomic force microscopy (AFM)
  amyloidogenic proteins, 174–176
  CD1d-glycolipid complex
    imaging parameters adjustment, 66–68
    principle of simultaneous topography and recognition imaging, 65–66
    THP1 Cells, 68–71
  hCD1d–GSL complexes
    AFM tips and solid substrates, 54
    ceramide lipid, 53
    iNKT TCR, 53
    isolated molecules and living THP-1, 58–62
    kinetic on-rate measurements, 62–63
    principle of single-molecule force spectroscopy, 54–58
  protein-DNA interactions (*see* Protein-DNA interactions)
  protein nanovisualization (*see* High-speed AFM, protein nanovisualization)
  sample and tip preparation methods
    CD1d transfected THP-1 cells, 52
    iNKT TCR tip functionalization, 50–51
    mica, CD1d immobilization, 52
    simultaneous topography and recognition imaging and instrumentation, 52–53
    THP-1 cell culture, 51–52
  single-molecule techniques
    application of, 153
    force-clamp spectroscopy, 154–155

A.F. Oberhauser (ed.), *Single-molecule Studies of Proteins*,
Biophysics for the Life Sciences 2, DOI 10.1007/978-1-4614-4921-8,

Atomic force microscopy (AFM) (*cont.*)
mechanical forces, 153–154
stretching and imaging individual protein (*see* Stretching and individual protein imaging)
Atomic force microscopy (AFM)-based SMFS
invention and uses, 5
physical principle and modes, 5–8

B
Bacillus circulans, 23
Bell's formula, 247
Beta tail domain (βTD), 79
B1 immunoglobulin binding domain of streptococcal protein G (GB1), 14
Biomembrane force probe (BFP), 78
Biotin-streptavidin, 80

C
Calmodulin (CaM), 11, 15
Catch bonds, integrin/ligand interactions
activation, 92–94
conformational changes, 83–84
dissociation kinetics, 77
force-dependent lifetimes, 84–86
heterodimeric transmembrane adhesive receptors, 78
$\alpha_L\beta_2$ /ICAM-1 catch bonds, 88–91
ligand-binding affinities, 79
molecular interactions, 77
plasma membrane, 80
receptor/ligand binding, 2D kinetics of, 80–82
single-molecule force techniques, 78
structural analysis, 86–88
Cell adhesion proteins
force-induced amyloid nanodomains
Als5p formation and propagation, 225–228
biological significance, 228–229
dual detection of single Als5p, 225, 226
formation in yeast, 228
molecular mechanism, 222
unfolding single Als5p
in vitro, 222–223
in vivo, 224
Cell surface sensors
AFM analysis, 214–215
membrane sensor, 214
sensor clustering
CVD, 219
microdomain, 219
sensosome, 221–222
Wsc1, 219–220
single sensors detection, 216
Wsc1, 216–218
Cellulase
cellulose hydrolysis, 135
cellulose Iα, 136
cellulose III_I, 136–138
TrCel7A and TrCel6A, 138–140
Chromatin structure
DNA assembly process, 108
levels of compaction, 107–108
native unfixed chicken erythrocyte, 111
NCP, 110–111
nucleosome
array, 108–109
dynamics of, 110
structural characterization, 111
Citrobacter freundii, 13
Contour length increase, ΔL_c, 7
Cysteine-rich domain (CRD), 219

D
7,8-Dihydrofolate (DHF), 17
Dihydrofolate reductase (DHFR), 17
Dimethyl sulfoxide (DMSO), 24
DNA repair system, 103–104
Dynamic processes, 121

E
Elastin-like-polypeptides (ELPs), 28
Enzyme catalysis
force-clamp spectroscopy
application, 156–157
disulfide bond reduction, 156
electronic feedback, 155
spurious interaction, 155
kinetics of, 149–150
single-molecule techniques
AFM, 153–155
optical tweezer, 152–153
sm-FRET, 150–151
structural techniques, 150
thioredoxin (*see* Thioredoxin)
Trx catalysis
ancestral enzymes, 163–165
computational reconstruction, 163
concentration, 161–162
disulfide bond reduction, 164
force-clamp assay, 157–158
thioredoxin phylogeny, 162

Epidermal growth factor (EGF), 79
Epi-fluorescence microcopy, 63
Extracellular matrix (ECM), 77
Ex vivo amyloid
  AFM topography, 199
  AL amyloidosis, 196
  extraction, 196, 198
  fibrillogenesis, 197

F
Fibronectin (FN), 78
Folded proteins
  barrier-activated process, 24
  chemical reactions, 25–26
  filamin domain, 25
  Monte Carlo models, 23
  osmolytes, 24
  polyubiquitin folding, 23
  structured proteins, mechanical stability of
    cohesin module, 10
    electrostatic forces, 17–18
    α-helical proteins, 11–12
    hydrophobic core, 14
    knots, 13
    length-clamp mode, 9
    ligand binding, 15–17
    mechanical clamp motif, 9
    mechanically resistant structures, 13–14
    neighboring strands, 14
    nonnative mechanical clamps, 12–13
    protein loops, 14–15
    solvents, 18–19
    β-structured proteins, 11
Force-clamp spectroscopy
  application of, 156–157
  disulfide bond reduction, 156
  electronic feedback, 155
  spurious interaction, 155
  Trx catalysis, 157–159
Force spectroscopy, 5, 182–184

H
Hand-over-hand movement
  mechanism, 129–130
  unidirectional, 126–128
Hessian matrix, 261–262
Highest possible imaging rate, 125
High-speed AFM, protein nanovisualization
  biomolecules and dynamic studies, 120–121
  cellulase movement (*see* Cellulase)
  dynamics of, 119
  instrumentation
    amplitude detector and feedback controller, 123–124
    cantilevers and OBD sensor, 122–123
    highest possible imaging rate, 125
    scanner and active damping, 124
    system design, 121–122
  myosin V (*see* Myosin V)
  proof-of-principle stages, 120
  rotorless F_1-ATPase (*see* Rotorless F_1-ATPase)
  surface topography, 119–120
Huntington's disease (HD), 32

I
Immunoglobulin (Ig), 78
Integrin/ligand interactions, catch bonds
  activation, 92–94
  conformational changes, 83–84
  dissociation kinetics, 77
  force-dependent lifetimes, 84–86
  heterodimeric transmembrane adhesive receptors, 78
  $\alpha_L\beta_2$ /ICAM-1 catch bonds, 88–91
  ligand-binding affinities, 79
  molecular interactions, 77
  plasma membrane, 80
  receptor/ligand binding, 2D kinetics of, 80–82
  single-molecule force techniques, 78
  structural analysis, 86–88
Intercellular adhesion molecule 1 (ICAM-1), 78
Intrinsically disordered proteins (IDPs)
  amyloid diseases, 27
  biochemical and traditional structural methods, 27
  computational algorithms, 27
  human conformational diseases, 27
  nonamyloydogenic IDPs
    amyloid β, 33
    N2B, 29
    PEVK, 28–29
    polyglutamine tracts, 32–33
    Sup35NM, 33
    α-synuclein, 31–32
    Tau, 34
    tropoelastin and elastin, 28
    VAMP2, 31
Invariant Natural Killer T (iNKT), 50

K

Kramers' theory, 57
Kymography, 184

M

Magnetic alternating current (MAC), 52
Maltose binding protein (MBP), 17
Mechanical stress
  compact denatured (CD) state, 240
  conformational transitions, 235
  coordinate system, 236
  denatured protein, 238
  enzymatic activity, 235
  folding–unfolding transition, 236, 239
  free energy barrier, 237
  individual protein domain
    coarse-grained models, 257
    computer simulations, 258
    Gaussian chain model, 258
    globular proteins, 257
    high mechanical stability, 258
    phantom polymer model, 259
    sawtooth force-extension curve, 256
    slipknot mechanism, 259
    SMD data, 256
  mechanical compliance, 239
  molecular jamming
    catch-bond phenomenon, 250
    degradation process, 251
    human ubiquitin hydrolase, 251
    slip-knot structural motifs, 250
  random polymer chain, 237
  sacrificial bond mechanism
    AFM pulling experiment, 252
    force *vs.* extension curve, 252–253
    Gaussian chain model, 254
    Hook's law, 254
    natural biomaterials, 252
    nonequilibrium pulling process, 256
    polyprotein chain, 255
    sawtooth pattern, 253
    spider capture silk/bone, 251
    unfolding process, 253
  ubiquitin, 240
  Zhurkov–Bell formula and extensions
    Arrhenius plot, 248
    coarser model, 243
    computer simulations, 249
    displacement-controlled experiment, 249
    force-induced changes, 248
    generalized inverses, 246
    harmonic theory, 249
    harmonic transition-state, 247
    Hessian matrix, 245
    Hookean spring, 247
    Kramers' theory, 244
    protein degradation, 242
    single-molecule pulling experiments, 250
    standard rate theories, 244
    Taylor series, 245
    thermodynamic equilibrium, 241
    transition state and initial state, 248
Metal ion-dependent adhesion site (MIDAS), 79
Methotrexate (MTX), 17
β_2-Microglobulin, 194, 196
Myosin V
  hand-over-hand movement
    mechanism, 129–130
    unidirectional, 126–128
  properties of, 125–126
  tension generation and foot stomp, 128–129

N

Nanomechanics
  cell adhesion proteins (*see* Cell adhesion proteins)
  cell surface sensors (*see* Cell surface sensors)
  of protein (*see* Protein nanomechanics)
Nicotinamide adenine dihydrogen phosphate (NADPH), 17
Nuclear magnetic resonance (NMR), 4
Nucleosome
  array, 108–109
  dynamics of, 110
Nucleosome core particles (NCP), 109–110
Nucleotide crystal structure, 130–131

O

Optical tweezer
  amyloidogenic protein, 176–177
  enzyme catalysis, 152–153

P

Photo-activated localization microscopy (PALM), 63, 65
Photon-flux force sensing, 180–181
Plexin-semaphorin-integrin (PSI), 79
Poly-L-Lysine (PLL), 52

Prion fibrils, 199
Protein-DNA complexes
  capabilities of, 100
  DNA repair system
    mismatches, 103
    MutS, 103–104
  dynamics of, 97
  site-specific
    EcoRII–DNA complexes, 102–103
    protective mechanism, 100–101
    SfiI complexes, 101
Protein-DNA interactions
  architectural protein
    chromatin structure (*see* Chromatin structure)
    SSB-DNA complexes, 105–107
  methodology
    instrumentation, 98–99
    sample preparation, 99–100
  principal mediators, 97
  protein-DNA complexes (*see* Protein-DNA complexes)
  time-lapse AFM analyses
    EcoP15I, 112
    flipping process, 112
    HS AFM analysis of EcoRII translocation, 112–113
    threading process, 113–114
Protein nanomechanics
  AFM-based SMFS
    invention and uses, 5
    physical principle and modes, 5–8
  force, 3
  ordered *vs.* disordered proteins, SMFS
    folded proteins (*see* Folded proteins)
    IDPs (*see* Intrinsically disordered proteins (IDPs))
    protein trinity, 9
    structure/function paradigm, 8
  in silico force spectroscopy
    all-atom MD, 37–38
    computer simulations, 34
    molecular dynamics, IDPs, 39
    SMD, 37
    structure-based/Go models, 38–39
  single-molecule *vs.* bulk biophysics, 3–4
  SMFS, 4–5
P-selectin glycoprotein ligand 1 (PSGL-1), 78

R
Red blood cell (RBC), 80
Root mean square distance (RMSD), 90
Rotorless F_1-ATPase
  α3β3 subcomplex
    ATP, 133–134
    nucleotide, 131–133
  properties of
    crystal structure, 130–131
    rotary motors, 130

S
SAP. *See* Serum amyloid P component (SAP)
Scanning force kymography, 175
Scanning probe microscopy (SPM), 5
Scanning tunneling microscope (STM), 5
Sensor clustering
  CVD, 219
  microdomain, 219
  sensosome, 221–222
  Wsc1, 219–220
Serum amyloid P component (SAP), 172
Single-molecule fluorescence resonance energy transfer (sm-FRET), 150–151
Single-molecule force spectroscopy (SMFS), 4–5
  bond rupture, 56–58
  cantilever, 55
  dissociation rate, 54
  irreversible process, 54
  loading rate, 56
  noncovalent bonds, 55
  single energy barrier, 54
  unbinding force, 56
Single-molecule techniques (SMT), 3
SSB–DNA complexes
  characterization of, 105–106
  homologous recombination, 105
  hybrid DNA, 106
  mica functionalization, 106
  69-tail-DNA, 107
Steered molecular dynamics (SMD), 37, 90
Stiffness calibration, 179–181
Stimulated emission depletion microscopy (STED), 65
Stretching and individual protein imaging
  adhesion and mechanosensing, 230–231
  AFM, 211–212
  elasticity, 212–213
  nanomechanics and nanoclustering
    adhesion proteins (*see* Cell adhesion proteins)
    surface sensors (*see* Cell surface sensors)

Stretching and individual protein imaging (*cont.*)
  unfolding, 213–214
Structure dynamics, 121
Surface plasmon resonance (SPR), 80
Synergetic hydrolysis
  conversion efficiency, 140
  TrCel6A and TrCel7A, 138–139

T
T cell receptor (TCR), 50
Thioredoxin
  bacterial Trx, 159, 161
  chemical mechanisms, 158–160
  disulfide bond reduction, 158–160
Time-lapse AFM analyses
  EcoP15I, 112
  EcoRII translocation, 112–113
  flipping process, 112
  threading process, 113–114
Topography and RECognition imaging (TREC), 49, 63–71
Transition state (TS) theory, 7
Transthyretin
  clinical forms, 192
  prevalence of, 192–193
  protofibrils, 194
  wild-type, 195
TrCel6A and TrCel7A, 138–139
Trx catalysis
  ancestral enzymes, 163–165
  computational reconstruction, 163
  disulfide bond reduction, 164
  enzyme concentration, 161–162
  force-clamp assay, 157–158
  thioredoxin phylogeny, 162

U
Unfolding force (F_U), 7, 12, 14, 15

V
von Willebrand factor (VWF), 78

W
Worm-like chain (WLC) model, 7
Wsc1
  sensor, 217–218
  spring constant, 218
  WLC model, 216

Z
Zhurkov–Bell formula and extensions
  Arrhenius plot, 248
  coarser model, 243
  computer simulations, 249
  displacement-controlled experiment, 249
  force-induced changes, 248
  generalized inverses, 246
  harmonic theory, 249
  harmonic transition-state, 247
  Hessian matrix, 245
  Hookean spring, 247
  Kramers' theory, 244
  protein degradation, 242
  single-molecule pulling experiments, 250
  standard rate theories, 244
  Taylor series, 245
  thermodynamic equilibrium, 241
  transition state and initial state, 248

MIX
Papier aus verantwortungsvollen Quellen
Paper from responsible sources
FSC® C105338
FSC
www.fsc.org

If you have any concerns about our products,
you can contact us on
ProductSafety@springernature.com

In case Publisher is established outside the EU,
the EU authorized representative is:
Springer Nature Customer Service Center GmbH
Europaplatz 3, 69115 Heidelberg, Germany

Printed by Libri Plureos GmbH
in Hamburg, Germany